METHODS IN MOLECULAR BIOLOGY™

Series Editor
John M. Walker
School of Life Sciences
University of Hertfordshire
Hatfield, Hertfordshire, AL10 9AB, UK

For further volumes:
http://www.springer.com/series/7651

In Silico Systems Biology

Edited by

Maria Victoria Schneider

The Genome Analysis Centre, Norwich Research Park, Norwich, United Kingdom

Humana Press

Editor
Maria Victoria Schneider
The Genome Analysis Centre
Norwich Research Park
Norwich, UK

ISSN 1064-3745 ISSN 1940-6029 (electronic)
ISBN 978-1-4939-6295-2 ISBN 978-1-62703-450-0 (eBook)
DOI 10.1007/978-1-62703-450-0
Springer New York Heidelberg Dordrecht London

Softcover reprint of the hardcover 1st edition 2013

Printed on acid-free paper

Humana Press is a brand of Springer
Springer is part of Springer Science+Business Media (www.springer.com)

Preface

I have the pleasure of introducing this edition of *in Silico Systems Biology: Methods and Protocols* in the successful *Methods in Molecular Biology* series. I must express my gratitude for the work done by the many contributors of this book. Their many efforts have made possible the realization of this project. Systems biology can now be considered an established and fundamental field in life sciences. It has allowed moving from the identification of molecular 'parts lists' for living organisms towards synthesizing information from different 'omics'-based approaches to generate and test new hypotheses about how biological systems work. To acquire a systems-level understanding of biology, neither experimental nor computational biology alone is sufficient. Parallel to the advances in the research front of systems biology, maturation in the approaches and actual analysis of the data and the resources/tools have also advanced. New generation of life scientists are able to find excellent text books in this field, as well as on the variety of flavours the term systems biology seems to acquire and its related disciplines. From experience in the nonprofit bioinformatics training sector it is clear that our days researchers in this field need to be able to master a complexity of tasks: (1) from handling and annotating networks, (2) getting acquainted and learning how to deal with network-based approaches to model, (3) investigating high throughput biological data, (4) understanding the formalisms and methods of existing correlation network models and gene pathways as well as how to model perturbations in the systems. All of these remain essential skills for which many have had no formal training. Scientists seek for opportunities to learn and share experiences upon. Based on this need, the idea of this book was born. It provides a practical set of chapters based often on actual materials used and develop for face-to-face training with examples and case studies. It centres on covering network biology and mathematical models of biological systems. The contributors are a magnificent set of experts and lead researchers in the field of systems biology. I am particularly grateful to Prof. Nicolas Le Novere, the inspiration behind the actual face-to-face courses as well as this book, in addition to Julio Saenz Rodrigues and all those that put time aside to offer useful and timely materials for this book. Not all contributors have been part of the training courses, but they have extended and enriched the contents to areas we had not covered previously, providing a comprehensive picture of the approaches and advances made in the field of in silico systems biology.

Norwich, UK *Maria Victoria Schneider*

Contents

Contributors

EMMANUEL BARILLOT • *Institut Curie, Paris, France; INSERM, U900, Paris, France; Mines ParisTech, Fontainebleau, France*
ANNA BAUER-MEHREN • *Stanford, CA, USA*
ERIC BONNET • *Institut Curie, Paris, France; INSERM, U900, Paris, France; Mines ParisTech, Fontainebleau, France*
BEN CALDERHEAD • *Department of Computing Science, University of Glasgow, Glasgow, Scotland*
LAURENCE CALZONE • *Institut Curie, Paris, France; INSERM, U900, Paris, France; Mines ParisTech, Fontainebleau, France*
VIJAYALAKSHMI CHELLIAH • *EMBL Outstation–European Bioinformatics Institute, Cambridge, UK*
DAVID COHEN • *Institut Curie, Paris, France; INSERM, U900, Paris, France; Mines ParisTech, Fontainebleau, France*
MANUEL CORPAS • *The Genome Analysis Centre, Norwich, UK*
DAVID CROFT • *EMBL Outstation–European Bioinformatics Institute, Cambridge, UK*
LUKAS ENDLER • *EMBL Outstation–European Bioinformatics Institute, Cambridge, UK*
MICHAEL EPSTEIN • *Department of Computing Science, University of Glasgow, Glasgow, Scotland*
ILLÉS J. FARKAS • *Statistical and Biological Physics Group of the Hungarian Academy of Sciences, Budapest, Hungary*
COLIN S. GILLESPIE • *School of Mathematics Statistics, Newcastle University, Newcastle upon Tyne, UK*
MARK GIROLAMI • *Department of Computing Science, University of Glasgow, Glasgow, Scotland*
ANDREW GOLIGHTLY • *School of Mathematics Statistics, Newcastle University, Newcastle upon Tyne, UK*
DAVID HENRIQUES • *Instituto de Investigaciones Marinas (CSIC), Vigo, Spain*
RAFAEL C. JIMENEZ • *EMBL Outstation–European Bioinformatics Institute, Cambridge, UK*
NICK JUTY • *The EMBL-European Bioinformatics Institute, Wellcome Trust Genome Campus, Hinxton, Cambridge, UK*
SARAH M. KEATING • *EMBL Outstation–European Bioinformatics Institute, Cambridge, UK*
TAMÁS KORCSMÁROS • *Department of Genetics, Eötvös Loránd University, Budapest, Hungary*
INNA KUPERSTEIN • *Institut Curie, Paris, France; INSERM, U900, Paris, France; Mines ParisTech, Fontainebleau, France*
CAMILLE LAIBE • *EMBL Outstation–European Bioinformatics Institute, Cambridge, UK*

AIDAN MACNAMARA • *EMBL Outstation–European Bioinformatics Institute, Cambridge, UK*

PABLO PORRAS MILLÁN • *EMBL Outstation–European Bioinformatics Institute, Cambridge, UK*

NICOLAS LE NOVÈRE • *Babraham Institute, Cambridge, Cambridgeshire, UK*

MÁTÉ PÁLFY • *Department of Genetics, Eötvös Loránd University, Budapest, Hungary*

DANIEL ROVERA • *Institut Curie, Paris, France; INSERM, U900, Paris, France; Mines ParisTech, Fontainebleau, France*

JULIO SAEZ-RODRIGUEZ • *EMBL Outstation–European Bioinformatics Institute, Cambridge, UK*

THOMAS SCHLITT • *Department of Medical and Molecular Genetics, King's College London, London, UK*

MARIA VICTORIA SCHNEIDER • *The Genome Analysis Centre, Norwich Research Park, Norwich, UK*

PROTEOMICS SERVICES • *EMBL Outstation–European Bioinformatics Institute, Cambridge, UK*

LUCIA SIVILOTTI • *Department of Computing Science, University of Glasgow, Glasgow, Scotland*

GAUTIER STOLL • *Institut Curie, Paris, France; INSERM, U900, Paris, France; Mines ParisTech, Fontainebleau, France*

TIBOR VELLAI • *Department of Genetics, Eötvös Loránd University, Budapest, Hungary*

ANDREI ZINOVYEV • *Institut Curie, Paris, France; INSERM, U900, Paris, France; Mines ParisTech, Fontainebleau, France*

Chapter 1

Defining Systems Biology: A Brief Overview of the Term and Field

Maria Victoria Schneider

Abstract

Here we provide a broad overview of the definition of the term "systems biology" as well as pinpoint specific events in biological research and beyond that are consistently cited to have contributed and led to the current science of in silico systems biology. Since there have been many reviews and historical accounts describing the term, it would be impossible to include all single references. However, we do attempt to provide a consensus vision of how the field has evolved and consequently the terminology that followed it. We also highlight the development and general acceptance, and use, of standards for model representations as being crucial to the continued success of the in silico systems biology field.

Key words Systems biology, In silico systems biology, Omics, SBML, Modeling

1 Introduction

Several reviews, specifically describing the semantic meaning and historical use of the term "systems biology" have been written. There is a general consensus that the term "systems biology" was first coined in the 1960s, when theoretical biologists began creating computer-run mathematical models of biological systems. Noble [1] is cited as the first to use the term in this context. Prior to Noble's connotation of the term, Hodgkin and Huxley [2], published a series of papers which described the initiation and propagation of a nerve impulse, using a set of Ordinary Differential Equations (ODEs), for which they were awarded the Nobel Prize in 1963. This mathematical approach to describe biological systems was then adopted by others to reconstruct the function of other body parts (e.g., reconstructing the electrical functioning of the heart [1]). The work of Harvey and Mendel has also been cited as having roots in this field [3, 4]. The actual origins of the term are a matter of debate amongst scientists. For instance, some regard it as originating from Claude Bernard about 150 years ago, or Norbert Wiener and Erwin Schrödinger, around 90 years ago [5].

Maria Victoria Schneider (ed.), *In Silico Systems Biology*, Methods in Molecular Biology, vol. 1021,
DOI 10.1007/978-1-62703-450-0_1,

Others will cite the development of the concept of homeostasis by Walter B. Cannon, who built upon Bernard's homeostasis concept, at the turn of the past century as a key factor towards the development of systems biology [6–8].

Mathematical models were extensively used to understand biological processes in life science, making their use in the field of systems biology an actual continuation of a long tradition as seen in many fields like genetics, physiology, biochemistry, evolutionary biology and ecology. Noble [9] still underlines the importance of "mathematical insight" being a compulsory component of the available and increasing computational power in order to develop the necessary theoretical framework required to deal with multi-level interactions.

However, regardless of a strong start the field did not prosper during the second half of the twentieth century. A leap forward took place at the beginning of the 90s when high-throughput tools, developed for the sequencing of the human genome, brought experimental scientists up to the speed of theoretical biologists. The widespread use of the Internet also made possible, for the first time, international collaborations and the sharing of the huge amounts of data.

"In silico Systems biology" herein refers to the actual modeling of the parts at the systems level [10]. The late 80s and the whole 90s saw a large influx of biological data largely driven by the human genome project [11, 12]. The need to actually decipher the information gained from this project went beyond pure genomics, leading to efforts in many related disciplines such as computational biology, therein systems biology. Advances in systems biology have also been driven by the assumed underlying predictive factor that models of systems would provide, therefore giving significant insights into the functionality of the systems as a whole, as well as an understanding of the individual different parts. These would also extend across species, since models from different species can be used to predict behavior of similar systems in humans which in turn can be applied to develop new medical remedies. The term "Systems Biology" has been defined and used also in the context (or even as a synonym) for the suffix-term OMICS (a suffix indicating the measurement of the entire set of a given level of biological molecules and information). There are several Omics [13], most commonly known are genomics, transcriptomics, proteomics and metabolomics. In silico systems biology as intended in this book relies on the definition of systems biology in terms of modeling and simulations of biological processes. Ultimately, the goal of developing the various omics technologies is to combine the data into iterative models in order to understand how the elements and their interactions together give rise to emergent properties of a system [14, 15].

The aim of modeling a biological system is to gain sufficient understanding such that the behavior of the system can be predicted. Simulations are utilized as a form of validation for virtual experiments, saving time and resources.

2 Standards Matter

A truly important advancement in the field of in silico systems biology was the understanding that a standard method for encoding a model would encourage interoperability between different modeling strategies. This could only be achieved with the set of clear standards. No generally accepted standards existed for developing models of biological systems prior to 2000. Models were developed according to individual tastes and trends within certain fields. Generally, existing models were specific with only their respective field in mind. The development of any standard needed to be versatile enough to accommodate different fields. The idea behind this being that by using these standards, established existing models can be integrated in order to develop larger more comprehensive ones. A huge step forwards in standardized modeling came with the adoption of the Systems Biology Markup Language [16, 17] and Cell Markup Language [18]; both were developed as XML based [19]. However, a hurdle that still needs to be overcome is that most applications store important data in application specific annotations [20].

There are several repositories that contain models of various formats including SBML (e.g., BioModels.net) and KEGG (Kyoto Encyclopedia on Genes and Genomes) (*see* **Note 1**). However, recently (*see* **Note 2** for link to the INBIOMEDconsortium) it was underlined how the challenge remains when it comes to reproducibility, reliability and verification of computational methods and their use. Evaluating algorithms/models in an unbiased manner is severely time consuming, journals themselves struggle with the issue of how to verify computational results, since ultimately this relies on being able to download and test the software used as well as the data.

3 In Silico Systems Biology: All Around Modeling

The properties of systems are the result of two important characteristics: (1) systems have a hierarchical structure, and (2) the structure is held together by numerous links to construct very complex networks. Systems were recognized by Woodger [21] as a hierarchy of organization. Higher organisms (most at least) start their life cycle as single cells, developing as multiple and diverse cells. This happens with a specific spatial distribution and temporal order. Likewise, tissues are organized complexes of cells. The recognition that many systems were constructed from hierarchies of organization represented an important advance in this field [22–24].

In the study of systems, one often starts with modeling of the individual components (e.g., protein networks, signal transduction pathways). Models are initially used in descriptive manner (e.g., involving an equation that shows the relationships between the proteins in the cell). Subsequently, graphical models visualizing the cell (often as very complicated flow charts or webs) are generated and the relationships between the participants shown using distance or color. The empirical part then starts by using model systems (such as yeast) to explore what happens at the organism level when the participants are disturbed. Often however, genetic perturbations (e.g., knock out genes) or environmental perturbations (e.g., give or take away certain kind of sugars) lead to observed data that cannot be explained by the actual model. Thus, new hypothesis are formulated to justify such discrepancies and the cycle starts again. Of course, the insights and advances achieved in studying yeast models can be extrapolated to better understand human cell models for example (e.g., scaling up the model for simple systems), bringing comparative genomics (*see* **Note 3**) as a powerful tool in systems biology. There seems to be a big favorite in the field when it comes to modeling approaches, led by differential equations (e.g., using a series of differential equations where parameters are chosen in order to predict the behavior of the system). There are of course other approaches, which would also reflect a better representation of a biological scenario (e.g., on/off switch process, where the best representation is a binary system instead of a differential equation, more adequate when modeling diffusion dependent processes for example). There are several reviews on modeling formalisms and the comparison of the features and their integration (e.g., Machado et al. [25]). Uhrmacher et al. [26] also discussed multi-level models, as opposed to most models in Systems Biology which can be located within the space that is spanned by three dimensions of modeling: continuous and discrete; quantitative and qualitative; stochastic and deterministic. Many modeling approaches are hybrid as they combine continuous and discrete, quantitative and qualitative, stochastic and deterministic aspects [27, 28].

An approach that has been used in in silico systems biology is to use a series of smaller models that overlap with each other to represent the biological tasks of a complex system [29]. How to translate this into the bench later is far from trivial. In order to be able to integrate experimentally the various levels of information (e.g., omics) it is vital that all the "omics" are integrated from the start to ensure the experimental design and variables measured are done so coherently as to allow real integration of data derived from a variety of experiments and labs. Merging data is one of the crucial steps researchers are striving for. Ideally, if all were to use the same standards as well as actually submit the data to (open) repositories, merging could be achieved by several (simple) databases pointing to the same experimental design. However, a critical flaw when it comes to integrating data (across the Omics) does not derive from the computational technical aspects, but from trying to integrate data from experiments that had no integrated design (e.g., trying to integrate data at the genomic level from study a with proteomics level from study b). The future seems to bring biology, computational science closer with nanotechnology (*see* **Note 4**) and microfluidics (*see* **Note 5**). Such technologies have the potential to enable researchers to measure small amounts of material in parallel, at the single-molecule level (*see* **Note 6**). Now that a common language for systems biology has been developed and is being accepted by the community, there is a concrete pathway where data sharing could truly happen. However, this will require not only those working in silico (e.g., the scientists that feel comfortable looking at data, writing programs, some numerics) but also those that know about the disease and environmental changes, that can truly aid the actual interpretation of what is observed, all to work together and make an effort to use and apply the standards developed.

4 Systems Biology: Significant Events and Where It's Heading

When looking at the historical developments around and leading to the generally recognized field of systems biology one could define two phases, one prior to the age of genomics and large-scale biology, another one starting from the advent of the technological developments in the actual production of data to the current dawn of High Throughput technologies. Figure 1 shows a time line pinpointing (some) of the significant events that contributed to the development of this field and its evolution in time till 2002 marked by Barabasi's contributions, based on several historical reviews written about this. The previous decade showed the foundation of new Institutes devoted to the study and research in systems biology (*see* **Note 7**). The recognition of the maturity of the field does not only transpire from the proliferation of actual institutions, departments, groups named after but also through the last 10 years of publications in terms of books (including text books) now available as well as trends at funding levels (Le Novère per comm.). The current pressures on the field now lie in how applicable the advances from in silico systems biology are towards the empirical research and contributions into revealing mechanisms and processes at a variety of levels: molecular, cellular and physiological. Yarden & Pines [30] pointed out the lack of numbers when it comes to examples of deeper molecular understanding of the

- 1885 • Claude Bernard, theory of the permanence of the milieu intérieur (later called homeostasis) due to integrated regulatory mechanisms
- 1926 • Smuts, criticise reductionism
- 1929 • Woodger, recognized that systems are a hierarchy of organization.
- 1932 • Cannon, recognized feedback controls and termed the process homeostasis.
- 1940 • Weiss criticise reductionism
- 1943 • Cori and Green, Protein kinases and phosphatases were identified first in animals
- 1948 • Weiner's initiating cybernetics and raising the profile of negative feedback owed much to Cannon (1932)
- 1950 • von Bertallanfy suggested all systems shared the common property of being composed of interlinked component underlying similarities in detailed structure and control design.
- 1952 • Hodgkin and Huxley integrated equations for the nerve impulse
- 1954 • Burnett and Kennedy, identified Protein kinases and phosphatases in animals
- 1956 • Williams publications ended mechanistically belief illustrating individual variations across a variety of levels
- 1956 • Umbarger as well as Yates and Pardee described molecular feedback
- 1965 • Toulmin and Goodfield, construction of clockworks leading to deterministic principles for living organisms
- 1966 • Crick noted that only in the cell are certain crucial nucleotide sequences constrained to act as a code. Marking where biology departs from physics and chemistry and enters into a systems perspective.
- 1968 • Polyani clarified the relationship between levels in a hierarchy, echoed in Weiss, 1973
- 1969 • Kuo and Greengard, detected the first second messenger kinase; Leake and Whyte independently recognised that systems are a hierarchy of organisation
- 1972 • Bateson pointed to the two-way interaction between genes and environment that had earlier been established by Schmalhausen(1949), Waddington (1953, 1957), and later by Rendel (1967)
- 1973 • Kacser and Burns developed unambiguous methods for measuring the control exerted by any particular enzyme.
- 1973 • Weiss noted much greater variation at lower levels output of individual pathways is more ordered within a system than expected from random operation of those pathways outside the system.
- 1977 • Waddington (1977) described manipulation of systems behaviour can only be sensibly accomplished by modifying many steps (recalls Bateson (1972)
- 1989 • van Roon et al., noted that protein composition varied
- 1990 • Ko et al., 1990 corroborates van Roon et al finding: protein composition does vary!
- 2002 • Barabasi and Buchanan independently, brought recognition that a common stable systems structure is represented by hubs and connectors.

Fig. 1 Illustrative chronological list of events and key researchers considered to have influenced and shaped the field of systems biology research

underlying pathogenesis to improve treatment of patients with cancer. They suggest the need for a constant dialogue between basic research and medical oncology to reach a sustained pipeline of novel drugs and ways to overcome acquired treatment resistance in patients under a systems biology framework. Trautman and Sekali [31] refer to systems biology as an emerging and promising research strategy that can be applied to vaccine development and the identification of new mechanisms and predictors of inactivated vaccine immunogenicity. Although the holistic idea, conceptual framework and application of mathematical modeling and simulation for representation and predicting living systems is not new, in terms of the actual impact in application of the discoveries in the field, systems biology is considered still a young field (InBIomedvision consortium *see* **Note 2**).

The last decade shows a clear active attempt to bring into real applications the findings and results obtained at the modeling level. Expected benefits from systems biology research include faster routes towards candidates for new drugs [32, 33], better diagnostics for diseases of animals and plants [34], increased ability to "design" products such as bio-compatible materials for bio-fuels [35, 36] and healthier foods [37].

Finally, there is a strong trend and potential future investments from research funding programs towards synthetic biology, intended here as both the re-design and fabrication of existing biological systems as well as the design and fabrication of biological components and systems that do not already exist in the natural world. Obviously, the advances in the field of in silico systems biology (whose focus has been on natural systems), would benefit from the application of engineering to study how to build artificial biological systems. One could not exclude synthetic biology from contributing to in silico systems biology, by providing insights into the ways parts of natural biological systems works, characterizing and simplifying them and using them as a component of a highly unnatural, engineered, biological system. Synthetic biology provides a complementary perspective from which to consider, analyse, and ultimately understand the living world.

5 The Next Chapters

This book reflects the topics and contents originally brought together for a practical course on in silico systems biology. The books also expands on topics and aspects not covered in the course, keeping always in mind to offer the information at a practical level with the idea to complement the several excellent resources that cover the concepts and background information of this field [38, 39].

5.1 Part I: Network Reconstruction and Visualization

This section presents three chapters. Chapter 2, from Schlitt provides an overview of the approaches to modeling gene regulatory networks, including examples of the use of different data sources. In Chapter 3, Bauer illustrates how to analyse the functional effect of sequence variations in the context of biological networks such as protein–protein interaction networks and signaling pathways by using Cytoscape. The focus on the practical aspect of the book is apparent by the use of a step-by-step case study examples by the authors. Porras (Chapter 4) provides a practical overview of many different tools and approaches used to build, represent and analyse biological networks by illustrating the full process using a practical example.

5.2 Part II: Network Analysis & Logical Modeling

This section starts with a contribution from MacNamara and Saez-Rodriguez, covering the overview of the various approaches most commonly used for modeling cell signaling network, using as illustrative example the MAPK cascade (Chapter 5). Cohen et al. provide a complete overview of Chapter 6, on how to get from a formulated biological hypothesis to the construction of a mathematical model. Bonnet et al. (Chapter 7) illustrate how to use a specific software package: Biological Network Manager software (BiNoM).

5.3 Part III: Mechanistic Modeling & Stochastic Simulations

This section starts with a contribution in Chapter 8 from Le Novère and Endler explaining how to use chemical kinetics to model biochemical pathways. Chaper 9, by Golightly & Gillespie provides an introduction to Stochastic simulations. Finally Chapter 10 presents and describes the BioModels Database as a repository of mathematical models of biological processes.

5.4 Part IV: Encoding Syntax & Semantics

This section is devoted to the so important advances in standards! Two chapters present SBML as a model exchange format in software applications (Chapter 11 by Keating & Le Novere and Chapter 12 by Juty et al.)

5.5 Part V: Bayesian Modeling in Systems Biology

One single (Chapter 13) but crucial contribution is made here on a Bayesian System Identification for Biological Pathway Modeling by Calderhead et al.

5.6 Part VI: Pathways and Data Integration Workflows

In Chapter 14, Croft very descriptively presents how to build models using Reactome pathways as templates. This is followed by Chapter 15, which illustrates using a metazoan signaling pathway to predict novel components by uniform curation protocol (by Pálfy et al.). The book concludes with a chapter from Jimenez and Corpas. This chapter aims to show, using a step by step pragmatic example, how to create and expand workflows and use web services. This is done in such a way as to be easily understandable by non-computational scientists or non-bioinformaticians.

It is foreseeable that this book will evolve in years to come by increasing the number of chapters covering new approaches but also illustrating case studies where the concepts herein presented have been applied. Sharing experiences, know-how, and actual workflows through case studies as presented in this book, remain effective ways to disseminate valuable information, which so many researchers have, but cannot share at a peer review level, as well as enabling many others to effectively choose and apply such knowledge.

6 Notes

1. URLs of three main repositories of models (formats including SBML and CellML):
   BioModels.net: http://biomodels.net/
   KEGG: http://www.genome.jp/kegg/
   CellML.org repository: http://www.cellml.org/
2. INBIOMEDvision: http://www.inbiomedvision.eu/index.html, particularly relevant this document entitled "Strategic Report for Translational Systems Biology and Bioinformatics in the European Union," link: http://www.inbiomedvision.eu/PDF/Report-TranslationalBioinformatics-FINAL.pdf
3. Comparative genomics defined herein as the ability to learn about complex systems by modeling simpler systems that have similar genetics.
4. Nanotechnology involves manipulating molecules smaller than 100 nm—the scale of viruses.
5. Microfluidics, commonly used in ink-jet printing, uses pumps and valves to transport nanoliter volumes of fluids through microchannnels in a tiny glass or plastic chip.
6. Emerging and evolving technologies are increasing in this area. Single cell sequencing high throughput techniques have a big potential, recently a technique called Strand-seq, which works by labeling newly synthesized DNA during replication selectively damaging these new strands in the daughter cells, so only the original parental strand gets sequenced has just been published [40].
7. Some of the major Institutes devoted to systems biology and their urls:
   - Institute for Quantitative Systems Biology at the University of Washington (Seattle, WA); https://www.systemsbiology.org/
   - The Systems Biology Institute in Japan http://sbi.jp/aboutSBI.htm
   - Institute for Genomics and Systems Biology, USA http://www.igsb.anl.gov/Chicago

- Institute of Molecular, Cell and Systems Biology (2010) in Scotland, Berlin Institute for Medical Systems Biology—MDC http://www.mdc-berlin.de/en/bimsb/index.html
- ETH—IMSB Institute of Molecular Systems Biology Note http://www.imsb.ethz.ch/
- The Simons Center for Systems Biology http://www.sns.ias.edu/csb
- Department of Systems Biology @HMS: https://sysbio.med.harvard.edu/
- FAS Center for Systems Biology: http://sysbio.harvard.edu/csb/and many more, see http://en.wikipedia.org/wiki/List_of_systems_biology_research_groups

Acknowledgement

I am grateful to Boye Gricar for useful comments on earlier and final versions of this chapter. Thanks to Nicolas Le Novère for inspiration. Big thanks to Jacqueline Dreyer, Nick Juty and Julio Saez-Rodriguez for useful comments to the final version of this chapter.

References

1. Noble D (1960) Cardiac action and pacemaker potentials based on the Hodgkin-Huxley equations. Nature 188:495–497
2. Hodgkin AL, Huxley AF (1952) A quantitative description of membrane current and its application to conduct and excitation in nerve. J Physiol II7:500–544
3. Auffray C, Noble D (2009) Origins of systems biology in William Harvey's masterpiece on the movement of the heart and the blood in animals. Int J Mol Sci 10:1658–1669
4. Auffray C, Imbeaud S, Roux-Rouquie M, Hood L (2003) From functional genomics to systems biology: concepts and practices. C R Biol 326:879–892
5. Saks V, Monge C, Guzun R (2009) Philosophical basis and some historical aspects of systems biology: from Hegel to Noble—applications for bioenergetic research. Int J Mol Sci 10 (3):1161–1192
6. Csete ME, Doyle JC (2002) Reverse engineering of biological complexity. Science 295:1664–1669
7. Cannon WB (1941) The body physiologic and the body politic. Science 93:1–10
8. Joyner MJ, Pedersen BK (2011) Ten questions about systems biology. J Physiol 589:1017–1030
9. Noble D (2010) Biophysics and systems biology. Philos Trans R Soc A 2010(368): 1125–1139
10. Selinger DW, Wright MA, Church GM (2003) On the complete determination of biological systems. Trends Biotechnol 21(6):251–254
11. Collins FS, Morgan M, Patrinos A (2003) The human genome project: lessons from large-scale biology. Science 300:286
12. Frazier ME, Johnson GM, Thomassen DG, Oliver CE, Patrinos A (2003) Realizing the potential of the genome revolution: the genomes to life program. Science 300:290
13. Schneider MV, Orchard S (2011) Omics technologies, data and bioinformatics principles. Methods Mol Biol 719:3–30
14. Jamers A, Blust R, De Coen W (2009) Omics in algae: paving the way for a systems biological understanding of algal stress phenomena? Aquat Toxicol 92(3):114–121
15. Gopalacharyulu PV, Lindfors E, Bounsaythip C, Kivioja T, Yetukuri L, Hollmén J, Orešic M (2005) Data integration and visualization

system for enabling conceptual biology. Bioinformatics 21(suppl 1):i177–i185
16. Hucka M, Finney A, Sauro HM, Bolouri H, Doyle JC, Kitano H, Arkin AP, Bornstein BJ, Bray D, Cornish-Bowden A, Cuellar AA, Dronov S, Gilles ED, Ginkel M, Gor V, Goryanin II, Hedley WJ, Hodgman TC, Hofmeyr J-H, Hunter PJ, Juty NS, Kasberger JL, Kremling A, Kummer U, Le Novère N, Loew LM, Lucio D, Mendes P, Minch E, Mjolsness ED, Nakayama Y, Nelson MR, Nielsen PF, Sakurada T, Schaff JC, Shapiro BE, Shimizu TS, Spence HD, Stelling J, Takahashi K, Tomita M, Wagner J, Wang J (2003) The systems biology markup language (SBML): a medium for representation and exchange of biochemical network models. Bioinformatics 19(4):524–531
17. Finney A, Hucka M (2003) Systems biology markup language: level 2 and beyond. Biochem Soc Trans 31(6):1472–1473
18. Cuellar AA, Lloyd CM, Nielsen PF, Bullivant DP, Nickerson DP, Hunter PJ (2003) An overview of CellML 1.1, a biological model description language. Simulation 79(12):740–747
19. Hedley WJ, Nielsen PMF, Hunter Seattle PJ (2000) XML languages for describing biological models. In: Proceeding of BMES 2000 (Biomedical Engineering Society Annual Meeting)
20. Joyce AR, Palsson BØ (2006) The model organism as a system: integrating 'omics' data sets. Nature 7:198–210
21. Woodger JH (1929) Biological principles: a critical study, 2nd edn. Routledge, London
22. Whyte LL, Wilson AG, Wilson D (eds) (1969) Hierarchical structures. American Elsevier, New York
23. Pattee HH (1973) Hierarchy theory: the challenge of complex systems. Brazilier, New York
24. Trewavas A (2006) A brief history of systems biology. Plant Cell 18(10):2420–2430
25. Machado D et al (2011) Modelling formalisms in systems biology. AMB Exp 1:45
26. Adelinde M Uhrmacher1, Daniela Degenring1, and Bernard Zeigler2 C Priami et al. (eds) (2005) Transactions on Computational Systems Biology, 3380:66–89
27. Bosl WL (2007) Systems biology by the rules: hybrid intelligent systems for pathway modeling and discovery. BMC Syst Biol 1:13
28. Fisher J, Piterman N (2010) The executable pathway to biological networks. Brief Funct Genomics 9(1):79–92
29. Laszlo A, Krippner S (1998) Systems theories: their origins, foundations, and development, ch. 3. In: Jordan JS (ed) Systems theories and a priori aspects of perception. Elsevier Science, Amsterdam, pp 47–74
30. Yarden Y, Pines G (2012) The ERBB network: at last, cancer therapy meets systems biology. Nat Rev Cancer 12(8):553–563
31. Trautmann L, Sekaly RP (2011) Solving vaccine mysteries: a systems biology perspective. Nat Immunol 12(8):729–731
32. Kitano H (2002) Computational systems biology. Nature 420(6912):206–210
33. Liu Y-Y, Slotine J-J, Baraba'si A-L (2011) Controllability of complex networks. Nature 473:167–173
34. Chen L-L, Chung W-C, Lin C-P, Kuo C-H (2012) Comparative analysis of gene content evolution in phytoplasmas and mycoplasmas. PLoS One 7(3):e34407. doi:10.1371/journal.pone.0034407
35. Mukhopadhyay A, Redding AM, Rutherford BJ, Keasling JD (2008) Importance of systems biology in engineering microbes for biofuel production. Curr Opin Biotechnol 19(3):228–234
36. de Jong B, Siewers V, Nielsen J (2012) Systems biology of yeast: enabling technology for development of cell factories for production of advanced biofuels. Curr Opin Biotechnol 23(4):624–630
37. Nookaew I, Gabrielsson BG, Holmäng A, Sandberg A-S, Nielsen J (2010) Identifying molecular effects of diet through systems biology: influence of herring diet on sterol metabolism and protein turnover in mice. PLoS One 5(8):e12361. doi:10.1371/journal.pone.0012361
38. Uri A (2006) An introduction to systems biology: design principles of biological circuits, 2nd edn. (Chapman & Hall/CRC Mathematical & Computational Biology). http://www.amazon.com/Introduction-Systems-Biology-Mathematical-Computational/dp/1584886420
39. Klipp E, Herwig R, Kowald A, Wierling C, Lehrach H (2005) Systems biology in practice: concepts, implementation and application. Wiley-VCH, Weinheim. ISBN 3-527-31078-9
40. Falconer E, Hills M, Naumann U, Poon SS, Chavez EA, Sanders AD, Zhao Y, Hirst M, Lansdorp PM (2012) DNA template strand sequencing of single-cells maps genomic rearrangements at high resolution. Nat Methods 9:1107–1112. doi:10.1038/nmeth.2206

Chapter 2

Approaches to Modeling Gene Regulatory Networks: A Gentle Introduction

Thomas Schlitt

Abstract

This chapter is split into two main sections; first, I will present an introduction to gene networks. Second, I will discuss various approaches to gene network modeling which will include some examples for using different data sources. Computational modeling has been used for many different biological systems and many approaches have been developed addressing the different needs posed by the different application fields. The modeling approaches presented here are not limited to gene regulatory networks and occasionally I will present other examples.

The material covered here is an update based on several previous publications by Thomas Schlitt and Alvis Brazma (FEBS Lett 579(8),1859–1866, 2005; Philos Trans R Soc Lond B Biol Sci 361(1467), 483–494, 2006; BMC Bioinformatics 8(suppl 6), S9, 2007) that formed the foundation for a lecture on gene regulatory networks at the *In Silico* Systems Biology workshop series at the European Bioinformatics Institute in Hinxton.

1 Introduction to Gene Regulatory Networks

The term *gene regulatory network* refers to regulatory relationships between genes. Genes are the units of heredity in living organisms and encode proteins or RNAs. On molecular level genes correspond to stretches of DNA. The activity of genes crucially depends on proteins, such as the transcription factors.

Unfortunately I cannot go here into the molecular detail that would be necessary to really understand transcription control, but there are excellent scientific textbooks on the topic such as [4]. I will describe here a very much simplified version of the transcription process.

Transcription factors are proteins that consist of a DNA binding domain and a transactivation domain. The DNA binding domain allows the transcription factors to recognize and bind specific stretches of DNA by recognizing a specific DNA sequence. These so-called binding sites are usually defined by short

Maria Victoria Schneider (ed.), *In Silico Systems Biology*, Methods in Molecular Biology, vol. 1021,
DOI 10.1007/978-1-62703-450-0_2, © Springer Science+Business Media, LLC 2013

(consensus) sequences that the DNA binding domain recognizes. A gene harbors a so-called promoter region which contains binding sites for the core transcription machinery and specific transcription factors. The core transcription machinery ensures that the mRNA is correctly made, while additional transcription factors determine the correct timing. Once a transcription factor has bound to the DNA it can modulate the activity of a nearby gene via its transactivation domain.

Let us look at a very simplified example (*see* Fig. 1a): one transcription factor might be enough to initiate the transcription of a gene. Once the transcription factor has bound to its binding site in the promoter its transactivation domain triggers the transcription machinery to produce the corresponding mRNA. This mRNA encodes a protein, it might be another transcription factor. Thus, after this transcription factor has been made, it will modulate the activity of its target genes. Many transcription factors control the activity of their own genes forming so-called feedback loops. This feedback might be inhibiting further transcription activity (negative feedback) or promoting transcription activity (positive feedback). By ignoring all genes not encoding transcription factors, we see that we can build a network of transcription factors. Obviously such a system is a crude abstraction of the real world: many proteins are necessary to modulate the activity of transcription factors, for example in response to different environmental conditions. And it is also obvious that many proteins are necessary to ensure the maintenance of the cell and the organism as a whole. A system as presented in Fig. 1a is a stark abstraction of the real world. Figure 1b illustrates the complexity and size of the eukaryotic transcriptional machinery. A large number of proteins working together in complexes are necessary to ensure the correct production of mRNA, which involves unwinding and opening the DNA double helix, production of the corresponding mRNA, and subsequent closing and winding up of the DNA helix. In addition to the core promoters there are regulatory elements that can be at more distant locations in relation to the gene, such as enhancers. These long range effects on the transcriptional activity of genes are thought to be due to the DNA folding back onto itself. Additional processes can be involved in the control of gene activity, for example large sections of a chromosome can be inaccessible depending on its acetylation and methylation status.

Living organisms, even relatively simple ones such as bacteria, contain a large number, often several thousands of genes. The activity of these genes needs to be tightly controlled and additional factors need to be taken into account. One example where tight control is essential is the control of cell cycle, whether a cell is growing or dividing (*see* Fig. 1c). For multicellular organisms factors such as cell identity—is it a liver cell or a kidney cell—are import as well as timing. It is important to keep in mind that at

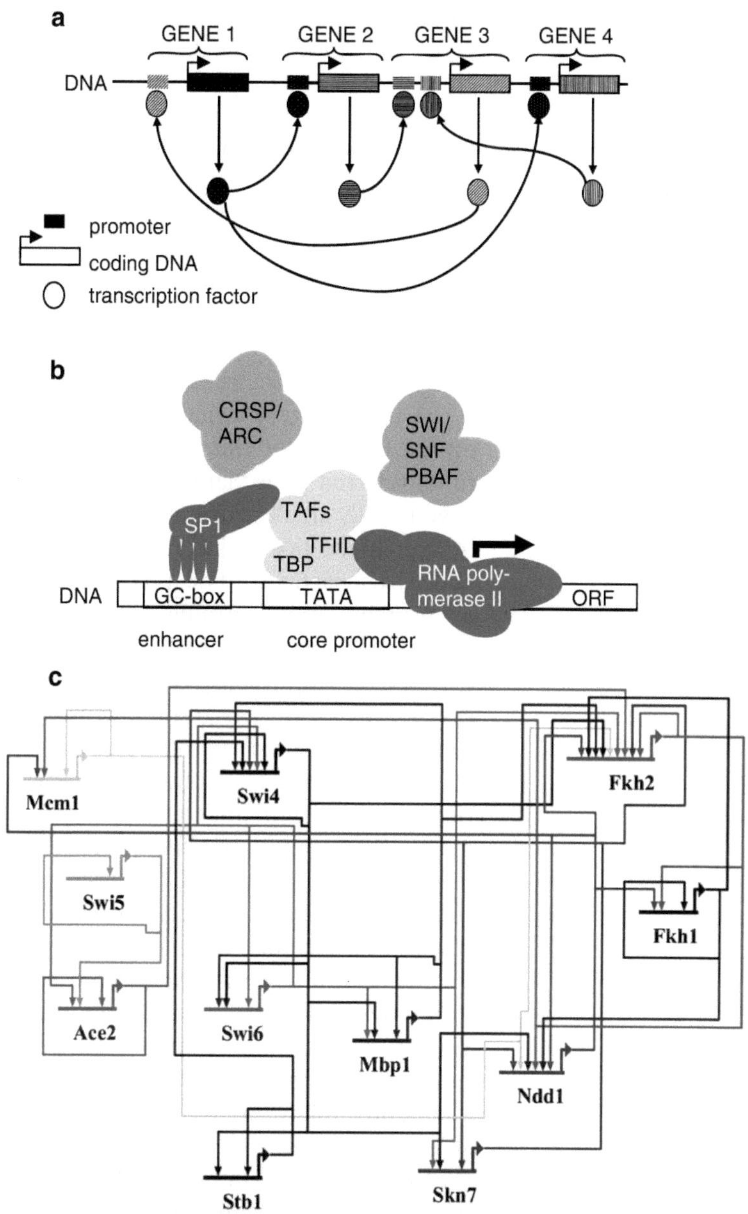

Fig. 1 (**a**) Simplified example for a gene regulatory network; it is assumed that all genes encode transcription factors and that one or two transcription factors are sufficient to control the activity of a gene; (**b**) the reality is more complex; each promoter involves a large number of proteins controlling the activity of RNA polymerase II (adapted from [112]). (**c**) Gene regulatory network model for yeast cell cycle control (adapted from [113])

each promoter represented in Fig. 1c, the whole transcriptional machinery described earlier (*see* Fig. 1b) needs to be active in a coordinated fashion involving probably hundreds of proteins.

While this complexity appears overwhelming, there have been early successes in understanding gene regulatory networks. Let us consider one example: the well-understood gene regulatory network of a simple model system: the bacteriophage lambda [5]. Bacteriophages are viruses that infect bacterial cells: lambda infects *Escherichia coli* cells. Upon infection the phage DNA is injected into the *E. coli* cell and a decision is made between two "life styles" [6]. The phage either integrates into the host genome and stays dormant, a process called *lysogeny.* Or, it triggers the production of new phage particles including the replication of its genome, ultimately leading to bursting the host cell and spreading of the new phage particles; this process is called *lysis*. If the conditions for the host cell deteriorate, lambda can switch from lysogeny to lysis, triggering the production of new virus particles ready to infect new host cells.

A series of very elegant experiments helped to understand the molecular basis of this complex decision making process, sometimes referred to as "lambda switch" [5, 7, 8]. Viruses and phages have usually small genomes; they are not independent organisms, but hitchhike host cells and modify their metabolism to produce new virus particles. The phage lambda genome encodes about 28 proteins, most forming parts of the phage capsule. There are some regulatory proteins that control the activity of other phage genes, most importantly the repressor *cI* and the activator *cro*. The decision between lysis and lysogeny crucially depends on the concentrations of these two molecules.

Modeling the gene regulatory network of phage lambda allows to study the system in detail and to predict the outcomes of experiments, for example, of gene deletions. The results of these predictions can be compared to the model and might elucidate shortcomings in our understanding of how the regulatory network works. Furthermore, if the model is in good agreement with experimental findings we might use the model to perform theoretical experiments that are not possible with the experimental system. We can, for example, compare the effects of all single gene deletions, double gene deletions, etc. We might as well be interested in asking how many different behaviors a particular network might be able to encode. Brazma et al. built a Finite State Linear Model (FSLM) of phage lambda [9, 10]. Without going into the details of the model I will only describe the major observations. The FSLM approach allows simulations of the model and it is possible to observe the systems' behavior, how the concentrations of the various molecules change over time. Using various parameter settings one can in principle observe two different kinds of dynamic "behaviour." One "behaviour" leads to the inactivation of all genes except lambda repressor *cI* after some initial burst of gene activity. This "behaviour" seems to correspond to lysogeny observed in phage lamba. The second "behaviour" leads to the inactivation of *cI* and a cascade of gene activations, corresponding to "lysis" in the phage lambda. Despite intensive simulations under many different parameter settings we were not able to discover a "behaviour" of the network model that was distinct from "lysis" and "lysogeny," suggesting that the network is very robust. But can we be sure that we did not miss a "behaviour?" And how many different behaviors can we obtain if we delete one gene, two genes, or manipulate some other parameters of the model?

A further problem that has had a lot of attention is the so-called "reverse engineering" of gene regulatory networks. Let's assume we work with a novel organism that has not been studied before. Given a list of genes and experimental measurements can we reconstruct the gene regulatory network? Obviously, this problem is far from trivial. While it might be possible to reconstruct "a" network, how do we know we found "the real" network? How do we know that the network predicted by us is complete and that it is the "correct" network? How many other networks would explain the data equally well and which experiments would allow us to identify the correct model?

Obviously these are important questions that can be addressed by the appropriate modeling system, but how do we choose the best modeling system?

Before getting into more detail, let us consider different approaches to gene network modeling that have been undertaken.

2 Approaches to Gene Network Modeling

What do we need to build a model for a gene regulatory network? In the first instance we need a list of the elements that make up the model. We could refer to this as a *parts list*. Actually, we could consider the parts list a, albeit very simplistic, model.

Once we obtained a parts list we need to elucidate the connections between the parts. We could call this the *architecture* or *topology* of the network. For a gene regulatory network we need to know which proteins control the activity of genes and which genes they control.

Often several proteins will influence the activity of a gene; we therefore need to find out how those proteins interact, if at all. Do several factors have to be there for the gene to be active (A "and" B) or is any one of the proteins sufficient (A "or" B). Which factors do act as activators or repressors? We call this level of description detail the "*logics* level."

So far, we did not consider if a gene is active or inactive. We just said "if A and B are present C is active," without making a statement on whether A or B actually are present. Therefore we also could not consider changes over time. Various *dynamic modeling* approaches have been developed that allow us to do precisely this to varying level of detail. Boolean networks allow only binary representations of gene activities, the states of the genes. Genes can be either "active" or "inactive"; time is represented by discrete time steps. It is obvious that these limitations have an impact on the level of detail we can obtain from a model. Differential equation models, on the other hand, represent continuous time and concentration changes with much more detail. So why not just always build a differential equation model? Because sometimes the level of detail is not necessary and often we do not have sufficient experimental measurements to fit all the parameters of a very detailed model. Therefore it is useful to explore a range of modeling approaches and choose the most appropriate depending on the application.

2.1 Parts List: Genes, Transcription Factors, Promoters, Binding Sites, . . .

Compiling a parts list is probably the first step for building any model in molecular biology. Often this is done implicitly, the parts list is extended when the model is growing. However, I believe that parts lists can be a valuable tool in understanding biological processes. It is useful to explicitly compile parts lists when building a model and to collect relevant information about the parts in an organized manner.

Such a parts list might, for example, be the result of a whole genome sequencing project where we subsequently annotate all open-reading frames and map them to genes already known in other organisms. Given such parts list we could start to compare between organisms and ask questions such as “How many genes are there?” or “Are there systematic differences in the distribution of gene functions between different organisms?” We can compare parts lists for a large number of different organisms and therefore derive information about evolutionary and environmental processes that might drive the functional composition of the genomes.

The comparison of parts lists from different organisms can be used to predict the presence or absence of particular gene regulatory elements, as well as signaling and metabolic pathways, see for example [11–13]. Obviously it is not trivial to find the corresponding proteins between different organisms, esp. if they are evolutionarily far apart. Confounding factors are gene loss and gain, for example, due to genome duplications, so that there will not be a 1:1 mapping between the different organisms [14].

In summary, parts lists provide a first impression of gene networks in different organisms and they are necessary before we continue to have a look at the network topology.

2.2 Architecture: A Graph Depicting the Connections of the Parts

Once we have established a parts list we are interested in describing the links between the different network elements. In gene regulatory networks we are, for example, interested which transcription factors regulate which genes. One indication for a regulatory relationship is the existence of a binding site for a transcription factor in the promoter region of a particular gene. Graphs are used to represent these relationships. Graph theory is a discipline within mathematics and computer science; its origins can be traced back to Leonhard Euler and his work on the walk across the Seven Bridges of Königsberg [15]. He found a mathematical proof showing that it is not possible to find a walk that crosses all seven bridges without

a graphical representation

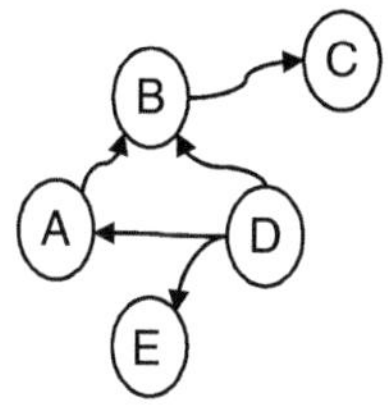

b edge list

A	B
B	C
D	E
D	B
D	A

c adjacency list

A	B
B	C
D	A, B, E

d adjacency matrix

	A	B	C	D	E
A	0	1	0	0	0
B	0	0	1	1	0
C	0	0	0	0	0
D	1	1	0	0	1
E	0	0	0	0	0

Fig. 2 Data structures for graphs. For the network depicted in (**a**) we show the corresponding edge list (**b**), the corresponding adjacency list (**c**), and the corresponding adjacency matrix (**d**)

using at least one bridge twice. Graphs allow the abstract representation of pairwise relationships between objects (or nodes) by edges (pairs of nodes). In graphical representations of graphs, the nodes are often represented by circles and the edges by lines (Fig. 2a). For the Königsberg problem, we could represent the different locations on the walk by the nodes and the bridges connecting the locations by edges. For gene regulatory networks genes are represented by nodes and regulatory relationships between the genes are represented by edges. The regulatory relationships are often directional, e.g., a particular transcription factor activates a particular gene but not vice versa. These directed relationships are usually represented by directed edges, sometimes called arcs (represented by arrows in graphical representations of graphs). The relationships might represent measurements of interaction strengths; in a graph these could be represented by weighted edges, where each edge is assigned a weight.

Several data structures are commonly used to store graphs in a computer. Graphs can be represented by a list of node pairs, for example an edge between nodes A and B would correspond to the pair of nodes (A, B). For undirected graphs we assume the order of the nodes in a node pair to be irrelevant, whereas for directed graphs the order of the nodes in a node pair is important. A simple data structure to represent a graph would be a table with two columns, column 1 containing the first node of an edge, column 2 containing the second node (*see* Fig. 2b). A weighted graph could be stored as a table with three columns, where the additional column would contain the weight. Other graph representations are commonly used, depending on the nature of the graph, for example the adjacency list, where a node in column 1 is followed by a list of all nodes connected to this node in column 2, *see* Fig. 2c. We can also use a matrix to represent a graph, where the columns c and rows r of the matrix represent the nodes and a matrix element $m(r_n,c_m)$ is 1 if there is an edge between the nodes and 0 if not (*see* Fig. 2d). To represent a weighted graph the matrix could contain real numbers instead of 0s and 1s. If 0 is a possible weight, then we would have to introduce an additional element NA that represents missing edges. High-throughput experiments often lead to a matrix summarizing the results of the experiment, such as gene expression microarray experiments lead to a gene (co-) expression matrix. This matrix can be converted in a graph representation either by applying a strict threshold above which genes are considered to be expressed, or by more sophisticated methods including the weighted network analysis method by Horvath et al. [16, 17] for which an R package is available [18].

There is no "best" representation of a graph; the optimal representation depends on the nature of the graph and the algorithm that we want to apply. Some graphs have very many edges and are almost fully connected; in this case the matrix representation might be memory efficient. If, however, the graph contains only a few edges relative to the number of nodes, adjacency lists are the better choice, since a matrix would mainly contain 0s. There are many tools that allow to analyze and visualize graphs [19]; one of the most popular tools for graph analysis in molecular biology is Cytoscape [20] for which a large number of specialized plugins exist (www.cytoscape.org).

2.2.1 Experimental Data

So, how do we obtain a graph representation for a gene regulatory network? Well, various experimental and computational methods allow to determine transcription factor binding sites in the genome. If we omit the distinction between genes and proteins, we can represent a gene regulatory network by using nodes to represent genes or their gene products (proteins) and directed edges to represent regulatory relationships. Only genes encoding transcription factors therefore would have outgoing edges, but all genes with known regulatory input would have incoming edges. If we want to distinguish between activators and repressors, we could use positive edge weights for activators while repressive interactions would be represented by edges with negative weights.

Transcription factor localizations can be identified experimentally. For example, individual binding sites can be detected using the DNAse I footprinting assay; proteins bound to the DNA protect it from degradation by DNAse I, therefore these regions can be analyzed further [21]. Another common experimental method is the "electrophoretic mobility shift assay" (EMSA) sometimes called "band shift assay" or "gel retardation assay"—DNA fragments that are bound by protein move slower in an electrophoretic gel than unbound fragments [22]. These methods allow fine mapping of individual binding sites, but are very labor intensive. High-throughput methods such as the ChIP-on-chip method allow the genome-wide detection of binding sites for a transcription factor [23], but the spatial resolution and signal quality is limited. Furthermore, assigning transcription factors to their target genes based on the genomic localization can be difficult due to the size of intragenic and intronic regions and long range effects of some transcription factors.

These experimentally determined sequences can be used to derive consensus sequences summarizing the sequences found to be bound by the transcription factor, for example as position specific scoring matrices (PSSMs) [24]. These consensus sequences are collected by databases such as Transfac [25] or JASPAR [26]. Computational methods have been developed allowing to screen genome sequences for putative transcription factor binding sites using PSSMs. To obtain a graph we would connect the genes encoding the transcription factors for which we used the PSSMs to their putative target genes. To decide if a regulatory relationship is an activation or repression we would need additional data.

Not for all transcription factors consensus binding sites are known, but it might be possible to identify short sequences that are overrepresented in the promoters of coexpressed genes by comparing those promoter sequences to the promoter sequences of all other genes [27]. This approach obviously depends on the availability of the sequences for many genes and their upstream regions. Such an approach was applied in a cell cycle study in *Schizosaccharomyces pombe*, where Rustici et al. showed that the presence or absence of consensus binding sites in the promoter regions corresponds to the cyclic expression pattern of the genes [28]. Genes with a peak expression at similar cell cycle stages often share similar sets of consensus binding sites.

However, the exact promoter regions are usually unknown and even the transcription start sites are only known for a few genes. For baker's yeast *Saccharomoyces cerevisiae* which has a relatively small genome with short intergenic regions, considering about 600–1,000 bp upstream of the translation start site (ATG) appears to be a good approximation for the promoter regions. In higher organisms like vertebrates the intergenic regions and thus the putative promoter regions are much larger than in yeast, therefore the identification of regulatory elements in the DNA sequence by computational means has turned out to be rather elusive. Some studies have focused on the computational analysis of higher-level organization of transcription factor binding sites in promoters, such as frequently occurring combinations of known binding sites [29], or restricted the search for regulatory elements to conserved sequence regions, identified by genome comparisons, a method often referred to as phylogenetic footprinting [30, 31]. However, phylogenetic footprinting does not always work, because the localization and the binding sites themselves are not always conserved.

While the existence of the binding site is necessary, this alone does not guarantee that the transcription factor will bind there—epigenetic modifications of the chromatin might make the binding site unaccessible or there might be steric competition with other DNA binding factors that block access to the binding site. Furthermore, many transcription factors need to be activated, for example by phosphorylation. Bansal and Califano developed a computational method to find post-translational interactions [32].

A very cost-efficient method to detect functionally related genes experimentally is to identify "synthetic lethals"—this method looks for double deletion mutants which are not viable. If single genes are deleted for example in baker's yeast, some deletions turn out to be lethal for the organisms. These genes are termed *essential genes*. Some deletions are not essential, but affect the cell, for example its appearance or ability to grow quickly, while deletions of other genes have no apparent effect. By mating yeast cells of strains carrying different deletions of non-essential genes it is possible to generate double deletion mutants. Again, it turns out that some cells survive without detectable effect, some are affected but survive, while some double deletions turn out to be lethal. The latter group is particularly interesting, because we know the single deletions were not essential, therefore, the double deletion, i.e., the combined effect of the two deleted genes, is responsible for the lethality. This could have various reasons, for example, the two genes are redundant; they have the same or similar function that is essential for the cell. If neither protein is present, the cell dies, but either protein is sufficient to ensure survival. By performing a large screen for synthetic lethal mutants in yeast Costanzo et al. were able to generate a large network of functionally related genes [33]. Unfortunately, it is challenging to interpret the biological meaning of the synthetic lethal network, as there might be many different reasons that lead to synthetic lethality. Furthermore, this method is more difficult to apply to multicellular organisms, where we can have tissue specific effects. Nonetheless, these data sets can provide an important resource when integrated with other types of experimental evidence described here.

We have considered different types of (large scale) experimental evidence that can be used to generate gene regulatory networks. Obviously an important approach for building gene regulatory networks is to read and collate evidence from literature, which can be of different nature and quality. For example Guelzim et al. compiled and analyzed a transcriptional regulatory network from literature for yeast [34].

Having compiled all these different data that can be represented as networks, why is it worthwhile to look at the network topology?

2.2.2 Why Study the Topology?

A large number of scientific publications have appeared since the mid-1990s examining the topology of many large networks of very different nature. These include the internet, the connections between the computers and servers as well as the connections (links) between websites, electrical power networks, social networks of humans as well as animals and various biological networks, to name a few. Intriguingly, all these networks seem to share topological features such as power-law degree distribution and small-world behavior. In a network with a perfect power-law degree distribution a log–log plot of the degree distribution follows a straight line over many orders of magnitude. Unfortunately for biological networks, the number of nodes is often limited. There are currently only about 22,000 known human genes, therefore in a human gene regulatory network we will not be able to generate a degree distribution plot for more than four orders of magnitude. It can be argued that for such "small" networks it is difficult to detect a power-law degree distribution confidently. However, it is obvious that the degree distribution of biological networks is different from "classical" Erdös and Rényi random networks where the degree distribution across the nodes follows a Poisson distribution [35]. We observe that in general in biological networks we find few nodes with very large degrees (often called "hubs") while most nodes have a very small degree. Another interesting observation is that in these networks the distance between any two nodes on average is smaller than in "classical" random networks. As for the reasons why these particular network structures arise is still the subject of active research and different hypotheses have been published. Here, we will not go into further detail but refer to the increasing literature on the topic (start for example with the review by de Silva [36]).

Nonetheless, this network structure has important consequences. Albert and Barabasi examined the consequences of removing either random nodes or highly connected nodes (hubs) on the average distance between all nodes and found that removing the hubs ("attack") had much more severe effects than removing random nodes ("failure"), thus this network structure seems to protect against random failure of network elements at the cost of increased sensitivity to attacks targeted against the hubs of the network [37, 38]. It has been speculated that biological networks are inherently modular and that these modules are reused during evolution [39]. Subsequently many groups developed methods for identifying modules or subnetworks in (biological) networks. Being able to identify modules would help in designing targeted experiments by minimizing side-effects from other modules [40]. When building a computational model of a biological system understanding the modular nature of the network would help to select the key parts of the biological system that are important for the functions of interest. Many methods to find network modules work "top-down" starting from the whole network and trying to break it down, e.g., [41, 42]. Uri Alon's group took a "bottom-up" approach. They started with enumerating all 13 possible connections between three nodes in a directed network. They then counted how often these network motifs occurred in networks of different origin. Comparing these results with the analysis of random networks allowed them to assess the complexity of connectivity in the original networks. They examined the number and size of potential feedback and feedforward loops in different networks and identified "network motifs" that can be seen as building blocks of complex networks [43, 44].

Genes that form hubs in the networks have been the focus of several studies. It appears that hubs in the protein network are more often encoded by essential genes than other proteins [45–51]. In 2004 Han et al. published a study where they examined the hubs in protein–protein interaction networks more closely [44]. Combining the interaction data with coexpression data they were able to identify two different groups of hubs: the party hubs and the date hubs [44]. The proteins directly connected to party hubs tend to be coexpressed, whereas the neighbours of date hubs are not. It appears that party hubs can interact with lots of proteins at the same time, whereas data hubs are able to bind to lots of different proteins sequentially. The main conclusion from this study is that we need to be aware that topological networks are a static representation of dynamical processes. Not all the interactions we see in a topological network are taking place at the same time, but might be active under different conditions, at different times, and at different localizations, within the organism but also within different cellular compartments. This does not only apply to the protein interaction network but also to gene regulatory networks, e.g., the localization transcription factors on the chromatin. The group of Rick Young performed chromatin-immunoprecipitation (ChIP) experiments for 203 transcription factors for yeast; 84 of these they tested under several different growth conditions. They found that some transcription factor binding patterns change considerably, depending on the growth conditions, suggesting a dynamic system of binding and releasing of transcription factors [52]. Some transcription factors bind their targets only under specific conditions (such as Msn2), others bind additional (Gcn4) or alternative targets (Ste12), whereas some show condition independent binding patterns (Leu3). Luscombe et al. tackled the dynamics of the topological networks by performing an analysis of a transcription factor localization network in combination with gene expression data for yeast [53]. By tracing back the putatively active transcription factors starting from actively expressed genes they found considerable differences of network activity depending on the particular conditions for the yeast culture [53].

We can therefore summarize here that studying the network topology has led to some important and at times surprising insights into the regulatory networks, despite its static nature. Elucidating the topology of the system one wants to study is an important step in deriving the scaffold for dynamic network models.

2.3 Logics: How Do Combinations of Regulatory Signals Interact?

Looking at a transcriptional regulatory network, we find that some genes have inputs from several transcription factors. We immediately see that in order to decide how the target gene behaves we need to know how the input signals are to be combined. For all genes in the model we need to define rules on how to map incoming signals to gene activity. Network models at this level of detail can probably be best described as "logical models." Please note, that at this level of description we still do not treat the networks as dynamic. All we compile is information of the format "if A and B are bound C is active." The group of Eric Davidson has studied the development of sea urchin in detail. In 1998 they published a study where they successfully dissected the interactions of transcription factors controlling the activity of one particular gene *Endo16* [54]. They were able to identify distinct seven modules in the upstream promoter region of *Endo16*. But more intriguingly they were able to express the rules governing the interactions between the transcription factors binding to the various modules by an "if-then-else" statement in the style of a programming language [54]. This stimulated further work into the analysis of the processes controlling the early development of sea urchin and culminated in a comprehensive representation of the underlying transcription regulatory network [55, 56]. To construct and analyze the transcription regulatory network the group generated a widely used software BioTapestry (www.biotapestry.org), an interactive tool for building, visualizing, and simulating genetic regulatory networks that allows to export models using the Systems Biology Markup Language (SBML) [57]. Logic network modeling has also been applied to signaling networks [58, 59] and the CellNetAnalyzer toolbox for Matlab is one of the tools that allow their analysis [60].

2.4 Dynamics: How Does It All Work in Real-Time?

Ultimately, we would like to obtain a model that describes changes of activities and concentrations over time. Various approaches have been developed and used to describe gene regulatory networks, see for example the review by de Jong [61].

2.4.1 Boolean Networks

Probably the most reductionist approach to gene regulatory networks are Boolean Network models, where genes can be in either of two states *on* or *off* and the state of a gene is determined by Boolean rules. A graphical representation often shows the genes as nodes and two types of directed edges (ending in an arrow for activating inputs or a perpendicular line for repressing inputs). Boolean networks can be represented in different ways, for example as a list of Boolean rules for all genes in the model, as a "truth" table mapping the current state to the following state (*see* Fig. 3). The state diagram shows the progression of the states and allows finding steady states and attractors (Fig. 3c); in this example there are two attractors, a steady state attractor (001) and a state cycle consisting of two alternating states (010–101). Probably the earliest work

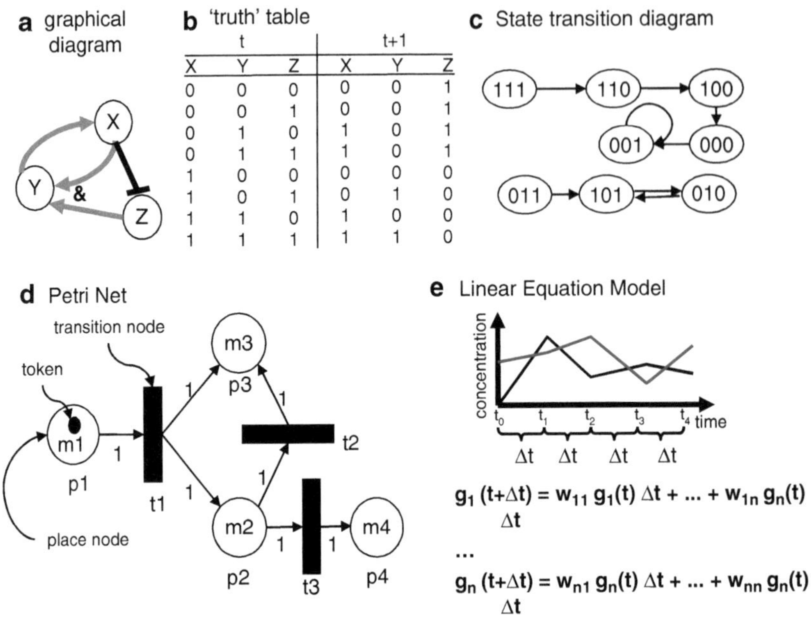

Fig. 3 (**a**) Graphical representation of a Boolean network model for three genes X, Y, and Z; (**b**) "truth" table describing how the next state will be determined by the current state; (**c**) state transition diagram represents the progression of states over time for the Boolean network depicted in (**a**); (**d**) in the Petri Net representation place nodes (*circles*) are denoted by *p*, transition nodes (*boxes*) by *t*, and token numbers by *m*. (**e**) linear equation model: concentration changes are approximated by linear equations; $g_1(t)$ concentration of gene g_1 at time *t*; Δt time difference between two time steps; w_{12} weighted influence of gene g_2 on concentration of gene g_1; $w < 0$ for repressors, $w > 0$ for activators, $w = 0$ if no influence

on transcription regulatory networks was published by Stuart Kauffman [62]. Lacking information on the structure of biological networks he worked with randomly generated Boolean networks and analyzed their properties using computational simulations [62]. In particular, he studied effect of varying the number of "genes" and inputs per gene on the lengths and number of state cycles. In his 1969 paper Kauffman speculated that the different state cycles in a network might correspond to steady states in biological networks representing different cellular differentiations [62]. Kauffman extended the study of regulatory networks and summarized his findings in an influential book [63]. Boolean networks remain an attractive option for modeling large biological networks, especially if most of the reaction kinetic constants are unknown [64–70]. BooleanNet, a software package for the simulation Boolean models of biological regulatory networks [71], is available from http://code.google.com/p/booleannet/.

An extension of Boolean networks are the asynchronous Boolean networks. While in the "traditional" Boolean networks all genes are updated at the same time (synchronously), asynchronous networks allow that genes are updated in a particular order or randomly at different times, see for example [72]. Rene Thomas developed a formalism for Boolean networks that distinguishes between the state and the image, where the image is initially a copy of the state vector that is updated as the individual genes are updated. Once all genes are updated the image vector becomes the new state vector and a new cycle begins [73, 74]. The group of Denis Thieffry are actively developing this approach [75] and provide with GINsim a graphical editor and simulation software for qualitative models of genetic regulatory networks [76] which is available from http://gin.univ-mrs.fr.

2.4.2 Petri Networks

Boolean networks use binary representations for gene activities and discrete time steps. However, some biological systems depend on the accumulation of biological substances, for example, the bacteriophage lambda lysis-lysogeny switch introduced earlier. A crucial step of the switch depends on the occupation of the so-called operator sites in the promoter regions of some genes by either *cro* or *cI*. *cI*, or lambda repressor, represses at low concentrations the production of *cro*; at higher concentrations it also represses the activity of its own gene in a negative feedback loop. Such a situation is not easily described in the Boolean formalism, as they do not generally represent different substrate concentrations. It would be possible to use several network elements to represent *cI* at different concentration levels, but this will make the model much less readable. If many processes in a biological system are concentration-dependent, it is therefore better to use a formalism that allows to represent different concentrations explicitly. Petri Nets provide such a formalism with discrete representations for both concentrations and time. Petri Nets were introduced by Carl Petri, originally to build models for chemical reactions [77].

Petri Nets consist of place nodes and transition nodes (Fig. 3d). In a model for biochemical reactions place nodes represent the molecules and transition nodes represent biochemical reactions. Concentrations or molecule counts are represented by marks or tokens on the place nodes; they represent the number of molecules of the respective molecule species. By convention in graphical representations place nodes are depicted by circles, transition nodes by boxes, and tokens as back dots. The transition rules define when a transition is taking place. For example, a particular transition rule might require one token as input and produces two different tokens as output (Fig. 3d). The transition rules therefore encode the stoichiometry of the biochemical reactions. Petri Net modeling approaches have been developed for over 40 years and there are plenty of tools to support building Petri Nets, a good overview provides the *Petri Nets World* website (http://www.informatik.uni-hamburg.de/TGI/PetriNets/). Similar to Boolean Networks, asynchronously acting transitions can add valuable insights compared to synchronous Petri Net models, but also add complexity in the analysis. There is a very readable book on Petri Net applications in Systems Biology [78]. In the same book

Banks et al. describe the modeling genetic regulatory networks using Petri Nets, more in particular, how to turn a Boolean network model into an asynchronous Petri Net model [79].

2.4.3 Linear Models

Moving from discrete to continuous representations of concentrations it is possible to use linear equations to represent gene regulatory networks [80, 81]. Assuming that every gene is influenced by the current level of activity of every other gene to some degree we use linear equations to describe concentration changes over time. To build a model we need to obtain suitable entries for the weight matrix, which correspond to the slopes of the linear functions. Positive entries mean activation and negative entries represent repression; the entries can be 0 if the expression of the corresponding genes is regulated independently. We can then express the activity of the gene at time $t + \Delta t$ by multiplying the concentrations at time t with the corresponding entries from the weight matrix (Fig. 3e). Linear equations are obviously an approximation to the concentration changes in real systems; however, it would be possible to choose a smaller Δt to achieve better approximation.

2.4.4 Differential Equation Models

In the extreme case, Δt is chosen to be infinitely small and that brings us into the realm of differential equation models. Differential equation models will be covered by other authors in this book in much more detail, therefore I will not go into much detail here. Using differential equation models, time and concentration changes are represented in a continuous manner; very detailed simulations are possible. Examples include the differential equation model by von Dassow et al. for the segment polarity network which is important for the correct formation of segments during the development of the *Drosophila melanogaster* fly [82] and the model for insulin-induced eukaryotic translation initiation by Lequieu et al. [83]. However, to achieve this level of detail one needs to obtain the correct values for many reaction constants and those values might not often be readily available. It is possible to estimate the best parameter values, but this might introduce a considerable source of error. Luckily there are methods that allow estimating the variability and uncertainty in ordinary differential equation models, see for example [84].

Systems of many differential equations are usually not solvable analytically; therefore one has to use numerical methods to find the solutions. In order to build differential equation models in Systems Biology there are some excellent tools available that support building the model as well as running simulations and finding numerical solutions such as Gepasi [85, 86], CellDesigner [87], and E-Cell [88].

2.4.5 Other Modeling Approaches

Other modeling approaches have been chosen to build models for particular systems. For example, McAdams et al. used an analogy to electrical circuits to build a model for lambda phage [89].

Bayesian approaches to build network models have been successful in predicting causal relationships between genes and have been applied to various biological systems [90–93]. They are a very attractive avenue, but often computationally expensive. Dana Pe'er has published a review that provides a good starting point to learn more about Bayesian network analysis [94].

2.4.6 Stochastic Models

So far, all modeling approaches we considered were deterministic. However, biological systems are often subject to stochastic behavior, especially if the number of molecules involved is small. Various adaptations to the previously described methods have been developed to study stochastic effects. For example, in stochastic Boolean networks, genes can randomly switch from *on* to *off* and vice versa [95]. Stochastic approaches to the modeling with Petri Nets have been developed as well [96–98]. McAdams et al. extended their electrical circuit model approach to include stochastic effects [99–101]. Stochastic simulations can be useful to study the properties of a range of networks if we lack the knowledge of biochemical details. For example, when designing a Petri Net model we might encounter transition nodes that compete for tokens at the same place node. It would therefore be necessary to decide, which transitions are treated preferentially, for example by examining binding constants for substrates to enzymes, but this information might not be available. Ruths et al. developed a stochastic approach where the transitions fire in random order [98]. They analyzed the overall behavior of a system after running large numbers of simulations [98]; an independent implementation of this method is available within Biolayout Express^{3D}, a software for visualization and analysis of network graphs, available at www.biolayout.org [102].

3 Conclusions

Many different biological systems are being studied and it is therefore not surprising that many different modeling approaches have been developed to capture and test our understanding of these biological systems. This introduction only scratches the surface of a dynamic area of research and only represents a personal view; nonetheless I hope it provides some orientation for newcomers to the exiting field of systems biology. There are excellent reviews covering the topics raised here in much more detail [36, 61, 66, 94, 103–110].

One cannot be a specialist in all methods and the task of identifying the best modeling approach for a particular system can be daunting. It is important to understand the principles of the modeling approaches and their requirements. There cannot be a one-size-fits-all approach, because the level of detail of knowledge differs vastly across the different biological systems. Even if we focus on gene regulatory processes only, there are well-studied transcription factors for which a wealth of data is available and others which have hardly been described. Picking the best model is rather an art than an exact science. But the development of methods and software tools is advancing rapidly; they are increasingly easier to use. Data exchange formats such as the SBML [111] help to migrate between tools and modeling approaches making it much less likely to get "locked in" to a particular modeling approach. This helps the biologists to keep their head free of the mathematical details and focus on the biological details.

Acknowledgements

I would like to thank Nicolas Le Novère, Julio Saez-Rodriguez, Vicky Schneider, and James Watson for organizing and running the excellent *In Silico* Systems Biology workshops. I would also like to thank my former supervisor Alvis Brazma for the introduction to gene network modeling and many interesting discussions.

References

1. Schlitt T, Brazma A (2005) Modelling gene networks at different organisational levels. FEBS Lett 579(8):1859–1866. doi:S0014-5793(05)00186-9[pii]10.1016/j.febslet.2005.01.073
2. Schlitt T, Brazma A (2006) Modelling in molecular biology: describing transcription regulatory networks at different scales. Philos Trans R Soc Lond B Biol Sci 361 (1467):483–494. doi:E686M06225352114 [pii]10.1098/rstb.2005.1806
3. Schlitt T, Brazma A (2007) Current approaches to gene regulatory network modelling. BMC Bioinformatics 8(suppl 6):S9. doi:1471-2105-8-S6-S9[pii]10.1186/1471-2105-8-S6-S9
4. Krebs JE (2009) Lewin's genes X, 10th edn. Jones and Bartlett, Burlington, MA
5. Ptashne M (1992) A genetic switch—phage lambda and higher organisms, 2nd edn. Cell Press & Blackwell Science, Oxford
6. Lederberg EM, Lederberg J (1953) Genetic studies of lysogenicity in Escherichia Coli. Genetics 38(1):51–64
7. Johnson AD, Poteete AR, Lauer G, Sauer RT, Ackers GK, Ptashne M (1981) Lambda repressor and cro–components of an efficient molecular switch. Nature 294(5838):217–223
8. St-Pierre F, Endy D (2008) Determination of cell fate selection during phage lambda infection. Proc Natl Acad Sci U S A 105 (52):20705–20710. doi:0808831105[pii] 10.1073/pnas.0808831105
9. Brazma A, Schlitt T (2003) Reverse engineering of gene regulatory networks: a finite state linear model. Genome Biol 4(6):P5
10. Ruklisa D, Brazma A, Viksna J (2005) Reconstruction of gene regulatory networks under the Finite State Linear Model. Genome Inform 16(2):225–236
11. Overbeek R, Begley T, Butler RM, Choudhuri JV, Chuang HY, Cohoon M, de Crecy-Lagard V, Diaz N, Disz T, Edwards R, Fonstein M, Frank ED, Gerdes S, Glass EM, Goesmann A, Hanson A, Iwata-Reuyl D, Jensen R, Jamshidi N, Krause L, Kubal M, Larsen N, Linke B, McHardy AC, Meyer F, Neuweger H, Olsen G, Olson R, Osterman A, Portnoy V, Pusch GD, Rodionov DA, Ruckert C, Steiner J, Stevens R, Thiele I, Vassieva O, Ye Y, Zagnitko O, Vonstein V (2005) The subsystems approach to genome annotation and its use in the project to annotate 1000 genomes. Nucleic Acids Res 33(17):5691–5702
12. Babu MM, Luscombe NM, Aravind L, Gerstein M, Teichmann SA (2004) Structure and evolution of transcriptional regulatory networks. Curr Opin Struct Biol 14(3):283–291
13. Madan Babu M, Teichmann SA (2003) Functional determinants of transcription factors in *Escherichia coli*: protein families and binding sites. Trends Genet 19(2):75–79
14. Koonin EV, Fedorova ND, Jackson JD, Jacobs AR, Krylov DM, Makarova KS, Mazumder R, Mekhedov SL, Nikolskaya

AN, Rao BS, Rogozin IB, Smirnov S, Sorokin AV, Sverdlov AV, Vasudevan S, Wolf YI, Yin JJ, Natale DA (2004) A comprehensive evolutionary classification of proteins encoded in complete eukaryotic genomes. Genome Biol 5(2):R7

15. Bornholdt S, Schuster HG (eds) (2003) Handbook of graphs and networks, 1st edn. Willey-VCH, Weinheim
16. Saris CG, Horvath S, van Vught PW, van Es MA, Blauw HM, Fuller TF, Langfelder P, DeYoung J, Wokke JH, Veldink JH, van den Berg LH, Ophoff RA (2009) Weighted gene co-expression network analysis of the peripheral blood from amyotrophic lateral sclerosis patients. BMC Genomics 10:405. doi:doi:1471-2164-10-405 [pii]10.1186/1471-2164-10-405
17. Zhang B, Horvath S (2005) A general framework for weighted gene co-expression network analysis. Stat Appl Genet Mol Biol 4: Article17. doi:10.2202/1544-6115.1128
18. Langfelder P, Horvath S (2008) WGCNA: an R package for weighted correlation network analysis. BMC Bioinformatics 9:559. doi:1471-2105-9-559[pii]10.1186/1471-2105-9-559
19. Suderman M, Hallett M (2007) Tools for visually exploring biological networks. Bioinformatics 23(20):2651–2659. doi:btm401 [pii]10.1093/bioinformatics/btm401
20. Cline MS, Smoot M, Cerami E, Kuchinsky A, Landys N, Workman C, Christmas R, Avila-Campilo I, Creech M, Gross B, Hanspers K, Isserlin R, Kelley R, Killcoyne S, Lotia S, Maere S, Morris J, Ono K, Pavlovic V, Pico AR, Vailaya A, Wang P-L, Adler A, Conklin BR, Hood L, Kuiper M, Sander C, Schmulevich I, Schwikowski B, Warner GJ, Ideker T, Bader GD (2007) Integration of biological networks and gene expression data using Cytoscape. Nat Protoc 2 (10):2366–2382
21. Galas DJ, Schmitz A (1978) DNAse footprinting: a simple method for the detection of protein-DNA binding specificity. Nucleic Acids Res 5(9):3157–3170
22. Garner MM, Revzin A (1981) A gel electrophoresis method for quantifying the binding of proteins to specific DNA regions: application to components of the Escherichia coli lactose operon regulatory system. Nucleic Acids Res 9(13):3047–3060
23. Orlando V (2000) Mapping chromosomal proteins in vivo by formaldehyde-crosslinked-chromatin immunoprecipitation. Trends Biochem Sci 25(3):99–104
24. Deonier RC, Tavaré S, Waterman MS (2005) Computational genome analysis: an introduction. Springer, Heidelberg
25. Matys V, Fricke E, Geffers R, Gossling E, Haubrock M, Hehl R, Hornischer K, Karas D, Kel AE, Kel-Margoulis OV, Kloos DU, Land S, Lewicki-Potapov B, Michael H, Munch R, Reuter I, Rotert S, Saxel H, Scheer M, Thiele S, Wingender E (2003) TRANSFAC: transcriptional regulation, from patterns to profiles. Nucleic Acids Res 31 (1):374–378
26. Bryne JC, Valen E, Tang MH, Marstrand T, Winther O, da Piedade I, Krogh A, Lenhard B, Sandelin A (2008) JASPAR, the open access database of transcription factor-binding profiles: new content and tools in the 2008 update. Nucleic Acids Res 36(Database issue): D102–D106. doi:gkm955[pii]10.1093/nar/gkm955
27. Brazma A, Jonassen I, Vilo J, Ukkonen E (1998) Predicting gene regulatory elements in silico on a genomic scale. Genome Res 8 (11):1202–1215
28. Rustici G, Mata J, Kivinen K, Lio P, Penkett CJ, Burns G, Hayles J, Brazma A, Nurse P, Bahler J (2004) Periodic gene expression program of the fission yeast cell cycle. Nat Genet 36(8):809–817
29. Werner T, Fessele S, Maier H, Nelson PJ (2003) Computer modeling of promoter organization as a tool to study transcriptional coregulation. FASEB J 17(10):1228–1237
30. Dickmeis T, Muller F (2005) The identification and functional characterisation of conserved regulatory elements in developmental genes. Brief Funct Genomic Proteomic 3 (4):332–350
31. Sauer T, Shelest E, Wingender E (2006) Evaluating phylogenetic footprinting for human-rodent comparisons. Bioinformatics 22 (4):430–437
32. Bansal M, Califano A (2012) Genome-wide dissection of posttranscriptional and post-translational interactions. Methods Mol Biol 786:131–149. doi:10.1007/978-1-61779-292-2_8
33. Costanzo M, Baryshnikova A, Bellay J, Kim Y, Spear ED, Sevier CS, Ding H, Koh JL, Toufighi K, Mostafavi S, Prinz J, St Onge RP, VanderSluis B, Makhnevych T, Vizeacoumar FJ, Alizadeh S, Bahr S, Brost RL, Chen Y, Cokol M, Deshpande R, Li Z, Lin ZY, Liang W, Marback M, Paw J, San Luis BJ, Shuteriqi E, Tong AH, van Dyk N, Wallace IM,

Whitney JA, Weirauch MT, Zhong G, Zhu H, Houry WA, Brudno M, Ragibizadeh S, Papp B, Pal C, Roth FP, Giaever G, Nislow C, Troyanskaya OG, Bussey H, Bader GD, Gingras AC, Morris QD, Kim PM, Kaiser CA, Myers CL, Andrews BJ, Boone C (2010) The genetic landscape of a cell. Science 327 (5964):425–431. doi:327/5964/425[pii] 10.1126/science.1180823
34. Guelzim N, Bottani S, Bourgine P, Kepes F (2002) Topological and causal structure of the yeast transcriptional regulatory network. Nat Genet 31(1):60–63
35. Erdös P, Rényi A (1960) On the evolution of random graphs. Publications of the Mathematical Institute of the Hungarian Academy of Sciences 17–61
36. de Silva E, Stumpf MP (2005) Complex networks and simple models in biology. J R Soc Interface 2(5):419–430
37. Albert R, Jeong H, Barabasi AL (2000) Error and attack tolerance of complex networks. Nature 406(6794):378–382. doi:10.1038/35019019
38. Albert R, Jeong H, Barabasi AL (2001) Correction: error and attack tolerance of complex networks. Nature 409(6819):542
39. Hartwell LH, Hopfield JJ, Leibler S, Murray AW (1999) From molecular to modular cell biology. Nature 402(6761 suppl):C47–C52
40. Schlosser G, Wagner GP (2004) Modularity in development and evolution, 1st edn. University of Chicago Press, Chicago
41. Enright AJ, Van Dongen S, Ouzounis CA (2002) An efficient algorithm for large-scale detection of protein families. Nucleic Acids Res 30(7):1575–1584
42. Frey BJ, Dueck D (2007) Clustering by passing messages between data points. Science 315(5814):972–976. doi:1136800[pii] 10.1126/science.1136800
43. Milo R, Shen-Orr S, Itzkovitz S, Kashtan N, Chklovskii D, Alon U (2002) Network motifs: simple building blocks of complex networks. Science 298(5594):824–827
44. Han JD, Bertin N, Hao T, Goldberg DS, Berriz GF, Zhang LV, Dupuy D, Walhout AJ, Cusick ME, Roth FP, Vidal M (2004) Evidence for dynamically organized modularity in the yeast protein-protein interaction network. Nature 430(6995):88–93
45. Park D, Park J, Park SG, Park T, Choi SS (2008) Analysis of human disease genes in the context of gene essentiality. Genomics 92(6):414–418. doi:10.1016/j.ygeno.2008.08.001
46. Zotenko E, Mestre J, O'Leary DP, Przytycka TM (2008) Why do hubs in the yeast protein interaction network tend to be essential: reexamining the connection between the network topology and essentiality. PLoS Comput Biol 4(8):e1000140
47. He X, Zhang J (2006) Why do hubs tend to be essential in protein networks? PLoS Genet 2 (6):e88. doi:10.1371/journal.pgen.0020088
48. Batada NN, Hurst LD, Tyers M (2006) Evolutionary and physiological importance of hub proteins. PLoS Comput Biol 2(7):e88. doi:06-PLCB-RA-0076R2[pii]10.1371/journal.pcbi.0020088
49. Hahn MW, Kern AD (2005) Comparative genomics of centrality and essentiality in three eukaryotic protein-interaction networks. Mol Biol Evol 22(4):803–806. doi: msi072[pii]10.1093/molbev/msi072
50. Yu H, Greenbaum D, Xin Lu H, Zhu X, Gerstein M (2004) Genomic analysis of essentiality within protein networks. Trends Genet 20 (6):227–231. doi:10.1016/j.tig.2004.04.008 S0168952504001015[pii]
51. Jeong H, Mason SP, Barabasi AL, Oltvai ZN (2001) Lethality and centrality in protein networks. Nature 411(6833):41–42
52. Harbison CT, Gordon DB, Lee TI, Rinaldi NJ, Macisaac KD, Danford TW, Hannett NM, Tagne JB, Reynolds DB, Yoo J, Jennings EG, Zeitlinger J, Pokholok DK, Kellis M, Rolfe PA, Takusagawa KT, Lander ES, Gifford DK, Fraenkel E, Young RA (2004) Transcriptional regulatory code of a eukaryotic genome. Nature 431 (7004):99–104
53. Luscombe NM, Babu MM, Yu H, Snyder M, Teichmann SA, Gerstein M (2004) Genomic analysis of regulatory network dynamics reveals large topological changes. Nature 431(7006):308–312
54. Yuh CH, Bolouri H, Davidson EH (1998) Genomic cis-regulatory logic: experimental and computational analysis of a sea urchin gene. Science 279(5358):1896–1902
55. Davidson EH, Rast JP, Oliveri P, Ransick A, Calestani C, Yuh CH, Minokawa T, Amore G, Hinman V, Arenas-Mena C, Otim O, Brown CT, Livi CB, Lee PY, Revilla R, Rust AG, Pan Z, Schilstra MJ, Clarke PJ, Arnone MI, Rowen L, Cameron RA, McClay DR, Hood L, Bolouri H (2002) A genomic regulatory network for development. Science 295 (5560):1669–1678
56. Davidson EH, Rast JP, Oliveri P, Ransick A, Calestani C, Yuh CH, Minokawa T, Amore G, Hinman V, Arenas-Mena C, Otim O, Brown CT, Livi CB, Lee PY, Revilla R, Schilstra MJ, Clarke PJ, Rust AG, Pan Z, Arnone MI,

Rowen L, Cameron RA, McClay DR, Hood L, Bolouri H (2002) A provisional regulatory gene network for specification of endomesoderm in the sea urchin embryo. Dev Biol 246 (1):162–190

57. Longabaugh WJ, Davidson EH, Bolouri H (2009) Visualization, documentation, analysis, and communication of large-scale gene regulatory networks. Biochim Biophys Acta 1789 (4):363–374. doi:S1874-9399(08)00162-4 [pii]10.1016/j.bbagrm.2008.07.014
58. Saez-Rodriguez J, Alexopoulos LG, Epperlein J, Samaga R, Lauffenburger DA, Klamt S, Sorger PK (2009) Discrete logic modelling as a means to link protein signalling networks with functional analysis of mammalian signal transduction. Mol Syst Biol 5:331. doi: msb200987[pii]10.1038/msb.2009.87
59. Samaga R, Saez-Rodriguez J, Alexopoulos LG, Sorger PK, Klamt S (2009) The logic of EGFR/ErbB signaling: theoretical properties and analysis of high-throughput data. PLoS Comput Biol 5(8):e1000438. doi:10.1371/journal.pcbi.1000438
60. Klamt S, Saez-Rodriguez J, Gilles ED (2007) Structural and functional analysis of cellular networks with Cell NetAnalyzer. BMC Syst Biol 1:2. doi:1752-0509-1-2[pii]10.1186/1752-0509-1-2
61. de Jong H (2002) Modeling and simulation of genetic regulatory systems: a literature review. J Comput Biol 9(1):67–103
62. Kauffman S (1969) Homeostasis and differentiation in random genetic control networks. Nature 224(215):177–178
63. Kauffman SA (1993) The origins of order, self-organization and selection in evolution. Oxford University Press, New York
64. Akutsu T, Miyano S, Kuhara S (2000) Inferring qualitative relations in genetic networks and metabolic pathways. Bioinformatics 16 (8):727–734
65. Albert R, Barabasi AL (2000) Dynamics of complex systems: scaling laws for the period of Boolean networks. Phys Rev Lett 84 (24):5660–5663
66. D'Haeseleer P, Liang S, Somogyi R (2000) Genetic network inference: from co-expression clustering to reverse engineering. Bioinformatics 16(8):707–726
67. Klamt S, Saez-Rodriguez J, Lindquist JA, Simeoni L, Gilles ED (2006) A methodology for the structural and functional analysis of signaling and regulatory networks. BMC Bioinformatics 7:56
68. Pandey S, Wang RS, Wilson L, Li S, Zhao Z, Gookin TE, Assmann SM, Albert R (2010) Boolean modeling of transcriptome data reveals novel modes of heterotrimeric G-protein action. Mol Syst Biol 6:372. doi: msb201028[pii]10.1038/msb.2010.28
69. Kauffman S, Peterson C, Samuelsson B, Troein C (2003) Random Boolean network models and the yeast transcriptional network. Proc Natl Acad Sci U S A 100(25):14796–14799. doi:10.1073/pnas.2036429100 2036429100[pii]
70. Kauffman S, Peterson C, Samuelsson B, Troein C (2004) Genetic networks with canalyzing Boolean rules are always stable. Proc Natl Acad Sci U S A 101(49):17102–17107. doi:0407783101[pii]10.1073/pnas.0407783101
71. Albert I, Thakar J, Li S, Zhang R, Albert R (2008) Boolean network simulations for life scientists. Source Code Biol Med 3:16. doi:1751-0473-3-16[pii]10.1186/1751-0473-3-16
72. Saadatpour A, Albert I, Albert R (2010) Attractor analysis of asynchronous Boolean models of signal transduction networks. J Theor Biol 266(4):641–656. doi:S0022-5193(10)00379-6[pii]10.1016/j.jtbi.2010.07.022
73. Thomas R (1973) Boolean formalization of genetic control circuits. J Theor Biol 42 (3):563–585
74. Thomas R (1991) Regulatory networks seen as asynchronous automata: a logical description. J Theor Biol 153:1–23
75. Naldi A, Carneiro J, Chaouiya C, Thieffry D (2010) Diversity and plasticity of Th cell types predicted from regulatory network modelling. PLoS Comput Biol 6(9):e1000912. doi:e1000912[pii]10.1371/journal.pcbi.1000912
76. Naldi A, Berenguier D, Faure A, Lopez F, Thieffry D, Chaouiya C (2009) Logical modelling of regulatory networks with GINsim 2.3. Biosystems 97(2):134–139. doi: S0303-2647(09)00066-5[pii]10.1016/j.biosystems.2009.04.008
77. Petri CA, Reisig W (2008) Petri net. Scholarpedia 3(4):6477
78. Koch I, Reisig W, Schreiber F (eds) (2011) Modeling in systems biology: the Petri net approach. Computational biology, 1st edn. Springer, London. doi:10.1007/978-1-84996-474-6
79. Banks R, Khomenko V, Steggles LJ (2011) Modeling genetic regulatory networks. In: Koch I, Reisig W, Schreiber F (eds) Modeling in systems biology: the Petri net approach, 1st edn. Springer, London. doi:10.1007/978-1-84996-474-6

80. D'Haeseleer P, Wen X, Fuhrman S, Somogyi R (1999) Linear modeling of mRNA expression levels during CNS development and injury. Pac Symp Biocomput 41–52 http://www.ncbi.nlm.nih.gov/pubmed/10380184
81. van Someren EP, Wessels LF, Reinders MJ (2000) Linear modeling of genetic networks from experimental data. Proc Int Conf Intell Syst Mol Biol 8:355–366
82. von Dassow G, Meir E, Munro EM, Odell GM (2000) The segment polarity network is a robust developmental module. Nature 406 (6792):188–192
83. Lequieu J, Chakrabarti A, Nayak S, Varner JD (2011) Computational modeling and analysis of insulin induced eukaryotic translation initiation. PLoS Comput Biol 7(11):e1002263. doi:10.1371/journal.pcbi.1002263 PCOMPBIOL-D-11-00832[pii]
84. Weisse AY, Middleton RH, Huisinga W (2010) Quantifying uncertainty, variability and likelihood for ordinary differential equation models. BMC Syst Biol 4:144. doi:1752-0509-4-144[pii]10.1186/1752-0509-4-144
85. Mendes P, Kell DB (2001) MEG (Model Extender for Gepasi): a program for the modelling of complex, heterogeneous, cellular systems. Bioinformatics 17(3):288–289
86. Mendes P (1997) Biochemistry by numbers: simulation of biochemical pathways with Gepasi 3. Trends Biochem Sci 22 (9):361–363. doi:S0968000497011031[pii]
87. Funahashi A, Matsuoka Y, Jouraku A, Morohashi M, Kikuchi N, Kitano H (2008) Cell Designer 3.5: a versatile modeling tool for biochemical networks. Proc IEEE 96 (8):1254–1265
88. Takahashi K, Ishikawa N, Sadamoto Y, Sasamoto H, Ohta S, Shiozawa A, Miyoshi F, Naito Y, Nakayama Y, Tomita M (2003) E-Cell 2: multi-platform E-Cell simulation system. Bioinformatics 19 (13):1727–1729
89. McAdams HH, Shapiro L (1995) Circuit simulation of genetic networks. Science 269 (5224):650–656
90. Friedman N, Linial M, Nachman I, Pe'er D (2000) Using Bayesian networks to analyze expression data. J Comput Biol 7 (3–4):601–620
91. Murphy K, Mian S (1999) Modelling gene expression data using dynamic Bayesian networks. Technical report, U.C. Berkeley, Department of Computer Science
92. Pournara I, Wernisch L (2004) Reconstruction of gene networks using Bayesian learning and manipulation experiments. Bioinformatics 20(17):2934–2942
93. Titsias MK, Honkela A, Lawrence ND, Rattray M (2012) Identifying targets of multiple coregulating transcription factors from expression time-series by Bayesian model comparison. BMC Syst Biol 6(1):53.doi:1752-0509-6-53 [pii]10.1186/1752-0509-6-53
94. Pe'er D (2005) Bayesian network analysis of signaling networks: a primer. Sci STKE 2005 (281):pl4
95. Shmulevich I, Dougherty ER, Kim S, Zhang W (2002) Probabilistic Boolean Networks: a rule-based uncertainty model for gene regulatory networks. Bioinformatics 18(2):261–274
96. Matsuno H, Doi A, Nagasaki M, Miyano S (2000) Hybrid Petri net representation of gene regulatory network. Pac Symp Biocomput 341–352 http://www.ncbi.nlm.nih.gov/pubmed/?term=10902182
97. Goss PJ, Peccoud J (1998) Quantitative modeling of stochastic systems in molecular biology by using stochastic Petri nets. Proc Natl Acad Sci U S A 95(12):6750–6755
98. Ruths D, Muller M, Tseng JT, Nakhleh L, Ram PT (2008) The signaling petri net-based simulator: a non-parametric strategy for characterizing the dynamics of cell-specific signaling networks. PLoS Comput Biol 4(2):e1000005. doi:10.1371/journal.pcbi.1000005
99. Arkin A, Ross J, McAdams HH (1998) Stochastic kinetic analysis of developmental pathway bifurcation in phage lambda-infected *Escherichia coli* cells. Genetics 149 (4):1633–1648
100. McAdams HH, Arkin A (1997) Stochastic mechanisms in gene expression. Proc Natl Acad Sci U S A 94(3):814–819
101. McAdams HH, Arkin A (1999) It's a noisy business! Genetic regulation at the nanomolar scale. Trends Genet 15(2):65–69
102. Theocharidis A, van Dongen S, Enright AJ, Freeman TC (2009) Network visualization and analysis of gene expression data using BioLayout Express(3D). Nat Protoc 4 (10):1535–1550. doi:nprot.2009.177[pii] 10.1038/nprot.2009.177
103. Albert R, Barabasi AL (2002) Statistical mechanics of complex networks. Rev Mod Phys 74(1):47–97
104. Barabasi A-L (2007) Network medicine—from obesity to the "diseasome". N Engl J Med 357(4):404–407. doi:10.1056/NEJMe078114
105. Dewey TG (2002) From microarrays to networks: mining expression time series. Drug Discov Today 7(20 Suppl):S170–S175

106. Friedman N (2004) Inferring cellular networks using probabilistic graphical models. Science 303(5659):799–805
107. Huber W, Carrey VJ, Long L, Falcon S, Gentleman R (2007) Graphs in molecular biology. BMC Bioinformatics 8(suppl 6):S8
108. Kitano H (2002) Computational systems biology. Nature 420(6912):206–210
109. Moore JH, Boczko EM, Summar ML (2005) Connecting the dots between genes, biochemistry, and disease susceptibility: systems biology modeling in human genetics. Mol Genet Metab 84(2):104–111
110. van Someren EP, Wessels LF, Backer E, Reinders MJ (2002) Genetic network modeling. Pharmacogenomics 3(4):507–525
111. Hucka M, Finney A, Sauro HM, Bolouri H, Doyle JC, Kitano H, Arkin AP, Bornstein BJ, Bray D, Cornish-Bowden A, Cuellar AA, Dronov S, Gilles ED, Ginkel M, Gor V, Goryanin II, Hedley WJ, Hodgman TC, Hofmeyr JH, Hunter PJ, Juty NS, Kasberger JL, Kremling A, Kummer U, Le Novere N, Loew LM, Lucio D, Mendes P, Minch E, Mjolsness ED, Nakayama Y, Nelson MR, Nielsen PF, Sakurada T, Schaff JC, Shapiro BE, Shimizu TS, Spence HD, Stelling J, Takahashi K, Tomita M, Wagner J, Wang J (2003) The systems biology markup language (SBML): a medium for representation and exchange of biochemical network models. Bioinformatics 19(4):524–531
112. Levine M, Tjian R (2003) Transcription regulation and animal diversity. Nature 424 (6945):147–151. doi:10.1038/nature01763 nature01763[pii]
113. Lee TI, Rinaldi NJ, Robert F, Odom DT, Bar-Joseph Z, Gerber GK, Hannett NM, Harbison CT, Thompson CM, Simon I, Zeitlinger J, Jennings EG, Murray HL, Gordon DB, Ren B, Wyrick JJ, Tagne JB, Volkert TL, Fraenkel E, Gifford DK, Young RA (2002) Transcriptional regulatory networks in *Saccharomyces cerevisiae*. Science 298 (5594):799–804

Chapter 3

Integration of Genomic Information with Biological Networks Using Cytoscape

Anna Bauer-Mehren

Abstract

Cytoscape is an open-source software for visualizing, analyzing, and modeling biological networks. This chapter explains how to use Cytoscape to analyze the functional effect of sequence variations in the context of biological networks such as protein–protein interaction networks and signaling pathways. The chapter is divided into five parts: (1) obtaining information about the functional effect of sequence variation in a Cytoscape readable format, (2) loading and displaying different types of biological networks in Cytoscape, (3) integrating the genomic information (SNPs and mutations) with the biological networks, and (4) analyzing the effect of the genomic perturbation onto the network structure using Cytoscape built-in functions. Finally, we briefly outline how the integrated data can help in building mathematical network models for analyzing the effect of the sequence variation onto the dynamics of the biological system. Each part is illustrated by step-by-step instructions on an example use case and visualized by many screenshots and figures.

Key words Cytoscape, Biological network, Biological pathway, Integration, Genomic information, Sequence variation, SNP

1 Introduction

The advent of high-throughput experimentation in research in the last decades has led to a better understanding of the biological processes within a cell. The identification and characterization of its molecular components became possible in a systematic way. In addition, the discovery of connections between each of these components promoted the reconstruction/identification of biological networks. These networks are crucial for understanding the ways in which the cells respond to external cues and how they communicate with each other. In this context, a biological network can circumscribe several types of biological processes including regulatory, metabolic, and signaling processes or protein–protein interactions (PPI). For simplicity, we will only distinguish between pathways, including regulatory, metabolic, and signaling processes, and PPI networks within this chapter.

Maria Victoria Schneider (ed.), *In Silico Systems Biology*, Methods in Molecular Biology, vol. 1021,
DOI 10.1007/978-1-62703-450-0_3, © Springer Science+Business Media, LLC 2013

A variety of pathway and interaction databases have emerged storing our knowledge on the biological processes (*see* refs. 1, 2 for an overview). The pathway and PPI databases contain manually curated information and provide cross-references to other biomedical databases such as Entrez Gene [3] or UniProt [4] and annotate entities and processes with gene ontology (GO) terms [5]. Moreover, most of the databases allow programmatic access to their data and support current pathway exchange formats such as BioPAX and SBML, which allow data visualization and analysis of the biological networks using dedicated tools such as Cytoscape [6]. Cytoscape is an open-source software for network visualization and analysis. Due to its plug-in environment, Cytoscape offers a vast variety of analysis tools ranging from basic functions, such as the automatic retrieval of pathways from pathway and PPI databases, to advanced network analysis, such as the topological analysis of the networks or network clustering.

The representation of biological networks in exchangeable, computer readable formats in which entities (genes, proteins) are represented by standard identifiers enables the integration of the pathways and PPI networks with genomic information such as gene expression data [7, 8], as well as information on single-nucleotide polymorphisms (SNPs) [9]. SNPs are the most frequent type of sequence variation between individuals and represent a promising tool for finding genetic determinants of complex diseases and understanding interindividual drug response. SNPs are currently used in genome wide association studies and pharmacogenomics studies. Once the SNPs associated with a disease phenotype are identified, the elucidation of their functional effect is a key factor for understanding the mechanisms underlying the disease. In this context, the genetic perturbation introduced by the SNP has to be considered at two levels. First, the SNP can affect the protein function directly, for example by affecting the protein stability or folding [10], second the SNP can have further consequences on interacting proteins/genes and downstream reactions in the pathways in which the affected protein participates. Hence, the functional effect of SNPs can be better appreciated if the evaluation is performed at the systems level.

In this chapter, we first briefly outline the extraction of information about the functional effect of SNPs from UniProt and the conversion into a format useful for integration with biological networks in Cytoscape. Second, we use Cytoscape to load and visualize biological pathways and PPI networks. Third, we integrate the information about the functional effect of SNPs with the biological networks loaded before. Here, we establish powerful visual mappings across these data. Then, we perform an advanced analysis using Cytoscape built-in functions as well as plug-ins to analyze the effect of perturbations introduced by the SNP(s) onto the network structure. Last, we outline how the integrated information can be used to build mathematical network models, which allow assessing

Table 1
Data sources of biological networks

Name	Contains	Link
HPRD	Human protein–protein interaction network	http://www.hprd.org/
Reactome	Human biological pathways	http://www.reactome.org/ReactomeGWT/entrypoint.html
BioModels	Curated mathematical network models	http://www.ebi.ac.uk/biomodels-main/

the functional effect of the SNP(s) onto the network dynamics. We guide the reader step by step through the integration process using an illustrative use case.

2 Materials

2.1 Software Requirements

Cytoscape version 2.8.2 can be downloaded from http://www.cytoscape.org/ including the following plug-ins:

- EnhancedSearch v.1.1.0 (ESP).
- AdvancedNetworkMerge v.1.1.2.

2.2 Attribute and Visualization Files

The attribute files needed for the integration of information about SNPs and mutations with biological networks, as well as the visual style file allowing a better visualization of sequence variations in the networks, can be downloaded from https://drive.google.com/folderview?id=0B0xCoXo3BVZjVTNQRG9FMl9yQUE&usp=sharing.

2.3 Biological Networks Repositories

A review about current pathway repositories and tools for their exploitation can be found in [1, 2]. In this tutorial, we use a human PPI network from HPRD and a pathway in BioPAX format from Reactome (*see* Table 1).

3 Methods

3.1 Extraction of Information on SNPs and Mutations from UniProt

In this section, we briefly outline how to extract information about SNPs and other sequence variations from UniProt and how to build Cytoscape node and edge attribute files for integrating this information with biological networks.

The Universal Protein Resource (UniProt) [4] is a centralized resource for protein sequences and functional information. The UniProt/SwissProt part of UniProt provides curated information on the functional and phenotypic effects of natural variations, including SNPs, as well as on mutations of protein sequences.

For several of these natural variants and mutants, it furthermore provides associations to disease phenotypes. Thus, it provides a comprehensive framework to extract information about the association of sequence variations and human diseases.

3.1.1 Generation of Node Attribute Files

Details of how to obtain information about the functional effect of sequence variations from UniProt are described in [9]. In brief, the procedure includes parsing of relevant information from UniProt and several steps of mapping UniProt protein identifiers to Entrez Gene identifiers and mapping SNPs to dbSNP identifiers. Additionally, within this tutorial, we map UniProt identifiers to HPRD identifiers to allow the integration of the information about sequence variations from UniProt with the PPI network from HRPD [11].

Cytoscape uses a very simple node and edge attribute file format and allows the import of attributes from tables. Such attribute tables in their simplest format are tab-delimited text files containing one column matching the primary identifier (or any other attribute) of the nodes in the network, such as the UniProt identifier of protein nodes, and additional columns containing the attributes one wants to map onto the nodes, such as the functional effect description of an SNP or the protein name.

In UniProt, the functional effect description is free text complicating the automatic integration of the effect of the sequence variation onto the reactions in the biological networks. Hence, we use text-mining tools to detect mentions of proteins and processes in the textual description that are likely being affected by the sequence variation. For instance for the protein MUC1 (P15941), there is a sequence variation of *an amino acid exchange from Y to N at position 1243*, which is annotated with the functional effect description that *greatly reduced binding to GRB2*. Hence, we use text mining to detect mentions of gene ontology (GO) terms, as well as proteins in this text. In this example, we detect the GO molecular function term *binding* (GO:0005488) and the *GRB2* (uniProtId:P62993). There are several tools available for that purpose including the Whatizit web service (http://www.ebi.ac.uk/webservices/whatizit/info.jsf) or the National Center for Biomedical Ontology (NCBO) Annotator [12], which allows the annotation of free text with terms of more than 250 biomedical ontologies in BioPortal [13].

We then merge all information about sequence variations from UniProt into one attribute file (*mutPoly_nodes.attr*). An overview of all attributes is given in Table 2. The two most important columns are *uniProtId* and *hprdId*, which contain the UniProt and HPRD identifiers, respectively, needed for the integration of the information about sequence variations with protein nodes in the networks. The other columns contain different kinds of information including the mapped Entrez Gene identifiers, disease

Table 2
Node and edge attributes

Attribute name	Attribute type	Description
hprdId	Node	HPRD identifier of the protein
uniProtId	Node	UniProt identifier of the protein
Mutagenesis	Node	List of the mutagenesis information contains the amino acid exchange, the sequence position, and the textual phenotypic description from UniProt
Polymorphism	Node	List of the natural variant/polymorphism information contains the amino acid exchange, the sequence position, the textual phenotypic description from UniProt, and if available an MIM identifier and the textual description of the disease association; if at the same position mutagenesis data is also available, this data is listed as a sub-list of the polymorphism
OMIM	Node (list attribute)	List of disease names and OMIM identifiers associated with the natural variant
dbSNP	Node (list attribute)	List of dbSNP identifiers
GObiolProcess	Node (list attribute)	List of GO biological process terms that are associated to the natural variant or mutant
GObiolProcessId	Node (list attribute)	List of GO biological process identifiers that are associated to the natural variant or mutant
GOmolFunction	Node (list attribute)	List of GO molecular function terms that are associated to the natural variant or mutant
GOmolFunctionId	Node (list attribute)	List of GO molecular function identifiers that are associated to the natural variant or mutant
GOcellComponent	Node (list attribute)	List of GO cellular component terms that are associated to the natural variant or mutant
GOcellComponentId	Node (list attribute)	List of GO cellular component identifiers that are associated to the natural variant or mutant
extUniProtIds	Node (list attribute)	List of UniProt identifiers that are associated to the natural variant or mutant
mutPolyFlag	Node	Required for the visual styles 1. Only mutagenesis information available 2. Only polymorphism information available 3. Mutagenesis and polymorphism information available but not on the same position 4. Mutagenesis and polymorphism information available on the same position

(continued)

Table 2 (continued)

Attribute name	Attribute type	Description
Edge	Edge	Edge identifier represented by the two HPRD protein identifiers of the interacting proteins delimited by the type of their interaction (pp) for protein–protein interaction
edge_mutPolyFlag	Edge	Flag useful for visualization purposes

associations, and GO annotations, among others. Since there can be multiple sequence variations per protein, we separate different annotations by the pipe symbol (|) enabling their import as list attributes (*see* Subheading 3).

3.1.2 Generation of Edge Attribute Files

In addition to the node attribute files, we generate edge attribute files for the integration of the functional effect of the sequence variants with the edges in the biological networks. The text-mining process described above allows the identification of protein mentions in the functional effect description provided by UniProt. Proteins mentioned in the functional effect description of sequence variations are likely being affected by the perturbation introduced by the sequence variant. Let us consider the same example as described above; for the protein MUC1 (P15941), there is a sequence variations which *greatly reduces binding to GRB2.* Using the text-mining tools described above, we identify the mention of GRB2. In this case the functional effect of the variation in the MUC1 protein affects the protein GRB2. Hence, when integrating this information with a biological network, we want to mark all edges between MUC1 and GRB2 because they might be affected by the sequence variation of MUC1 causing a reduced binding to GRB2. Therefore, we can use the columns *hprdId* defining the HPRD identifier of the protein and *extUniProtIds* containing the proteins, which are likely affected by sequence variations of the protein, to create an edge attribute file. Similarly to node attribute files, the edge attribute file contains the identifier of the edge (which generally consists of the primary identifiers of the two nodes defining the edge) and some additional columns representing the edge attributes. For each protein and sequence variation pair, multiple proteins can be affected. Hence, we parse the *extUniProtIds* column and create a new edge attribute for each unique combination. We represent the PPIs (edges) using the HPRD [11] protein identifiers because we want to integrate the genomic information with the HPRD PPI network in this tutorial. Nevertheless, for other PPI networks or pathways, edge attribute files can be generated in a similar manner using the specific protein identifiers. The file *mutPoly_PPI_edges.attr* is provided, containing the edge attributes for the next step of the tutorial. An overview of all attributes is given in Table 2.

3.2 Loading Biological Networks into Cytoscape

Let us start with a human PPI network from HPRD [11]. To download the HRPD PPI network, go to http://www.hprd.org/download and download the latest HPRD FLAT FILES containing protein–protein interactions in tab-delimited file format as well as information about posttranslational modifications, tissue expression, and subcellular localization, among others. In this tutorial, we are working with HPRD Release 9 of 2010.

3.2.1 Import a Protein–Protein Interaction Network from HRPD

We first load the HPRD PPI network (BINARY_PROTEIN_PROTEIN_INTERACTIONS.txt) into Cytoscape. For this purpose, use the Cytoscape built-in function: *File→Import→Network from Table*. It is important to use HPRD identifiers as primary identifiers because they are unique for each protein (*see* Fig. 1). Also, it is important to use the default interaction type (pp) since our edge attribute file requires this type of interaction (*see* **Note 1** for more information). The resulting network contains 9,673 nodes (proteins) connected by 39,204 edges (interactions). For a better representation apply a layout algorithm. We recommend the Organic layout to get the first idea of the network structure (*Layout→yFiles→Organic*) (*see* **Note 2**).

3.2.2 Import a Biological Pathway from Reactome

Next, we import a pathway from Reactome. Go to http://www.reactome.org/ and search for pathways (restrict your search to pathways only) related to ATM (Q13315) in Reactome and download the *ATM-mediated response to DNA double-strand break (Homo sapiens)* pathway in BioPAX2 format. Import the pathway into your Cytoscape session using the Cytoscape built-in function *File→Import→Network (Multiple File Types)*. Rename the network to *double-strand break repair* (right click on the network in the *Networks* tab of the *Control panel)*. Save your Cytoscape session.

3.3 Integrate Information on SNPs/Mutations with the PPI Network

In this section, we integrate the information about sequence variations with the PPI network and pathways loaded in the previous section. Also, we use built-in functionalities of Cytoscape to query, filter, and analyze the networks.

3.3.1 Working with Node Attributes

First, we import the information about sequence variations (*mutPoly_nodes.attr*) into the PPI network as node attributes (*see* **Note 3**). We also import the *mutPoly.props* as *Vizmap Property File* and select the mutPoly_PPI visual style to visualize SNPs/mutations on the protein nodes in the PPI network. For this purpose, use the Cytoscape built-in function *File→Import→Vizmap Property File* (*see* **Note 4**). Now, let us explore the different node attributes. For this purpose, use the

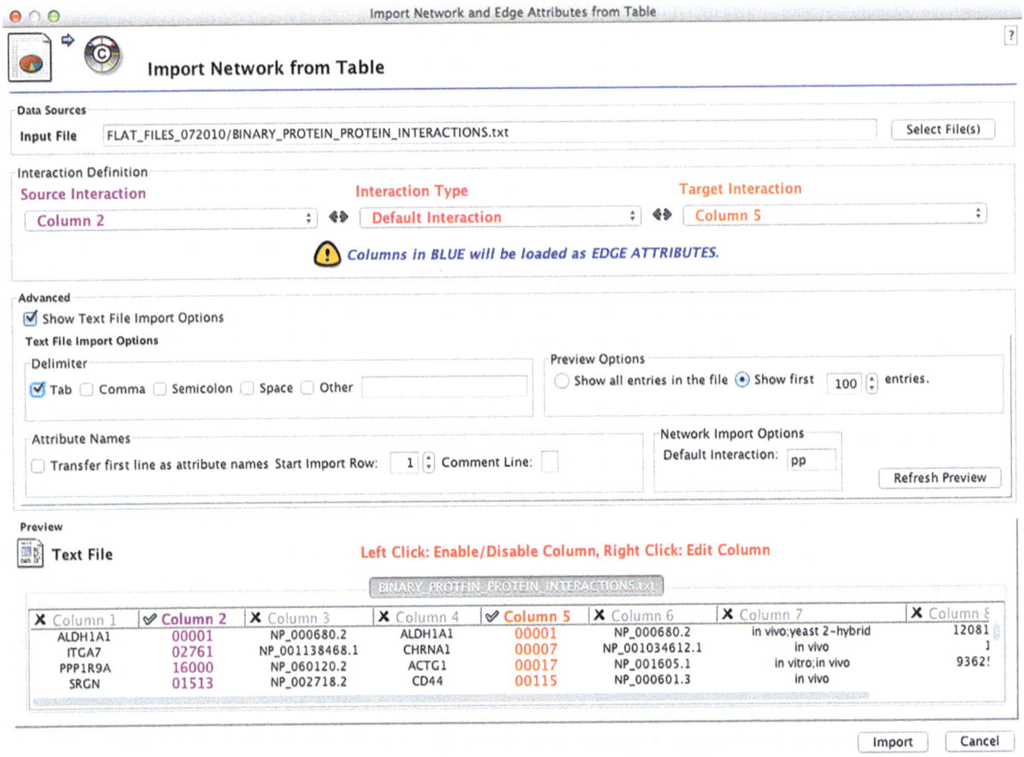

Fig. 1 Cytoscape dialogue to import a network from a table/text file. Import the HPRD PPI network into Cytoscape. Columns 2 and 5 contain the necessary information to load the PPI network. The other columns are not selected

Select Attributes icon in the *Data panel* and select some attributes such as *uniProtId, geneSymbol, mutagenesis, polymorphism, dbSNP, GObiolProcess, GObiolProcessId, and OMIM*. Then, select some nodes and check the values of the selected attributes (*see* Fig. 2).

Now, we want to select those proteins for which there is a sequence variant associated with "Crohn disease." For this purpose, we can use the built-in search for node attributes (*see* **Note 5**). There are three proteins associated with "Crohn disease": SLC22A4, NOD2, and IL10. Check the information about the SNPs for these proteins by inspecting the *polymorphism* node attribute. Figure out which polymorphism is associated with the disease, has a functional effect description, is annotated to a dbSNP identifier, and has an annotation to the GO term *transport* (*see* **Note 5**). We can also ask how many proteins in total have annotations to SNPs/mutations in the PPI network. How many of these SNPs/mutations are associated to disease (*see* **Note 6**)?

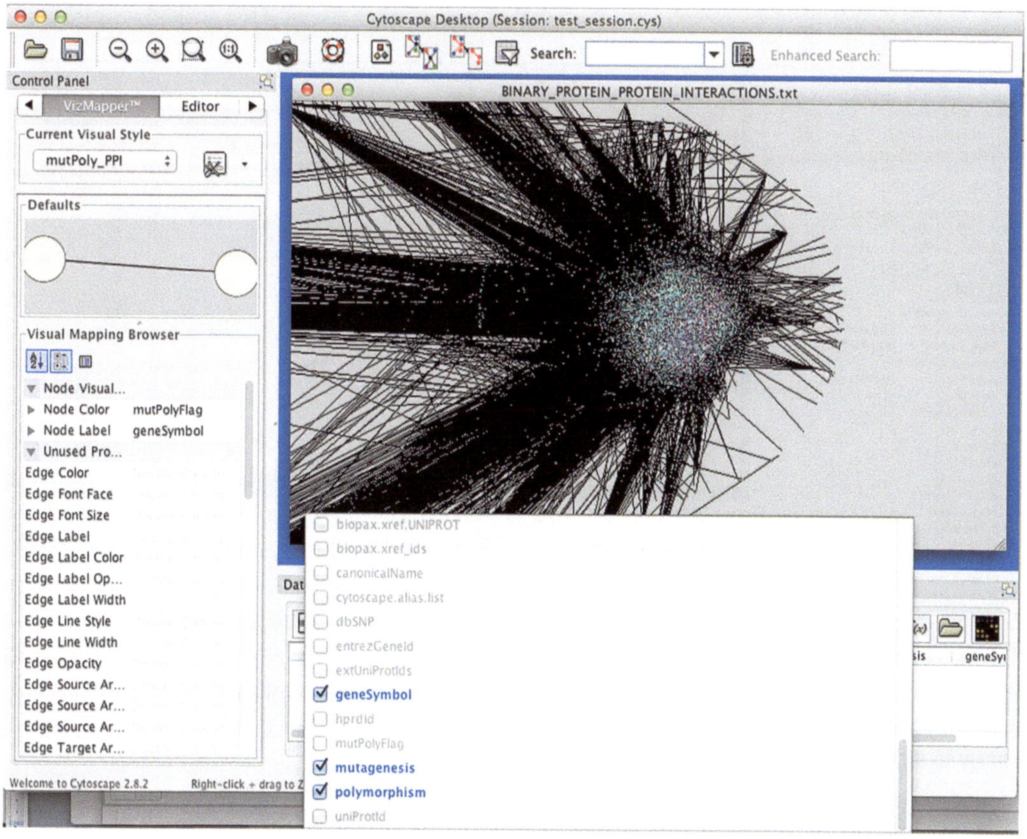

Fig. 2 Cytoscape screenshot of the HPRD PPI network. The visual style mutPoly_PPI is selected to visualize SNPs/mutations on the protein nodes in the network. *Purple* nodes only contain information on mutagenesis experiments; *turquoise* depicts nodes for which only polymorphism data exists. Some nodes have data for both mutagenesis and natural variant either at different sequence positions (*pink*) or at the same position (*light purple*). Several node attributes (geneSymbol, mutagenesis, and polymorphism) are selected and are therefore displayed in the node attribute browser

3.3.2 Working with Edge Attributes

In the previous section we have imported attributes for the nodes (proteins) in our PPI network and used them to select a set of proteins sharing specific properties. Now, we are using edge attributes to study the effect of the SNPs or mutation onto the interactions (edges) in the network. For this purpose, we first import edge attributes similarly as we did for the node attributes.

We import the *mutPoly_PPI_edges.attr* file as edge attributes and select the *mutPoly_PPI_edgeFlag* visual style in the *VizMapper* to highlight edges between proteins that are potentially affected by the SNPs/mutations (*see* **Note 7**). Figure 3 shows how the edge attributes are visualized in Cytoscape. The red edges denote those being potentially affected by a sequence variant.

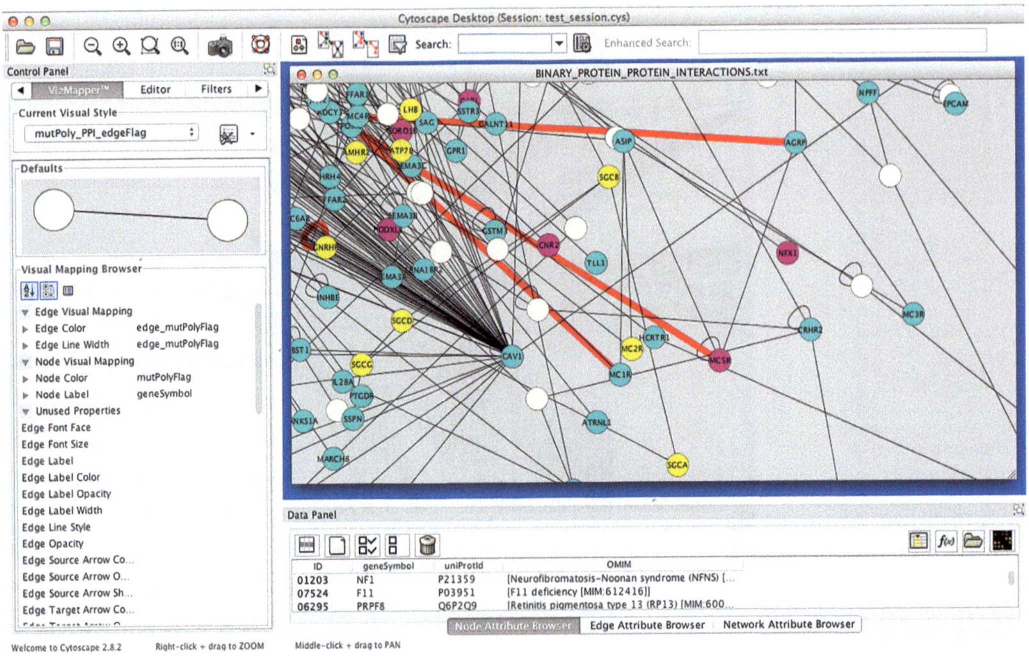

Fig. 3 Cytoscape screenshot of the HPRD PPI network. Nodes are colored according to the type of annotation available; edges are colored in red if there is a sequence variant for one of the interacting proteins, which is likely to affect their interaction

3.4 Exploring Disease-Associated Sequence Variations: Use Case Ataxia

In this section, we are focusing on a specific use case—ataxia. Ataxia is a neurological sign consisting of a lack of muscle coordination during voluntary movements, such as walking or picking up objects, due to cerebellar dysfunction. There are several ataxia subtypes including spinocerebellar ataxias (SCAs), episodic ataxias (EAs), ataxias telangiectasias (ATs), Friedrich ataxia (FA), ataxia–oculomotor apraxia (AOA) syndrome, and sensory ataxic neuropathy dysarthria and ophthalmoparesis (SANDO). Several human inherited ataxias share clinical and pathological features, and in 2006, Lim et al. used a PPI network constructed of proteins related to ataxias to study if the phenotypic similarities can be explained by some shared molecular mechanisms. They used the PPI network to understand the pathogenic mechanisms common for this class of neurodegenerative disorders and for identifying candidate genes for inherited ataxias [14]. In this section, we analyze the human HPRD PPI network and a pathway from Reactome in context with mutation/SNPs that are associated to ataxias to gain insight into the underlying mechanisms.

3.4.1 Creating Subnetworks Related to Ataxia

In order to start, we need to extract those proteins being associated with ataxia. For this purpose, use the enhanced search plug-in (*see* red circle in Fig. 4) to search for ataxia. Extract the proteins and build a subnetwork containing them and their interactions.

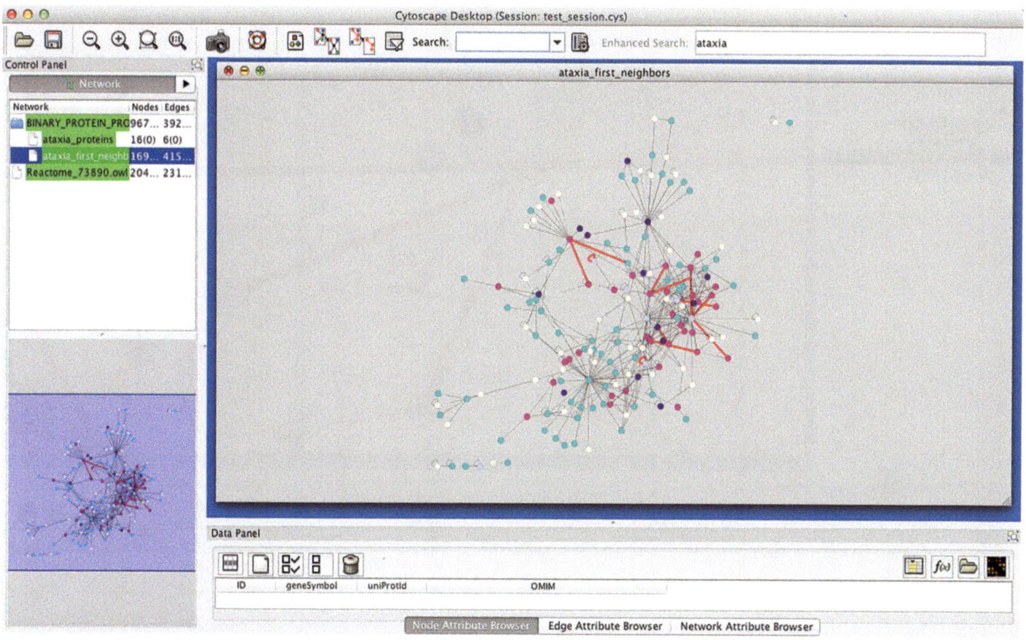

Fig. 4 Subnetwork of the 16 ataxia proteins and their neighbors. Most ataxia proteins (12) are interacting directly or indirectly in a large connected component. The enhanced search plug-in has been used to search all attributes for the term "ataxia" (see *upper right part*)

How many proteins do you find? Make use of the function *Select→Nodes→First Neighbors of Selected Nodes* in Cytoscape to extend the ataxia protein network, and check if the ataxia proteins are directly or indirectly connected in the resulting network. How many ataxia proteins appear in the largest connected component, and what does that imply (*see* **Note 8**)?

To better identify proteins being associated with ataxia, we can create a string attribute called *ataxia_protein* and fill it with "ataxia_protein" for the 16 ataxia proteins (*see* **Note 9**). Select the visual style mutPoly_PPI_ataxiaFlag. This adds a red circle around the ataxia protein nodes. You can also create your own visual style using the attributes or modify the provided styles. Now, find the ataxia protein for which there is a mutation or polymorphism affecting *DNA binding*. Is the sequence variation associated to the disease (*see* **Note 9**)?

3.4.2 Analyzing the Functional Effect of Sequence Variations Related to Ataxia

It is known that aprataxin (APTX) forms a complex with XRCC1 and XRCC4, which are partners of DNA ligase III and IV, respectively. This complex is involved in the DNA repair mechanism. Is there any sequence variation of aprataxin protein that could have an effect on DNA repair mechanism, and what does that imply (*see* Fig. 5 and **Note 10**)?

Now, we want to find the ataxia protein for which a sequence variation is affecting the molecular function *kinase activity*. This protein is highly connected (a hub node in the ataxia_first

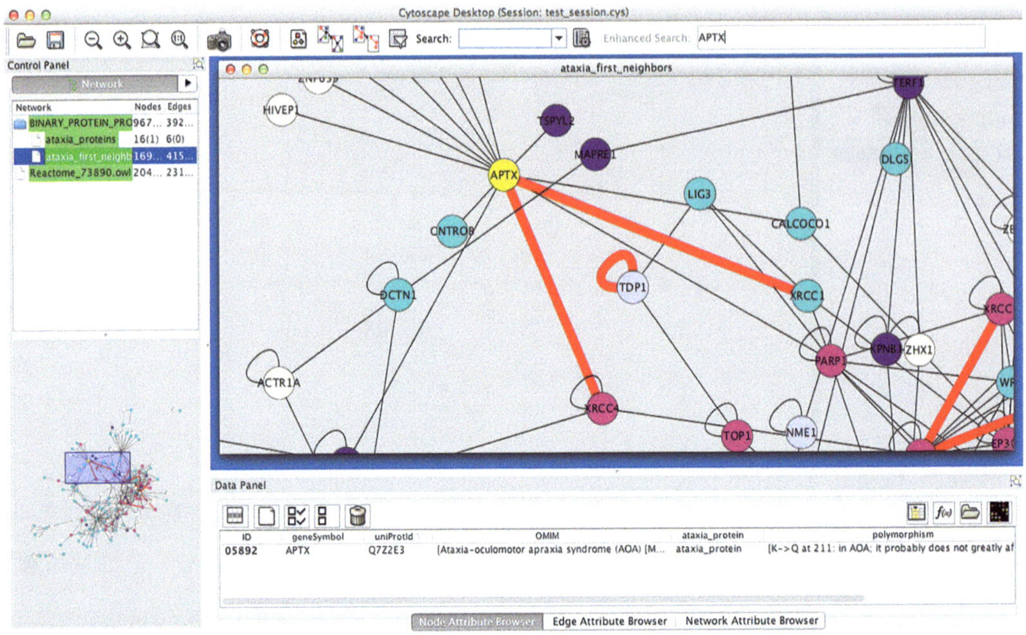

Fig. 5 PPI subnetwork of ataxia-related proteins and their first neighbors. Aprataxin interacts with XRCC1 and XRCC4. Both interactions are colored red indicating that they could be affected by a sequence variation of APTX. The complex of aprataxin, XRCC1, and XRCC2 plays an important role in the DNA repair mechanism

neighbors network). How many connections to other proteins does it have? Is there any neighbor protein also associated to ataxia? Use the visual styles to highlight the proteins (*see* Fig. 6 and **Note 11**).

3.4.3 Analyzing the "ATM-Mediated Response to DNA Double-Strand Break" Pathway

First, we need to import the *mutPoly_nodes.attr* node attributes into the BioPAX pathway we loaded before and saved as *double-strand break repair*. Also, select the visual style *mutPoly_BioPAX* to visualize the nodes for which information on SNPs and mutagenesis experiments exists. Furthermore, we import the ataxia flag attributes from the *ataxia.attr* and select the visual style *mutPoly_BioPax_ataxiaFlag*. Then, we apply the layout *Hierarchic* (*see* Fig. 7). Make sure to use the right identifiers for the mapping (*see* **Note 12**).

Find the proteins in this pathway that are phosphorylated by ATM. Next to ATM, are there any other ataxia proteins in this pathway (*see* **Note 13**)?

3.5 Incorporating the Effect of Perturbation into Dynamic Network Models

Once the data integration is accomplished, it is of interest to determine the effect of different perturbations onto the dynamics of the biological processes in which the affected protein are involved. Here, the perturbations are the functional effects of SNPs or mutations on the activity of the proteins. This task is usually performed by laborious and time-consuming review of the literature. We propose that the integration of the data on the functional effect of sequence variations

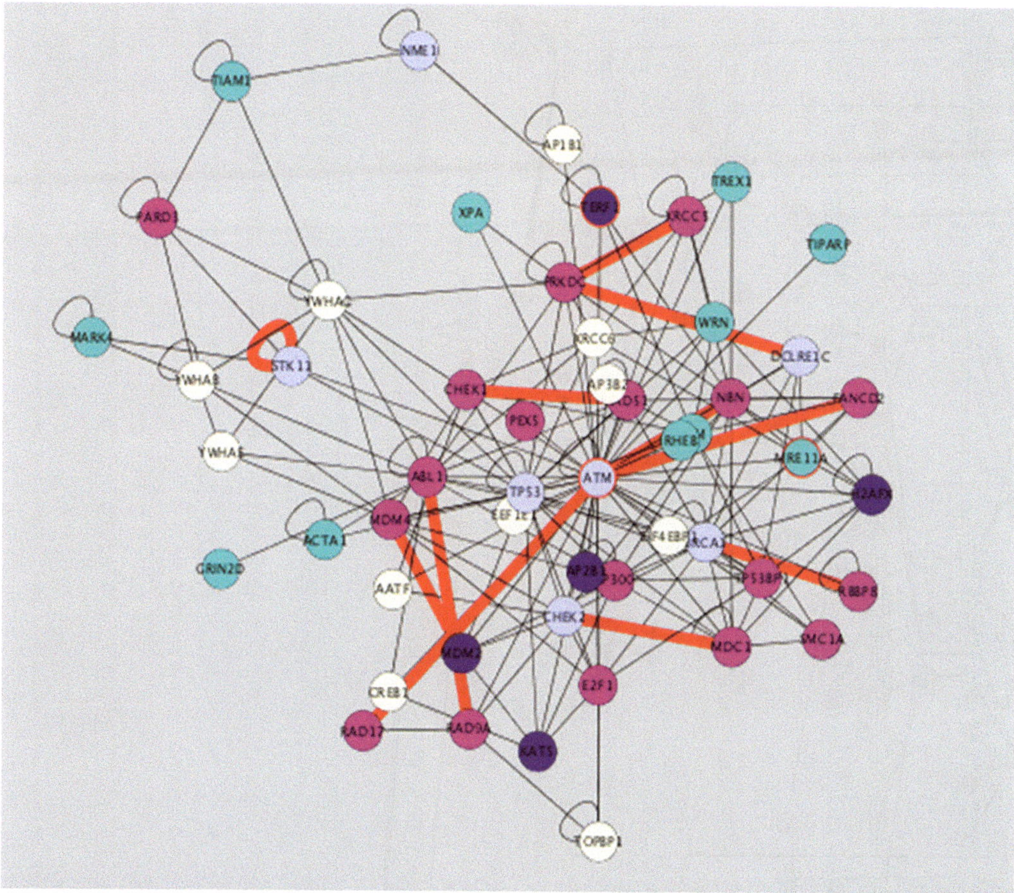

Fig. 6 PPI subnetwork of the ataxia protein ATM and its first neighbors. There are two more proteins related to ataxia (MRE11A and TERF1)

from UniProt with biological networks can aid in the evaluation of different perturbations on the dynamics of a model. To illustrate how this can be achieved, we use a model of the EGFR signaling network in MCF-7 cells [15]. We follow the same steps as described in the tutorial published in BMC Bioinformatics [9]. In brief, we first load a mathematical network model of the EGFR signaling in the SBML format into Cytoscape (similarly to loading pathways from Reactome) and then integrate information about sequence variations by loading the *mutPoly_nodes.attr* file. We then select a variation of interest; here we choose the mutation of serine 218 in the MEK1 kinase, which leads to protein inactivation. Next, we evaluate the effect of this sequence variation onto the dynamics of the signaling network by modifying the mathematical model accordingly (for details see [9]). The described strategy allows the evaluation of different conditions observed in experimental settings or in nature in predictive mathematical models. This is of particular interest, if the sequence variation is known to be associated with a disease, since the

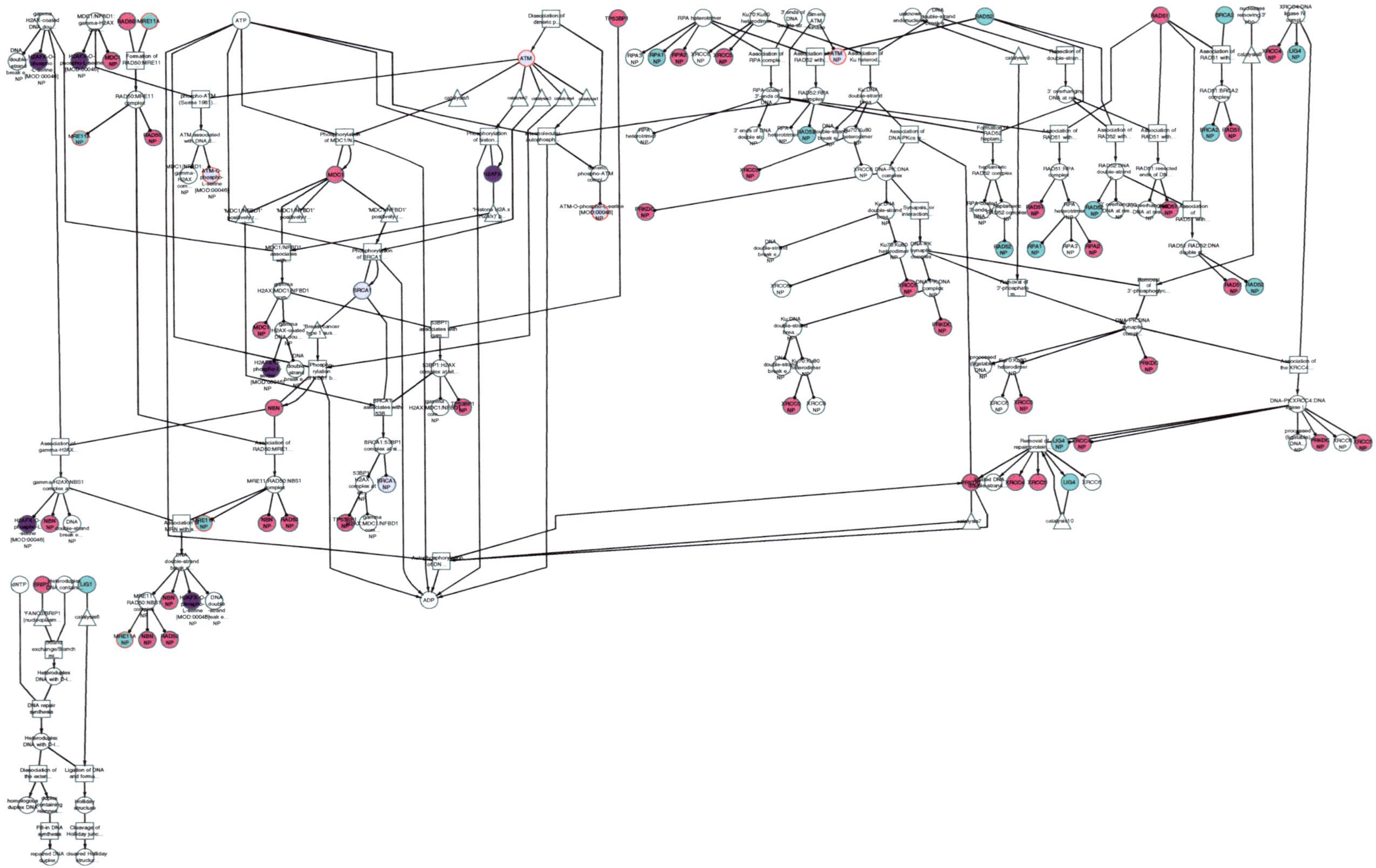

Fig. 7 ATM-mediated response to DNA double-strand break pathway in BioPAX format. Protein nodes for which SNPs or mutagenesis information is available are colored. *Red circles* around protein nodes denote that there is an SNP for this protein being associated with ataxia

model might help in understanding the mechanisms underlying the disease. Moreover, the approach can be used to analyze the effect of different mutations in the same model for the analysis of the polygenic character found in several complex diseases.

3.6 Conclusions

In this tutorial we have presented a strategy for the integration of information on SNPs and other sequence variations with biological pathways and PPI networks using Cytoscape. We want to stress that the presented workflow can be used to integrate any kind of genomic information with biological networks if the information is represented in a Cytoscape readable format such as a simple table. We showed how pathways and PPI networks are loaded into Cytoscape and how the information about sequence variations can be imported as node and edge attributes. In many illustrative use cases we explained the application of Cytoscape's built-in functions and plug-ins for powerful visualization and for the analysis of the effect of the sequence variations onto the network and pathway structures. Furthermore, we outlined how the integrated data can be used to study the functional impact of sequence variations onto the dynamics of the biological processes.

4 Notes

(All results might vary depending on the Cytoscape, HPRD, and Reactome version used):

1. Tab-delimited network representation: Use the readme file (readme.txt) provided by HPRD for more information. The PPI network is represented as tab-delimited text file (see BINARY_PROTEIN_PROTEIN_INTERACTIONS.txt). Each row represents an interaction between two proteins. Columns 1 and 4 represent the gene symbols of the genes encoding the interacting proteins. Columns 2 and 5 contain their HPRD identifier and columns 3 and 6 their RefSeq identifier, respectively. The other columns contain additional information about the interaction, such as the experimental evidence for the interaction and the PubMed identifiers of the original publication reporting the interaction. Such additional information regarding the interaction (edge) could be imported as edge attributes. However, we only import the raw interactions. Note that the first row represents an interaction of the protein ALDH1A1 with itself.
2. Import a network from text file: Use the *Show Text File Import Options* to customize the import. Uncheck the *space delimiter* box if you want to load edge attributes correctly while importing the PPI network. We suggest to first load the network and to add any attributes either for the nodes (proteins) or the edges (interactions) in subsequent steps. Make sure that only

those columns you want to import are actually colored. Disable any column you do not want to import by clicking on them. For a correct import use column 2 as *Source Interaction* and column 5 as *Target Interaction*. In case something goes wrong with the import, just delete the network by right clicking onto the network in the *Networks* tab of the *Control panel,* and repeat the import. Once the import is completed, the network is visualized and the *Nodes, Edges* tab in the left panel shows the number of nodes/edges in brackets. For a better representation of your network, you can apply any built-in layout algorithm. If nothing is known about the structure of the network, the Organic layout *Layout→yFiles→Organic* gives a good first overview.

3. Node attribute import: Use the *file→Import→Attribute from Table* function, and select the *mutPoly_nodes.attr* file. Use *Show Text File Import Options* to customize the import. Select only the tab delimiter and check the *Transfer first line as attribute names* to transfer the column header as attribute names. Note that these names will be used throughout your Cytoscape session as names for the attributes (e.g., to search for specific attributes). Hence, always assign informative names to the attributes. Some of our attributes, *mutagenesis, polymorphism,* all *GO* attributes, *extUniProtIds, OMIM,* and *dbSNP,* contain lists of elements (for an overview of all attributes applied, see Table 2). To correctly import them as lists, we need to configure the import for each attribute separately. Click on each of the list attributes with the right mouse to open the dialogue for the attribute import. Select *List of Strings* as attribute and the pipe symbol (|) as *List Delimiter* (*see* Fig. 8) before clicking *OK.* Repeat this step for each of the list attributes. Then, deselect the *Show Text File Import Options* and select *Show Mapping Options.* Here, it is crucial that the *Key column in the annotation file* and the *Key Attribute for the network* match. In this example, we want to map the *hprdId* column in the annotation file with the ID attribute on the network. Always check your mapping in the *Key Attributes* window in the lower right side, and only click on *Import* if the format of both columns matches (*see* Fig. 9). In case something goes wrong with the import, just delete all attributes (using the *Delete Attributes* icon in the *Data panel)* and repeat the import.

4. Visual style import: The *mutPoly.props* file contains information for different visual styles and can be imported into Cytoscape using the built-in function *File→Import→Vizmap Property File.* Alternatively, we can create visual styles within Cytoscape (for more information see http://wiki.cytoscape.org/Cytoscape_User_Manual/Visual_Styles). To select a visual style, go to the *VizMapper* tab in the *Control panel* on the left side.

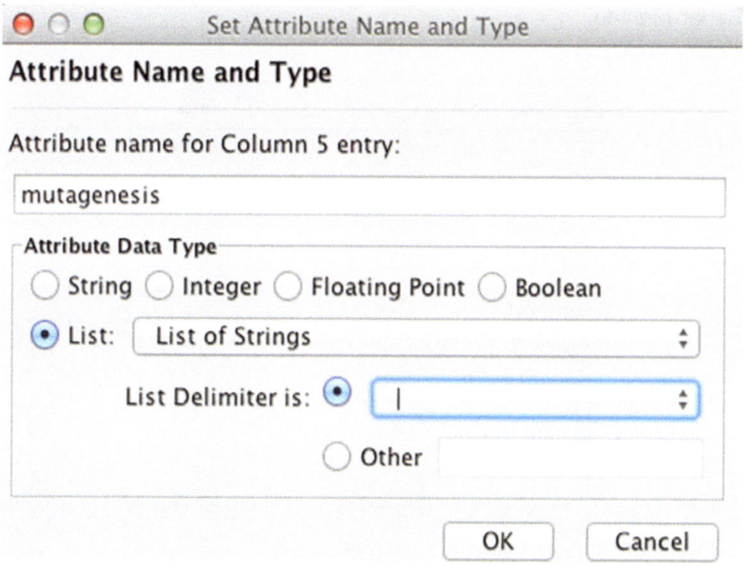

Fig. 8 Cytoscape dialogue to import node attributes as lists

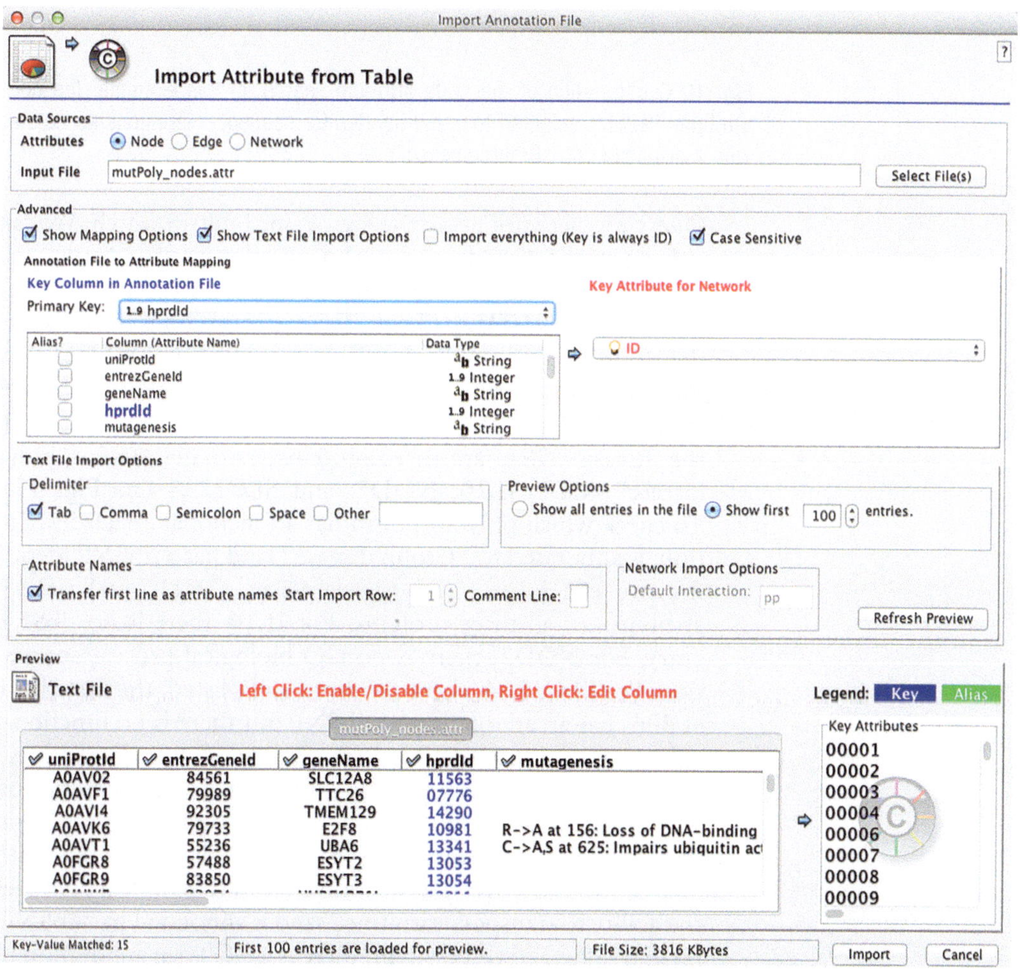

Fig. 9 Cytoscape dialogue to import node attributes. The mapping file contains column headers, which are imported as attribute names. HPRD identifiers (*hprdId* column) are mapped onto the primary identifiers (*ID*) in the network. The *Key Attributes* window (*lower right*) allows to check the mapping

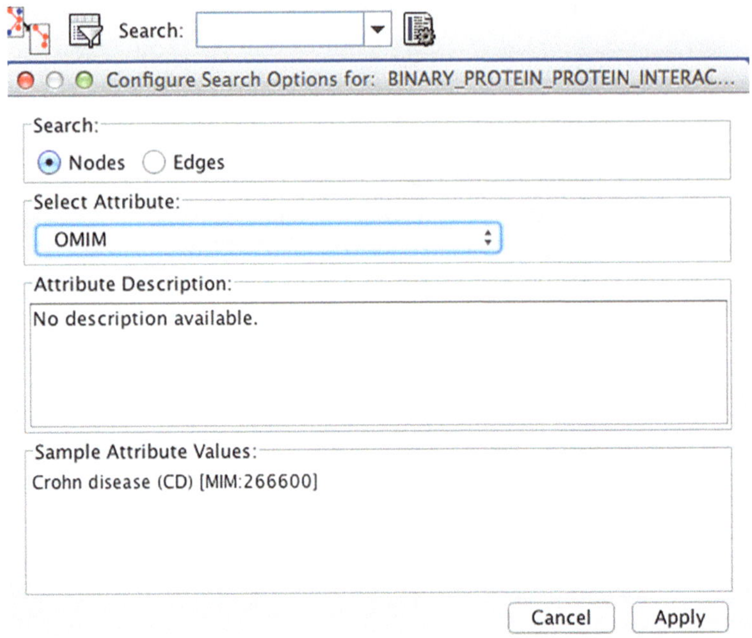

Fig. 10 Configuration of the node attribute search. In this example, the node attribute *OMIM* is selected to allow queries for sequence variations associated with a disease (e.g., "Crohn disease")

5. Query the network using the Cytoscape built-in search: We can either use the Cytoscape built-in function to search the network restricted by the node attribute *OMIM* or create a filter based on the *OMIM* attribute. The easier option is to configure the Search Options and select the attribute OMIM (*see* Fig. 10). This attribute contains disease information for proteins for which an SNP is known to be associated with the disease. There are three proteins associated with "Crohn disease" (CD), IL10, NOD2, and SLC22A4 (*see* Fig. 11). To check which polymorphism has a functional effect description, is annotated to "Crohn disease," and has a dbSNP identifier, we can inspect the *polymorphism, OMIM,* and *dbSNP* attributes of all three proteins. For IL10, there is no dbSNP entry. For NOD2, there is no GO biological process annotation. For SLC22A4, there are three SNPs listed: the first (I→T at 306) has an annotation to dbSNP but there is no functional effect description and it is not associated with "Crohn disease," the second (G→E at 462) has a functional effect description and an annotation to dbSNP but is not associated with "Crohn disease," and the third (*L→F at 503*) is associated with "Crohn disease," has the functional effect description "reduces the ability to transport carnitine," and is annotated to "dbSNP: rs1050152." For this protein, there is a GO term annotation to "transport" with the identifier "GO:0006810" (see the attributes

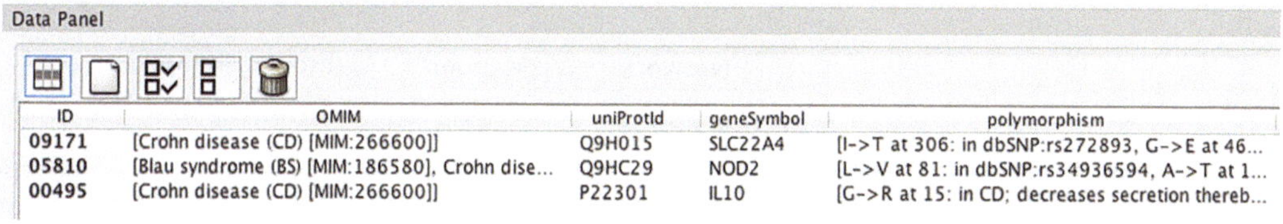
Data Panel

ID	OMIM	uniProtId	geneSymbol	polymorphism
09171	[Crohn disease (CD) [MIM:266600]]	Q9H015	SLC22A4	[I->T at 306: in dbSNP:rs272893, G->E at 46...
05810	[Blau syndrome (BS) [MIM:186580], Crohn dise...	Q9HC29	NOD2	[L->V at 81: in dbSNP:rs34936594, A->T at 1...
00495	[Crohn disease (CD) [MIM:266600]]	P22301	IL10	[G->R at 15: in CD; decreases secretion thereb...

Fig. 11 The three proteins, SLC22A4, NOD2, and IL10 for which an SNP is known to be associated with "Crohn disease." The *polymorphism* attribute contains information about which sequence variation is annotated to "Crohn disease"

GObiolProcess and *GObiolProcessId*). If you have difficulties visualizing the node attribute polymorphism for this protein, copy all or one entry into a text editor (right mouse click opens the dialogue).

6. Using the Cytoscape filter: To see how many proteins have SNP annotations, we apply the Cytoscape built-in filter using the node attribute *mutPolyFlag*. Go to the *Filters* tab, create a new filter (*Option→Create New Filter*), and name it *SNP_filter*. Add the *node.mutPolyFlag* attribute to the filter, and if not done already, pull the slider to the right to include all values between 1 and 4. The attribute is empty if there is no information on SNPs/mutations for this protein. Hence, by applying this filter we select all nodes for which the *mutPolyFlag* is between 1 and 4. In total, there are 6277 proteins with a sequence variant annotation. To figure out how many of these proteins have an SNP that is also associated with a disease, we can extend our filter using the *OMIM* node attribute. Now the filter contains two constraints: the *mutPolyFlag* must have a value between 1 and 4 and the *OMIM attribute* must not be empty. We can now apply the filter (*see* Fig. 12). There are 1,044 proteins for which this filter applies; thus for 1,044 proteins an SNP is known to be associated with a disease.
7. Import edge attributes: The edge attribute import from text is very similar to the import of node attributes (*see* **Note 3**). Use the *File→Import→Attributes from* Table function. Be careful to select *Edge* as the attribute type and to *transfer the first line as attribute names*. Moreover, check that the correct identifiers are mapped in the *Key Attributes* window in the lower right part of the import window (*see* Fig. 13).
8. Create subnetworks: First, we search the PPI network for proteins related to "ataxia" using the enhanced search plug-in (*see* Fig. 5). Once, the ataxia proteins are selected, use the Cytoscape built-in function *File→New→Network→From Selected Nodes, All Edges* to create a subnetwork, and name it "ataxia_proteins." Also apply the *Organic* layout. In total, there

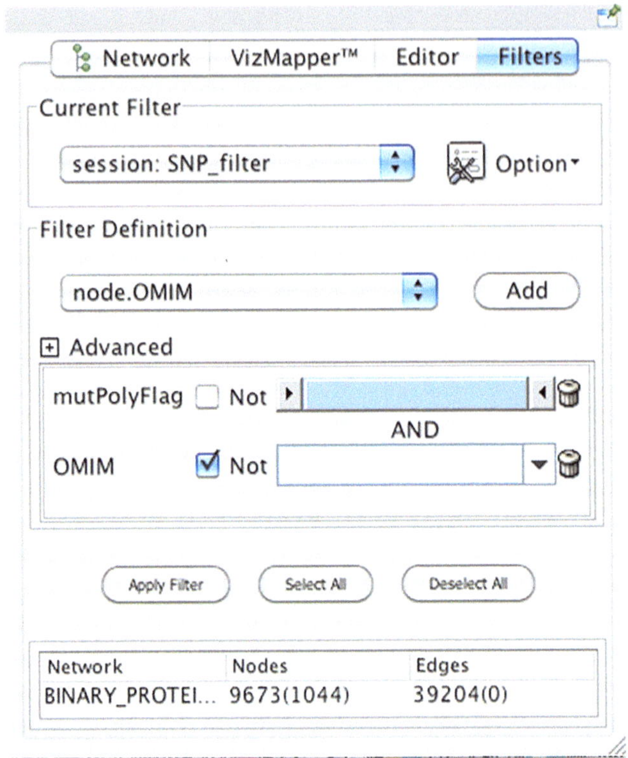

Fig. 12 Cytoscape filters with Boolean logic. A filter named SNP_filter is displayed, which filters for proteins that have an SNP/mutation (attribute *mutPolyFlag* between 1 and 4) and are also associated with a disease (attribute *OMIM* not empty)

are 16 proteins related to different ataxia subtypes. Some of the proteins are connected with other ataxia proteins, e.g., ATM, TERF1, and MRE11A. Second, we select all 16 ataxia proteins in the large PPI network (e.g., by applying again the enhanced search using "ataxia" as search string) and use the Cytoscape built-in function *Select→Nodes→First Neighbors of Selected Nodes* and then the *File→New→Network→From Selected Nodes, All Edges* to extend the ataxia protein network by the first layer of neighbors. We name this subnetwork with 169 nodes and 415 edges "ataxia_first_neighbors." We analyze the network using the *Plugins→Network Analysis→Analyze Network* function provided by the NetworkAnalyzer plug-in to study the network properties (treat the network as undirected). The analysis tells us that the network contains five connected components in total. We do not save or visualize the results.

9. Create node attributes: We can also create attributes within Cytoscape manually, instead of loading them into the networks. The easiest way to create the attribute *ataxia_protein*

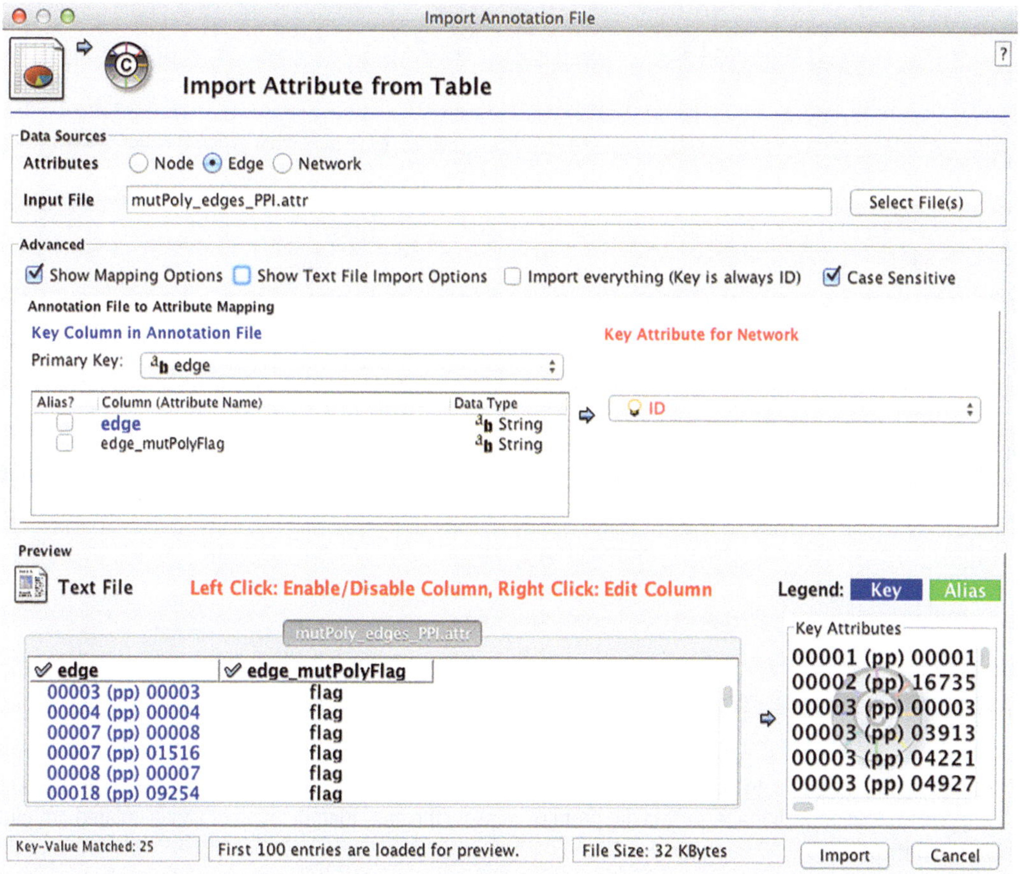

Fig. 13 Cytoscape dialogue to import edge attributes. The first line of the attribute file is used as attribute names

manually is to select all nodes in the small subnetwork only containing the 16 ataxia proteins. Then, we create a new string attribute (second icon on the left in the *Data panel*), name it *ataxia_protein*, and use the *Attribute Batch Editor* (first icon on the right in the *Data panel*) to fill it with the string "ataxia_protein" (*see* Fig. 14). The *ataxia_protein* attribute is now empty for all proteins except the 16 ataxia proteins, note that the attribute is set globally, and also ataxia proteins in the large PPI network have this attribute now. Next, we apply the visual style *mutPoly_PPI_ataxia_flag*. This adds a red circle to all ataxia proteins (those for which the *ataxia_protein* node attribute contains "ataxia_protein"). In the ataxia_first_neighbors network, we can see that most (12) ataxia proteins are in the largest connected component and there are four small connected components, each consisting of only two proteins where one is an ataxia protein. The fact that most ataxia-related proteins are either connected directly or indirectly

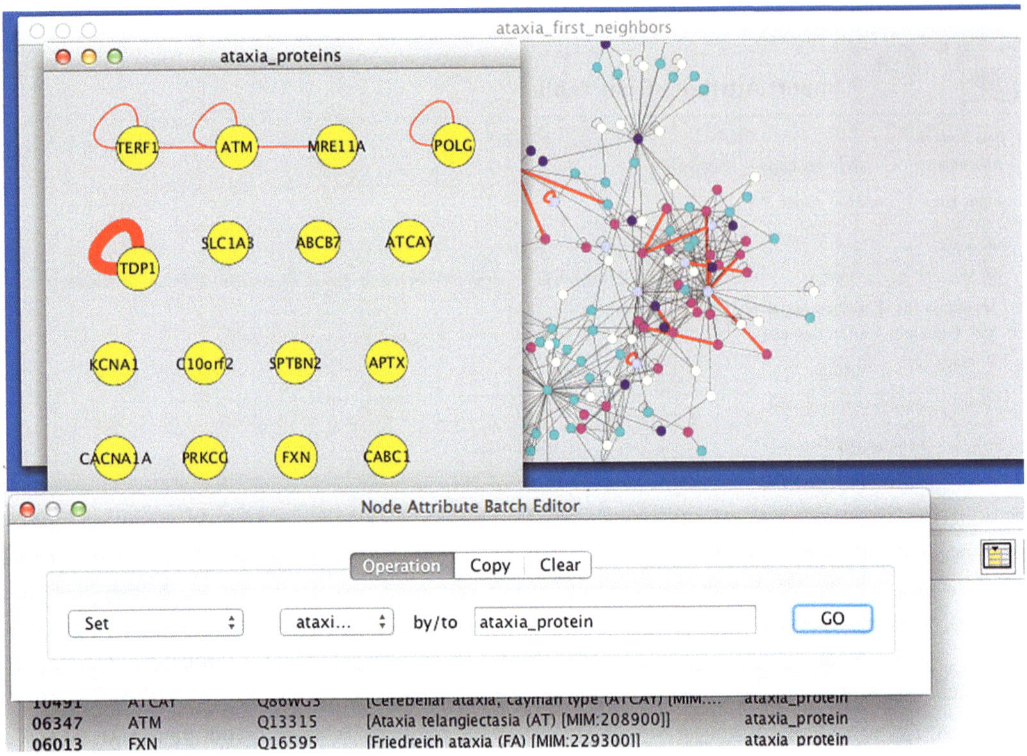

Fig. 14 Batch node editor to fill an attribute for selected nodes. In this example, the new node attribute *ataxia_protein* is filled with the term "ataxia_protein" to allow the easy identification of ataxia-related proteins in the PPI networks

through their neighbors and therefore form the large connected component suggests that similar mechanisms might be associated with the diverse ataxia types. The different forms of ataxia can be caused by mutations in different genes, which in turn lead to disruption of the same/similar processes represented by the large connected component. This finding is in good agreement with the research on ataxias by Lim et al. [14].

To search for an ataxia protein for which an SNP is affecting DNA binding, we use the Cytoscape built-in search on the node attribute *GOmolFunction* and search for *DNA binding* in the small network consisting of 16 ataxia proteins. The protein aprataxin with *geneSymbol* APTX and *uniProtId* Q7Z2E3 is associated with the AOA syndrome [MIM: 208920]. There is a polymorphism in this protein at position 277 (V→G), which is associated with the disease and which "abolishes DNA-binding and enzymatic activity towards Ap, MIM: 208920." If you have difficulties visualizing the node attribute polymorphism for this protein, copy all or one entry into a text editor (right mouse click opens the dialogue).

10. Analyzing the functional effect of sequence variations: We manually inspect the polymorphism and mutagenesis attributes of APTX and find that there is a mutagenesis experiment (R→A at 43) with the functional effect description "Impairs interaction with XRCC1 and XRCC4." If this sequence variation disrupts the interaction of APTX with XRCC1 and XRCC4 and the complex of the three proteins plays a role in DNA repair mechanism, then it is likely that the sequence variation has an effect on DNA repair. Note it is not known that this sequence variation is associated with the disease; nevertheless the disease-related mutations could have a similar effect and also affect DNA repair mechanism.
11. Analyzing the functional effect of sequence variations: We again use the Cytoscape built-in search using the node attribute *GOmolFunction* and search for *kinase* activity. In total, six proteins have sequence variations that affect *kinase activity* in the ataxia_first_neighbors network. But there is only one ataxia protein (ATM with *uniProtId* Q13315) for which there is mutagenesis data available about the "loss of kinase activity" for a mutation of D→A at position 2870. Interestingly, there is also an SNP at the same position related to ataxia, even though the amino acid exchange is different (polymorphism, D→N at 2870; in dbSNP rs55798854, MIM, 208900). In the ataxia_first_neighbors network, ATM has 52 neighbors, of which two are also related to ataxia (TERF1 and MRE11A). Use the *Select→Nodes→First Neighbors of Selected Nodes* function to create a subnetwork of first neighbors of ATM, name it "ATM_first_neighbors," and use the visual style *mutPoly_PPI_ataxia_flag*.
12. Import node attributes to a BioPAX pathway: The steps to import node attributes onto a pathway in BioPAX format are the same as for importing node attributes onto a PPI network (compare Note 3). Again, it is crucial to map the correct identifiers in the attribute file with the node attribute in the network. The BioPAX format contains a lot of information about the nodes (proteins, complexes, etc.) and edges (enzymatic reactions, formation of complexes, etc.). The node attributes, which are generated by the import of a BioPAX pathway, all contain the prefix "biopax." Hence, for a correct integration of the node attributes, we need to use the *biopax.xref.UNIPROT* attribute in the pathway (*see* Fig. 15) and map with the *uniProtId* attribute in the *mutPoly_nodes.attr* file. Also, we again need to modify the import for all list attributes such as *polymorphism* and *mutagenesis* (*see* Table 2).
13. Analyzing a BioPAX pathway: In the BioPAX format, triangles represent catalysis reactions. Use the Hierarchic layout (*Layout→yFiles→Hierarchic*). Now, select the protein ATM

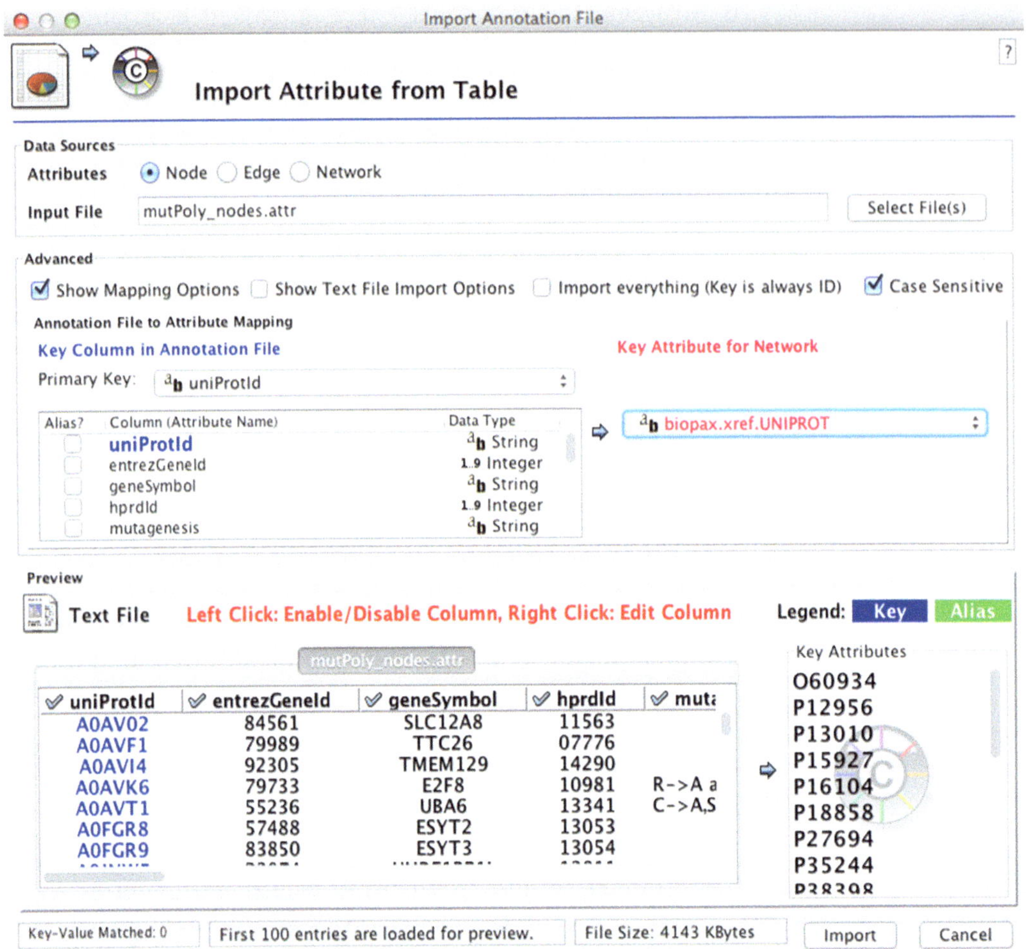

Fig. 15 Cytoscape dialogue to import node attributes onto a BioPAX pathway. The uniProtId column in the attribute file is mapped onto the node attribute *biopax.xref.UNIPROT* in the pathway

using the Cytoscape built-in search for the *biopax.xref. UNIPROT* Q13315. Then, use the *Select→Nodes→First Neighbors of Selected Nodes* function twice and create a new subnetwork. Apply the *Hierarchic* layout to the small subnetwork. The network contains five phosphorylation events catalyzed by ATM (see five triangles and the resulting phosphorylation events). Check the *biopax.entity.SHORT-NAME* attribute of these phosphorylation events. The proteins being phosphorylated are H2AX, MDC1/NFBD1, BRCA1, and NBS1. To inspect which other proteins are associated with ataxia in this pathway, we use the visual style *mutPoly_BioPAX_ataxiaFlag*, which marks all ataxia proteins (Note: you either need to fill the *ataxia_protein* attribute for this purpose or import the ataxia.attr node attributes file). Next to ATM, there is another protein associated with ataxia (MRE11A). We can also use the Cytoscape search using the *ataxia_protein* attribute to search for ataxia-associated proteins in the pathway.

Acknowledgements

This tutorial was inspired by the study of integrating information about sequence variations with biological networks published in 2009 [9]. First of all, I want to thank Laura I. Furlong who designed that study and who helped in developing a similar tutorial. I also want to thank Ferran Sanz for helpful discussions and Frank Wallrapp for reading the manuscript and testing the tutorial. Moreover, I want to thank Nicolas Le Novère who invited me to teach this tutorial in the EMBO/FEBS course and especially Vicky Schneider who invited me to write this chapter.

References

1. Bauer-Mehren A, Furlong LI, Sanz F (2009) Pathway databases and tools for their exploitation: benefits, current limitations and challenges. Mol Syst Biol 5:290
2. Bader GD, Cary MP, Sander C (2006) Pathguide: a pathway resource list. Nucleic Acids Res 34:D504–D506
3. Maglott D, Ostell J, Pruitt KD, Tatusova T (2006) Entrez Gene: gene-centered information at NCBI. Nucleic Acids Res 35:D26–D31
4. Apweiler R, Bairoch A, Wu CH, Barker WC, Boeckmann B et al (2004) UniProt: the Universal Protein knowledgebase. Nucleic Acids Res 32:D115–D119
5. Ashburner M, Ball CA, Blake JA, Botstein D, Butler H et al (2000) Gene ontology: tool for the unification of biology. Nat Genet 25:25–29
6. Shannon P, Markiel A, Ozier O, Baliga NS, Wang JT et al (2003) Cytoscape: a software environment for integrated models of biomolecular interaction networks. Genome Res 13:2498–2504
7. Cline MS, Smoot M, Cerami E, Kuchinsky A, Landys N et al (2007) Integration of biological networks and gene expression data using Cytoscape. Nat Protoc 2:2366–2382
8. Emig D, Cline MS, Lengauer T, Albrecht M (2008) Integrating expression data with domain interaction networks. Bioinformatics 24:2546–2548
9. Bauer-Mehren A, Furlong L, Rautschka M, Sanz F (2009) From SNPs to pathways: integration of functional effect of sequence variations on models of cell signalling pathways. BMC Bioinformatics 10:S6
10. Wang Z, Moult J (2001) SNPs, protein structure, and disease. Hum Mutat 17:263–270
11. Mishra GR, Suresh M, Kumaran K, Kannabiran N, Suresh S et al (2006) Human protein reference database—2006 update. Nucleic Acids Res 34:D411–D414
12. Jonquet C, Shah NH, Musen MA (2009) The open biomedical annotator. Summit on Translat Bioinforma 2009:56–60
13. Noy NF, Shah NH, Whetzel PL, Dai B, Dorf M et al (2009) BioPortal: ontologies and integrated data resources at the click of a mouse. Nucleic Acids Res 37:W170–W173
14. Lim J, Hao T, Shaw C, Patel AJ, Szabó G et al (2006) A protein-protein interaction network for human inherited ataxias and disorders of Purkinje cell degeneration. Cell 125:801–814
15. Birtwistle MR, Hatakeyama M, Yumoto N, Ogunnaike BA, Hoek JB et al (2007) Ligand-dependent responses of the ErbB signaling network: experimental and modeling analyses. Mol Syst Biol 3:144

Chapter 4

Visualization and Analysis of Biological Networks

Pablo Porras Millán

Abstract

The study of the interactome—the totality of the protein–protein interactions taking place in a cell—has experienced an enormous growth in the last few years. Biological networks representation and analysis has become an everyday tool for many biologists and bioinformatics, as these interaction graphs allow us to map and characterize signaling pathways and predict the function of unknown proteins. However, given the size and complexity of interactome datasets, extracting meaningful information from interaction networks can be a daunting task. Many different tools and approaches can be used to build, represent, and analyze biological networks. In this chapter, we will use a practical example to guide novice users through this process. We will be making use of the popular open source tool Cytoscape and of other resources such as : the PSICQUIC client to access several protein interaction repositories and the BiNGO plugin to perform GO enrichment analysis of the resulting network.

Key words Interactome, Protein–protein interactions, Databases, Network analysis, PPI networks, Cytoscape, PSICQUIC, GO enrichment analysis

1 Introduction

The advent of high-throughput methodologies for protein–protein interaction (PPI) detection that has taken place in the last years has resulted in an explosion of data and aims to systematically uncover the totality of molecular interactions that take place within a cell, what is known as the "interactome." Protein–protein interactions (PPIs) are the driving force behind most—if not all—cellular processes. Thus, the detection, representation, and analysis of PPIs have gained popularity in the scientific community, as its study is of seminal importance to build an integrated and comprehensive view on how cellular processes work. In this chapter we will show one of the multiple ways in which PPI networks can be represented and analyzed. We will discuss the limits and advantages of such approach, going from obtaining the interaction data from public databases and then representing it with the open source software Cytoscape, to finally functionally annotating and analyzing the dataset using the Gene Ontology.

Maria Victoria Schneider (ed.), *In Silico Systems Biology*, Methods in Molecular Biology, vol. 1021,
DOI 10.1007/978-1-62703-450-0_4, © Springer Science+Business Media, LLC 2013

PPI information is an invaluable resource that can be accessed through a number and variety of databases that curate, represent, and make the data available for the scientific community both manually and programmatically. Proper data representation entitles the use of unique, stable identifiers, controlled vocabularies, and cross-referencing to other types of resources such as the UniProt database [1] or the Gene Ontology [2]. However, this variety of resources also entails a significant amount of redundancy and a lack of homogeneity in data representation (i.e., different types of identifiers can be chosen to represent a gene or protein and it is often not straightforward to map from one type of identifier to another). Initiatives such as the International Molecular Exchange (IMEx) consortium guidelines [3], part of the Human Proteome Organization-Proteomics Standards Initiative (HUPO-PSI), aim to standardize the level of detail that needs to be captured in order to accurately represent an interaction. Members of the IMEx consortium, such as IntAct [4], MINT [5], or DIP [6], represent PPIs following these guidelines. These databases also aim to curate non-overlapping spaces of the interactome, with the goal to improve the coverage in the representation of PPI data. Nevertheless, heterogeneity in PPI data repositories is still a problem and "secondary" databases (also called "metadatabases") such as UniHI [7] or the MPIDB [8] aim to solve this by incorporating and clustering data from several other "primary" databases—such as those cited above—instead of curating their own data. Another integrative approach, even more powerful, is the one taken by the Proteomics Standard Initiative Common QUery InterfaCe (PSICQUIC) [9], a querying tool that enables common access to a large number of repositories containing PPI and pathway information datasets, including both primary and secondary databases.

As we stated before, PPIs depicted in the form of graphical networks are used as maps of the interactome where other types of information can be integrated in order to accurately describe cellular events. Cytoscape [10] is an open source software platform written in Java that is widely used by researchers for network representation and analysis. It features customizable options for network representation and it is relatively easy to use, but the reason behind its popularity and arguably its most powerful feature is the variety of plugins that have been developed for it. The plugins give the user the means to perform sophisticated analysis, to provide elaborated representation features, or to integrate complementary information to networks loaded in Cytoscape. New plugins are constantly developed for Cytoscape and if a researcher needs to perform a very specific type of analysis and he can write Java, he can create his own plugin and easily integrate it following the Cytoscape team specifications. In this chapter, we will make use of a plugin that enables access to the PSICQUIC query tool directly to Cytoscape and generate a PPI network. We will use the BiNGO plugin to perform GO enrichment analysis.

Making sense of PPI networks can be a daunting task, given the size and complexity of the information represented in them. The correct visualization of the network is the first step of the process and its importance should not be underestimated. We will use a part of this chapter to provide you with the basic knowledge you need to improve the visual features of a network as represented in Cytoscape. Apart from that, the study of the network topology, the identification of highly connected clusters within a network, and the integration of external annotations and additional information (such as, for example, expression profiles or subcellular localization information of the proteins represented in the network) are examples of the approaches that can be taken to tackle the complexity underlying these representations. Using these strategies allows for the production of meaningful biological maps, but has its limitations. Apart from the problem of the incompleteness of the interactome mappings and the presence of significant amounts of false positives, the topological nature of biological networks is still not well understood and the extension of the annotations characterizing the proteins that take part in them is far from comprehensive (*see* refs. 11, 12 for more information on the subject).

Nevertheless, there are certain resources that have allowed for meaningful analysis of PPI networks. One of the most popular resources for protein annotation is the Gene Ontology (GO) [13], a controlled vocabulary of terms that describe gene product characteristics in its three branches of ontology that represent three different aspects of the gene product biology: biological process, cellular component, and molecular function. GO terms are assigned both via manual curation and using computational approaches based on sequence similarity and common ancestry, for example. They are widely used to annotate large protein datasets and they are invaluable to characterize unknown regions of the interactome. In order to identify which annotations best describe a large list of proteins—either alone as list or as part of an interaction network—GO enrichment analysis has become a must in most works facing PPI network analysis. There are several plugins in Cytoscape that can help performing this analysis directly in a represented network. In this chapter we will briefly describe the main features of one of them: the very popular and simple BiNGO plugin [14].

2 Objectives

With the present tutorial you will learn the following skills and concepts:

1. To build a molecular interaction network by fetching interaction information from a public database using the PSICQUIC client through its app in the open source software tool Cytoscape.
2. To load and represent that interaction network in Cytoscape.
3. The basic concepts underlying network analysis and representation in Cytoscape: the use of attributes, filters, and plugins.
4. To integrate and make use of quantitative proteomics data in the network.
5. To add Gene Ontology annotation to a protein interaction network.
6. To use the BiNGO Cytoscape plugin to identify representative elements of GO annotation and learn more about the biology represented in the network.

3 Materials

3.1 Software Requirements

Cytoscape version 2.8.3 (downloadable from www.cytoscape.org) including the BiNGO v. 2.44 plugin (www.psb.ugent.be/cbd/papers/BiNGO) and the PSICQUIC Universal Client v. 0.31 plugin (*see* Subheading 7 for installation instructions).

3.2 Additional Files

The files you need to follow this tutorial can be found in www.ebi.ac.uk/~pporras/SpringerProtocolsBook/.

4 Methods

4.1 Introduction to Cytoscape

Cytoscape 2.8.3 is an open source, publicly available network visualization and analysis tool (www.cytoscape.org) [10]. It is written in Java and will work on any machine running a Java Virtual Machine, including Windows, Mac OSX, and Linux. We will use version 2.8.3 of Cytoscape in this tutorial. At the beginning of 2013, a new version of Cytoscape was released (3.0). However, the migration of many plugins to the new version was not completed at the time this chapter was written. The BiNGO plugin was not yet available for 3.0, so we will stick to 2.8.3 in this tutorial. In case you want to use Cytoscape 3.0, you can use ClueGO as an alternative to perform GO enrichment analysis and you will not have to install the PSICQUIC Plugin, since it is built-in the new version.

Cytoscape is widely used in biological network analysis and it supports many use cases in molecular and systems biology, genomics, and proteomics:

1. It can import and load molecular and genetic interaction datasets in several formats.
   - In this tutorial, we will import a molecular interaction network fetching data from IMEx-complying databases, such as IntAct or MINT, using the Cytoscape PSICQUIC plugin.
2. It can make effective use of several visual features that can effectively highlight key aspects of the elements of the network.
   - We will use node and edge attributes to represent quantitative proteomics data and interaction features.
3. It can project and integrate global datasets and functional annotations.
   - We will make use of resources such as the Gene Ontology to annotate the interacting partners in our network.
4. It has a wide variety of advanced analysis and modeling tools in the form of plugins that can be easily installed and applied to different approaches.
   - The BiNGO plugin will be used to perform GO enrichment analysis and try to identify the functional modules underlying our network.
5. It allows visualization and analysis of human-curated pathway datasets such as Reactome or KEGG.

4.2 Dataset Description

In order to easily illustrate the concepts discussed in this tutorial, we are going to follow a guided analysis example using a dataset from a work published by König et al. Our working dataset is going to be a list of proteins coming from a quantitative proteomic analysis of the "kinome" (the totality of the protein kinases encoded by the human genome) of regulatory and effector T cells [15]. The authors use immobilized unspecific kinase inhibitors to purify kinases from both regulatory and effector T cells and then use iTRAQ™ labeling to differentially label the proteins obtained from each one of these cell types. This way, they obtain a set of 185 kinases that can be identified in T cells with a high confidence. The relative abundance of such kinases in regulatory vs. effector T cells was calculated using the iTRAQ-based quantification in combination with a MS-devise-specific statistical approach called iTRAQassist and a RF value is given for each kinase in the list. We are going to use the list of kinases plus the RF values to find out which proteins are known to interact with these kinases and in which processes are they known to have a role.

4.3 Generating an Interaction Network Using the PSICQUIC Plugin in Cytoscape

We are going to generate a protein interaction network that will help us identify the biological functions associated with those kinases identified in both regulatory and effector T cells. To do this, we will find out which proteins are interacting with the ones represented in the dataset as stored in some of the different molecular interaction databases that comply with the IMEx guidelines [3]. Here is a list of the databases that we will use:

1. IntAct (www.ebi.ac.uk/intact): One of the largest available repositories for curated molecular interactions data, storing PPIs as well as interactions involving other molecules [4]. It is hosted by the European Bioinformatics Institute.
2. MINT (http://mint.bio.uniroma2.it/mint): MINT (Molecular INTeraction database) focuses on experimentally verified protein–protein interactions mined from the scientific literature by expert curators [5]. It is hosted in the University of Roma.
3. MatrixDB (http://matrixdb.ibcp.fr): Database focused on interactions of molecules in the extracellular matrix, particularly those established by extracellular proteins and polysaccharides [16]. The data in MatrixDB comes from their own curation efforts, from other partners in the IMEx consortium and from the HPRD database. It also contains experimental data from the lab of professor Ricard-Blum in the Institut de Biologie et Chimie des Protéines in the University of Lyon, where it is hosted.
4. DIP (http://dip.doe-mbi.ucla.edu/dip): DIP (Database of Interacting Proteins) is hosted in the University of California, Los Angeles, and contains both curated data and computationally predicted interactions [17].
5. I2D (http://ophid.utoronto.ca/i2d): I2D (Interologous Interaction Database, formerly OPHID) integrates known, experimental (derived from curation), and predicted PPIs for five different model organisms and human [18]. It is hosted in the Ontario Cancer Institute in Toronto.
6. InnateDB (www.innatedb.com): InnateDB is a database of the genes, proteins, experimentally verified interactions, and signaling pathways involved in the innate immune response of humans, mice, and bovines to microbial infection [19]. Regarding their PPI datasets, they come both from their own curation and from integrating interaction data from other databases.

We will use the Protemics Standard Initiative Common QUery InterfaCe (PSICQUIC) importing plugin that can be found in Cytoscape (named as PSICQUICUniversalClient v. 0.31 if you look for it in the plugin installation wizard). PSICQUIC is an effort from the HUPO Proteomics Standard Initiative (HUPO-PSI, www.hupo.org/research/psi/) to standardize the access to molecular interaction databases programmatically, specifying a standard web service with a list of defined accessing methods and a common query language that can be used to search from data in many different databases. You will learn more about PSICQUIC further ahead in this chapter, but if you want to have more information, check their Google Code website at http://code.google.com/p/psicquic/ or have a look at the Nature Methods publication where the client is described [9]. PSICQUIC allows you to access data from many different databases, but we will limit our search to those resources that comply with the IMEx consortium curation rules (www.imexconsortium.org/curation) as listed before (*see* **Note 1**).

1. Open the file "TableS1_mapped.xlsx." This is an updated version of Supplementary Table 1 the König et al. publication in which each kinase has been mapped to their UniProtKB (www.uniprot.org) accession numbers (*see* **Note 2**).
2. Open Cytoscape and go to "File" → "Import" → "Network from Web Services." In the window that will appear, select the "PSICQUIC Universal Web Service Client" option from the "Data source" drop-down menu. To search for the interactions in which the proteins from your list are involved, you just have to paste the list of the UniProt AC identifiers in the query box and click "Search" (*see* **Notes 3** and **4**).
3. You will get a dialog window with the total number of interactions found by PSICQUIC among the different databases (or "services") that the client can access and you will be asked if you want to create a network out of them. Click "Yes" and then a list of services with the amount of interactions found for each one of them will show up.
4. For the selection of the source of our interactions, we will stick to just IMEx-complying datasets. You should get interactions from IntAct, DIP, I2D-IMEx, InnateDB-IMEx, MINT, and MatrixDB, among other resources that store predicted interactions or pathways or are just not IMEx-compatible. We will ignore these to avoid problems while merging the data from the different repositories. Notice that some databases, such as I2D or InnateDB, identify a subset of their interactions as "IMEx-complying." The number of interactions found for each database changes with time, because they are constantly updated. Select just the IMEx-complying datasets we mentioned before in the "Import?" column and then click "OK."
5. You will get yet another dialog box from which you will have a list of your databases of choice and the option to merge the results from them or just have them in separated networks. Click "Merge" and the "Advanced network merge" assistant will pop up.
6. Now the "Advanced Network Merge" assistant will open up. Select the networks you want to merge (in our case, all of them except the "PSICQUIC Search Results..." one) and then click on the "Advanced Network Merge" menu to select the

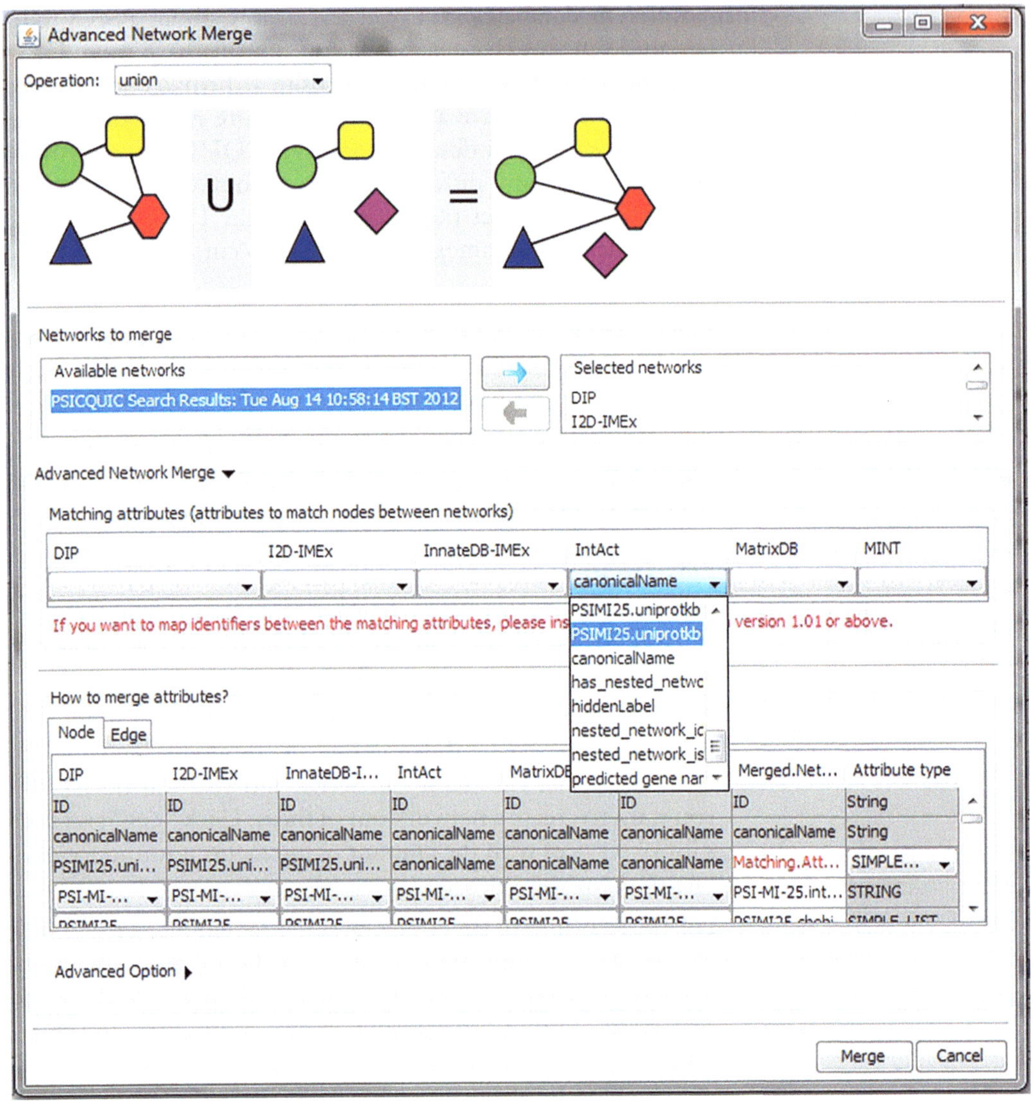

Fig. 1 "Advanced network merge" menu in Cytoscape 2.8.3. Notice the drop-down menus for each of the networks you select that allow you to choose which attribute will be used as an identity reference for each node when performing the merge

identifier you will use as a common ID for the merge. In our case, we are merging protein–protein interaction information and we will use UniProtKB ACs as our primary identifier. You will see a drop-down menu appearing for each network you select to be merged (*see* Fig. 1). In each drop-down menu you will find a list of the "attributes" that each node or edge of the network is assigned during the import. We will talk more about attributes later, for now, just select the attribute "PSICQUIC25.uniprotkb.top" in each menu. This attribute contains the UniProtKB AC for each node, so the merging can proceed properly.

7. Finally, several networks will be created by the PSICQUIC client plugin. The first one is just a graphical representation of the different resources that were associated with your query, named "PSICQUIC Query Results..." and the time and date of your query. Then a different network will be created for each of the resources that were accessed by PSICQUIC and will be named accordingly. The final one will be called "Merged.Network" and is the one we will use for our analysis. The networks will look like a grid of squares (nodes) connected by many lines (edges). We will learn how to make sense of it in the following sections of the tutorial.
8. Finally, since Cytoscape can be tricky (and buggy) and you don't want your precious time to be wasted, save your session (go to "File" → "Import," click on the floppy icon up left or just press "Ctrl + s"). A piece of advice: do this every time you want to try something new with Cytoscape, since going back to your initial file is sometimes not possible and you can waste a lot of time re-doing a lot of work!

4.4 Representing an Interaction Network Using Cytoscape

Finding a meaningful representation for your network can be more challenging than you might expect. Cytoscape provides a large number of options to customize the layout, coloring, and other visual features of your network. This tutorial does not aim to be exhaustive in exploring the capabilities of Cytoscape; we just want to give you the basics. More detailed information and basic and advanced tutorials for Cytoscape can be found in their documentation page: www.cytoscape.org/documentation_users.html.

Now we will learn how to use the basic tools that Cytoscape provides to manage the appearance of your network and make the information that it provides easier to understand.

1. If it is the first time you use Cytoscape, have a look at the user interface and get familiar with it. The main window displays the network (all the network manipulations and "working" will be visualized in this window). The lower-right pane (the Data Panel) contains three tabs that show tabulated information about node, edge, and network attributes. The left-hand pane (the Control Panel) is where navigation, visualization, editing, and filtering options are displayed.
2. By default, Cytoscape lays out all the nodes in a grid, so that is why your network is looking so ugly. You can change the layout going to "Layout" → "Cytoscape Layouts." There is a wide range of different layouts that will help displaying certain aspects of the network, like which proteins have a large number of interaction partners (the so-called "hubs"). Give some of

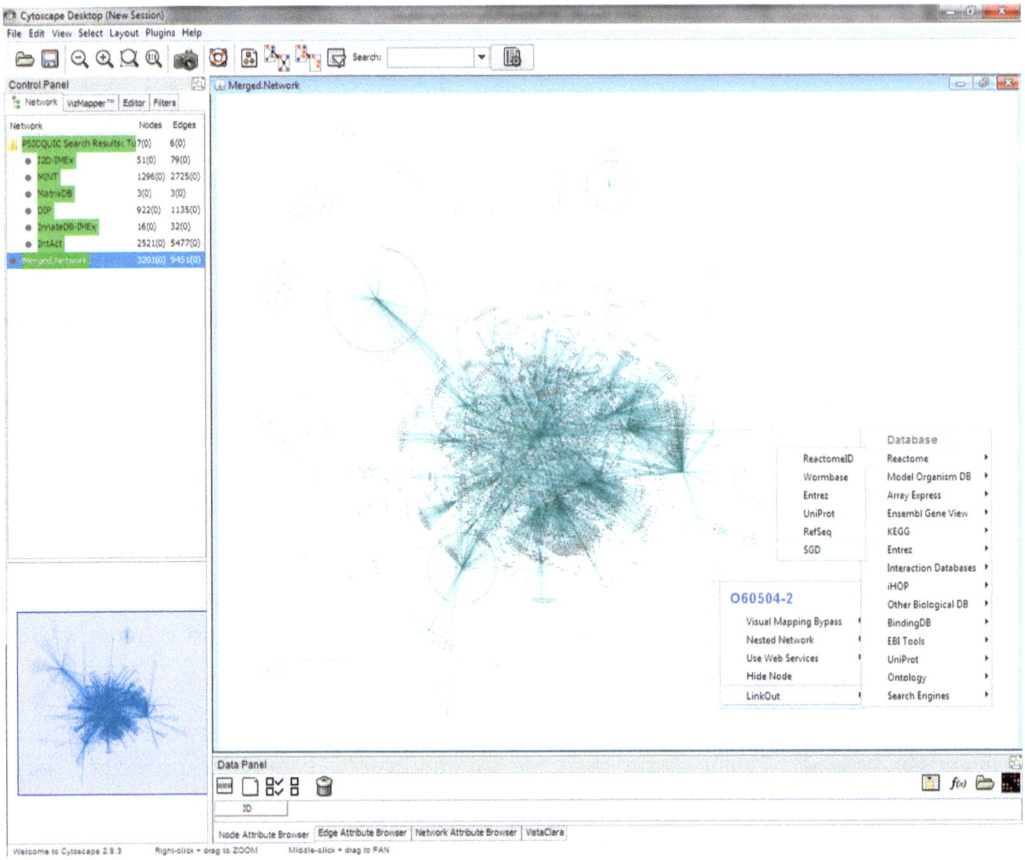

Fig. 2 Network visualization using the organic layout in Cytoscape 2.8.3. If you right-click a single node, a menu with different options will appear, and you will be able to select the "LinkOut" tool and perform further searches for the information concerning that particular node

them a try and stick to the one you prefer, like the "organic" layout shown in Fig. 2.

3. If you right-click on a node in the network representation, a small menu will open where you can see some representation options and the "LinkOut" tool (*see* Fig. 2, right-hand side). This tool allows you to quickly perform a web search for the ID of the node in question in a variety of databases and resources.
4. **Save your session** when you are happy with a layout and have tried the "LinkOut" tool.

Exercise: Find a layout that sorts your network nodes by the number of interactions that each one of them has.

4.4.1 Filtering with Edge and Node Attributes

In network graphs, interacting partners are represented as **nodes**, which are objects represented as circles, squares, plain text ... that are connected by **edges**, the lines depicting the interactions. All information referred to an interacting partner or an interaction must then be loaded in Cytoscape as a node or an edge **attribute**. An attribute can be a string of text, a number (integer or floating point), or even a Boolean operator and can be used to load information and represent it as a visual feature of the network. For example, a confidence score for a given interaction between two participants represented as nodes can be represented as the thickness of the edge connecting those nodes. Attributes can be created and loaded directly in Cytoscape using the "Create New Attribute" icon on the top of the Data Panel and then values can be added using the "Attribute Batch Editor" icon (*see* Fig. 5 as a reference for icons). The attributes can also be imported from data tables defined by the user or from external resources, as we will see later, and directly imported with the network from different network formats, as we will see right now.

Because we have used the PSICQUIC client, the information we took from the different PPI databases will be represented complying with the PSI-MI-2.5 tabular format (*see* **Note 5**), so the fields requested by the format will be loaded as attributes and we can start making use of them right away.

1. Let's have a look at the attributes that have been loaded with our network. First, select all the nodes and edges of the network.
2. Have a look at the Data Panel below the main window. By default, you should be in the Node Attribute Browser tab. So far, you can only see one column "ID" which corresponds to the identifier that Cytoscape uses for each node.
3. Click on the "Select All Attributes" icon in the Data Panel. All the attributes that have been loaded from the XGMML file will now be visible in a tabular format.
4. As you can see, there is a large number of attributes (some of them redundant, due to the merging of networks) and it is difficult to read the table. You can also select and load only those that you want to show by clicking the "Select Attributes" icon in the Data Panel. Choose the following node attributes to be displayed and try to figure out their meaning:
   (a) Predicted gene name
   (b) PSI-MI-25.uniprotkb
   (c) PSI-MI-25.uniprotkb.top
   (d) PSI-MI-25.taxid
   (e) PSI-MI-25.taxid.name
5. If you right-click on the node attributes in the table that appears below, you can perform a "Search [your term] on the web" in a similar way you do when you right-click on the nodes represented in the network and perform a "LinkOut" search.

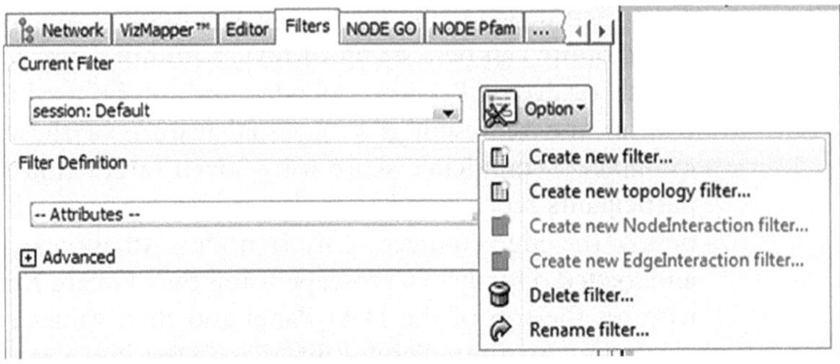

Fig. 3 "Filters" tab in Cytoscape 2.8.3. The "Option" box allows you to choose to create a new filter of different types and rename or delete an existing one. Filtering details are entered using the "Filter definition" box

6. Now go to the "Edge Attribute Browser" tab and do the same with the following edge attributes:
   (a) PSI-MI-25.interaction detection method
   (b) PSI-MI-25.interaction detection method.name
   (c) PSI-MI-25.interaction type
   (d) PSI-MI-25.interaction type.name
   (e) PSI-MI-25.source database
   (f) PSI-MI-25.source database.name
   (g) PSI-MI-25.author
   (h) PSI-MI-25.pubmed
   (i) PSI-MI-25.ConfidenceScore.author-score/mint-score/intact-miscore (*see* **Note 6**)

Let's make use of some of these attributes. Sometimes, homolog proteins coming from different species are used to perform interaction experiments. For this reason there are a number of "human-other species" interactions in the databases. Now we will use the "PSIMI25.taxid" node attribute to produce a human proteins-only network.

1. In the Control Panel, go to the "Filters" tab (*see* Fig. 3).
2. Choose "Create new filter" in the "Option" menu and give your filter a name (e.g., "human only").
3. Go to the "Filter definition" section. In the "Attributes" drop-down menu, choose the attribute you want to use for filtering. In this case, we will use the node attribute "PSIMI25.taxid." Select it and click "Add."
4. A search bar/drop-down menu called "PSIMI25.taxid" will appear where you can select the attribute value that you want to use. This attribute stores NCBI taxonomy identifiers for the species origin of each protein in the network. The code for human is "9606," write it down in the search bar and then click "Apply filter."

5. The nodes that bear the "9606" attribute will be then selected and highlighted in the network. Combinations of different attributes can be applied by using the "Advanced" menu in the Filter definition box.
6. Now generate a new network containing only human proteins by going to "File" → "New" → "Network" → "From Selected Nodes, All Edges." Alternatively, you can click the quick "Create new network from selected nodes, all edges" button at the top off the session window.
7. **Save your session.**

Exercise: Multiple methodologies can be used for PPI detection, each method entailing its own strength and weaknesses and none of them being perfect, since every PPI detection approach must be considered artefactual to some degree (several reviews on the subject are recommended in the Subheading 7 at the end of the chapter). Nevertheless, sometimes you want to look at interactions found with a particular methodology. Use edge attributes to create a network in which all the interactions have been found using the "two hybrid" method.

4.4.2 Integrating Quantitative Proteomics Data: Loading Attributes from a User-Generated Table

In order to load large amounts of information associated with the proteins in our network, it is often useful to import user-defined tables containing external data that can complement the network analysis. In our particular case, we will make use of the differential expression values that are given in Supplementary Table 1 of our selected publication in order to highlight the proteins that are enriched either in regulator or in effector T cells. Since no interaction information was extracted from the original article, the information we put in will be exclusively node-centric (no edge annotations) and can be loaded in the form of a user-produced node attributes table (*see* **Note 7**).

1. Open the "Table1_mapped.xlsx" file. This is an adaptation of the Table 1 in the original article. Have a look at the different fields and figure out what is represented in each column.
2. In Cytoscape, go to "File" → "Import" → "Attribute from table (text/MS Excel)…." The "Import Attribute from Table" wizard will pop up.
3. Select the attributes file in the "Data Sources" section and be sure to check the mapping and text file import options from the "Advanced" section while performing the import. It is important that you import the first line of the table as attribute names and that you choose the primary key for the attribute that will map with the key attribute in the network. In this case, the primary key in the attribute file will be "UniProt_AC" and the attribute you want to map to in the network is "PSI-MI-25.

uniprotkb.top." Both fields are populated with UniProtKB ACs, as can be seen in the "Preview" section.

4. In the "Preview" section you can choose which fields to import as new attributes in our network. Have a look and leave out the "Name (UniProt) [1]" attribute, since it would be redundant with the "predicted gene name" we got already in the network. Click "Import" to finish the process.
5. Finally, show the new node attributes in the "Data panel" using the "Select Attributes" icon in the Data Panel. Notice that only the proteins that were part of the original proteomics dataset from the paper have values in the newly imported attributes.
6. **Save your session**.

4.5 Using the Visual Representation Features of Cytoscape: VizMapper

After having integrated the quantitative proteomics information from the publication in the form of node attributes, we can use the visual editor of Cytoscape, VizMapper, to represent this information in our network in a meaningful way. This tool opens many representation possibilities, so we will just give an example to learn the basics.

1. Go to the "VizMapper" tab in the "Control Panel." Click on the "Options..." icon to create a new visual style and give it a name.
2. Click on the "Defaults" panel and select some default values for the node and edge colors, shapes, and size that make them easy to see. Don't use a big size (over 30) nor green or red as colors, since we are going to use them later on.
3. We are going to show the confidence with which the proteins in our dataset were identified with mass-spectrometry and whether they were over- or down-represented in the regulatory T cells with respect to the effector T cells. We will use the size of the nodes to represent confidence and node color for over- and under-representation.
4. In the "Visual Mapping Browser," look for "Node Color" first. Double-click and choose "Differential expression (RFmedian-Treg/Teff) [6]" as reference and "Node Color" and "Continuous Mapping" as "Mapping type" option. A graphical interface will appear and you can select how the node color will change between two reference colors (green and red, for example). Pick your favorite colors and have a look at the representation.
5. Now you can try to use the "Absolute differential expression (RFmedian-Treg/Teff) [7]" for "Node height" and "Node width." This way, the relative enrichment of a given kinase in one cell type or the other becomes even more evident. Use the "Continuous mapping" option again. You can try to use other mapping options and see what happens.

6. Now you have a representation in which we can easily differentiate between the original protein dataset, in which quantitative proteomics data has been integrated and represented, and its interactome context as given by PSICQUIC.
7. **Save your session**.

Exercise: Try to create a sub-network to see how the proteins that are over-represented in regulator T cells are connected (use the >1.5 cut-off that the authors use in the publication). Make use of filters and the "Create new networks for selected nodes, all edges" function.

4.6 Adding Annotation to a Network: Loading GO Annotations with Cytoscape

Protein interaction networks can be used as backbones in which to set up the elements of new pathways or functions; but in order to be able to do that, we need to have access to information about the elements of the network. We can make use of the functional annotation that is associated to genes and proteins to enrich our network with such information. One of the most important resources that annotate genes and proteins is the Gene Ontology (GO) project [2], which provides structured vocabulary terms for describing gene product characteristics (*see* **Note 8**).

Every GO annotation is associated to a specific reference that describes the work or analysis supporting it. The evidence codes indicate how that annotation is supported by the reference. For example, annotations supported by the study of mutant varieties or knock-down experiments on specific genes are identified with the IMP (Inferred from Mutant Phenotype) code. All the annotations are assigned by curators with the exception of those with the IEA code (Inferred from Electronic Annotation), which are assigned automatically based in sequence similarity comparisons. See http://geneontology.org/GO.evidence.shtml for more information about evidence codes.

The PSICQUIC plugin might have a red exclamation mark by it, stating that it has not been verified to work with this particular version of Cytoscape. Do not worry about it, we have tried it and it works.

First we will learn how to map GO terms, along with some general gene and protein annotation, to our interaction network. The objective is to bring some information to the nodes that were added from PSICQUIC, where little more than the name and a set of identifiers is given.

1. Go to "File" → "Import" → "Ontology and annotation…." This will open the "Import Ontology and Annotation" wizard (see screenshot in the next page).
2. In the "Data Source" section, select the "Annotation" file from the drop-down menu. In our case, we need the gene

association file for *Homo sapiens*. For the "Ontology" drop-down menu, select to import "Gene Ontology full."

3. Select the "Show mapping options" tick box in the "Advanced" section. As in the node attributes import, select the appropriate field as "Primary Key" in the Annotation file by checking the "Preview" section. In this case, the one to select is "DBObject_ID." The "Key Attribute" for the network is again "PSI-MI-25.uniprotkb.top."
4. In the "Preview" section, have a look at the information you are about to import as node attributes and figure out the meaning of the different fields. Click "Import" when you are done.
5. Go to the "Data Panel" and select the new node attributes "annotation.GO BIOLOGICAL_PROCESS," "annotation. GO CELLULAR_COMPONENT", and "annotation.GO MOLECULAR_FUNCTION" to be shown.
6. Click on one of the cells showing any of these three attributes and you will get a menu from which you can see all the GO terms associated with each protein as a list. As it happens with nodes and normal node attributes when you right-click on them. From each term a menu will show up allowing you to copy one or all the terms associated to that protein or to perform a search with the LinkOut tool.
7. **Save your session.**

4.7 Analyzing Network Annotations: GO Enrichment Analysis

As we have seen, we have incorporated annotation in the form of GO terms to the proteins in our network, but it is difficult to interpret and access that information when we try to analyze more than a few nodes, due to both the amount of information and its level of detail. Some of the terms will be redundant as well and distributed through many of the proteins represented in our list or network. GO enrichment analysis aims to figure out which terms are over- or under-represented in the population, thus extracting the most important biological features that can be learned from that particular set of proteins.

There are a couple of important considerations to make before doing any GO enrichment analysis, so we will briefly comment on them.

To start with, you will need to have solid knowledge about the biological and experimental background of the data you are analyzing to draw meaningful conclusions. For example, if you analyze a list of genes that are over-expressed in a lab cell line, you have to be aware that cell lines are essentially cancer cells that have adapted to live in Petri dishes. You will find a lot of terms related to negative regulation of apoptosis, cell adhesion, or cell cycle control; but that just reflects the genetic background your cells have.

It is also important to take into account that certain areas of the gene ontology are more thoroughly annotated than others, just because there is more research done in some particular fields of biology than in others, so you have to be cautious when drawing conclusions. GO terms are assigned either by a human curator that performs manual, careful annotation or by computational approaches that use the basis of manual annotation to infer which terms would properly describe uncharted gene products. They use a number of different criteria always referred to annotated gene products, such as sequence or structural similarity or phylogenetic closeness. The importance of the computationally derived annotations is quite significant, since they account for roughly 99% of the annotations that can be found in GO. If, nevertheless, you do not want to use computationally inferred annotations in your analysis, they can be filtered out by excluding those terms assigned with the evidence code "IEA" (Inferred from Electronic Annotation). Most analysis tools support this feature.

Finally, another factor that will make the analysis of GO annotation challenging is the level of detail and complexity you can reach when annotating large datasets. GO terms can describe very specific processes or functions—what is called "granularity"—and it is often the case that even the result of a GO enrichment analysis is way too complex to understand due to the large number of granular terms that come up. In order to solve this problem, specific sets of GO annotation that are trimmed down in order to reduce the level of detail and the complexity in the annotation are provided by GO or can be created by a user in need of a specific region of the ontology to be "slimmed." Check www.geneontology.org/GO.slims.shtml to learn more about them. Apart from that, some tools, such as ClueGO [20], give the option to cluster together related terms of the ontology, highlighting groups of related, granular terms together.

There are a number of tools that allow to perform this analysis using a list of genes or proteins as input, such as the DAVID Web Service [21] (see http://david.abcc.ncifcrf.gov/) or the previously mentioned ClueGO. We will present here the use of a simple tool that can use networks as an input and that make use of the visualization capabilities of Cytoscape to help the interpretation of the analysis: the BiNGO plugin.

4.8 Using BiNGO for Functional Annotation

In order to perform network-scale ontology analysis, we are going to use the BiNGO tool (www.psb.ugent.be/cbd/papers/BiNGO), a Cytoscape plugin that annotates proteins (nodes) with gene ontology (GO) terms and then performs an enrichment analysis [14]. BiNGO works by providing an answer to this basic question:

"When sampling X proteins (test set) out of N proteins (reference set; graph or annotation), what is the probability that x or more of these proteins belong to a functional category C shared by n of the N proteins in the reference set."

The main advantage of BiNGO with respect to other enrichment analysis tools is that it is very easy to use and it can be complemented with the basic network manipulation and analysis tools that Cytoscape offers. It also can provide its results in the form of a network that can be further manipulated in Cytoscape, a feature that eases the analysis, and it can be used in combination with its sister tool PiNGO [22], which can be used to find candidate genes for a specific GO term in interaction networks. On top of that, it is relatively light-weight when it comes to usage of computer resources and it can be run with reasonable speed in any desktop computer. On the negative side, it is not as customizable and does not offer as many visualization options as the more advanced tool ClueGO, for example.

1. Before you start, take into account that the Gene Ontology is updated continuously and both the ontologies and the annotations that are loaded by default in BiNGO are usually out of date. You should download the most updated version of the ontology file, which holds the structure and relationships between GO terms, from www.geneontology.org/GO.downloads.ontology.shtml. Get the full ontology file (OBO 1.2 version) and save it as "gene_ontology_ext.obo." The annotation file, holding list of proteins that are annotated for specific terms grouped by organism, must also be updated and can be downloaded from www.geneontology.org/GO.downloads.annotations.shtml. Save the file corresponding to human as "gene_association.goa_human."
2. As a starting point, we will apply the BiNGO analysis to the whole dataset, in order to see an overview of all the processes over-represented in this network. Subsequent analyses may then focus on sub-sets of the network, using a view suitable to pick out functional modules. Select all the nodes in the network.
3. To start BiNGO, go to "Plugins" → "Start BiNGO 2.44." Do this only once: Cytoscape will not stop you from opening multiple copies of the BiNGO setup menu (which will lead to confusion and chaos!).
4. The BiNGO setup screen will now appear. There are several operations you need to perform in this screen:
   (a) Name the fraction of the network you are going to analyze in the text box "Cluster name."
   (b) We will take the standard significance level and statistical analysis options for this exercise. For a detailed comment on these options, you might want to have a look at the BiNGO User Guide that can be found in their website: www.psb.ugent.be/cbd/papers/BiNGO/User_Guide.html.

(c) We want to know which terms are over-represented in the network with respect to the whole annotation, so we leave the corresponding categories as they are.

(d) Under "Select ontology file" choose the Gene Ontology file "gene_ontology_ext.obo" using the "custom" option in the drop-down menu.

(e) Under Select namespace select "Biological Process."

(f) Under Select organism/annotation choose the "gene_association.goa_human" file.

(g) The "Discard the following evidence codes" box allows you to limit the analysis discarding annotations that are given based on a specific evidence code (*see* **Note 9**).

(h) If you want to save the results of the analysis, mark the check-box and choose a path to save your files.

(i) Finally, press the "Start BiNGO" button.

5. You will receive a warning saying, "Some category labels in the annotation file are not defined in the ontology." The warning refers to identifiers that are not properly mapped in the GO reference file by BiNGO. There might often be a small discrepancy between the identifiers provided in the interaction network and those found in the GO reference file (when using isoforms, for example). Ignore this warning and click OK.

6. The GO terms found are displayed in two ways. The first is a table of GO terms found; the second is a directed acyclic network in which nodes are the GO terms found and directed edges link parent terms to child terms.

7. The table displays the most over-represented terms sorted in with the smallest p-values on top. In this table we see a list of GO terms (with their names and GO-IDs) and the uncorrected p-value and corrected p-value. Apart from that, total frequency values and a list of corresponding proteins (listed under the title "genes") are listed for each term. You can visualize which nodes have been significantly annotated under the listed terms by selecting the terms and then using the "Select nodes" button. Since the list is sorted just by p-value, many general terms (less descriptive terms) rise to the top of the table, making it difficult to see the more specific terms that are more useful. If you clicked the "save" option in the BiNGO setup window, then this table is already saved to file. If not, then you will need to copy and paste these results into an Excel file (or similar). The data in this table is not saved as part of a Cytoscape session file and you will lose this data if you do not save it separately.

8. The other representation of the results is a graphical depiction of the enriched GO terms in the form of a network. Each node is a GO term, and GO terms are linked by directed edges

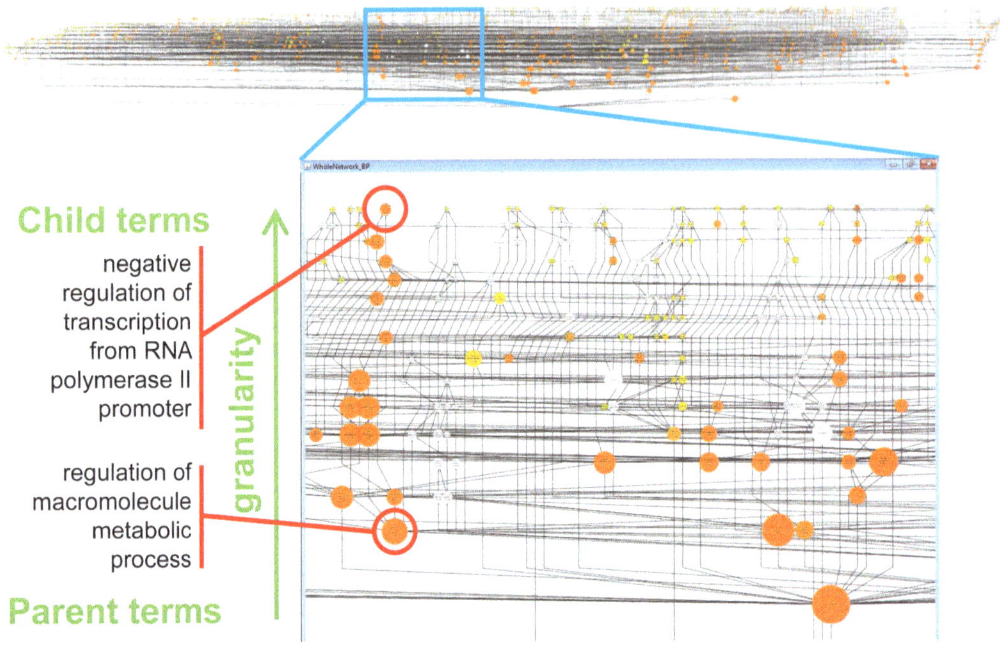

Fig. 4 Hierarchical nature of GO as seen with a BiNGO analysis result. After applying the "Hierarchical" layout we can see how granular children terms are placed at the top section of the graph, while parent, generic terms take their place at the bottom (root) part of the network

representing parent-to-child relationships. Nodes are colored by *p*-value (a small window depicting the legend is also produced) and the size of each node is proportional to the number of proteins annotated with that term. The default layout is less easy to read, but we may take advantage of one of Cytoscape's tools to provide a user-friendlier representation.

9. Make sure the graphical representation of the BiNGO results is selected. Choose "Layouts" → "Cytoscape Layouts" → "Hierarchical layout." Gene ontologies are a directed acyclic graph: Cytoscape utilizes this topology to organize the BiNGO results graph so that more specific and informative terms float to the top, while general, less informative terms sink to the bottom. You want to focus on orange-colored terms that branch-up the graph to find significantly enriched functions, as shown in Fig. 4. Navigating through this view provides a more useful impression of what biological processes are present in this network. When you find a term of interest, you may look it up in the table to see what proteins in the network were annotated with that term.
10. **Save your session** (*see* **Note 10**).

Exercise: A final test to put together what you have learnt about GO annotation.

1. Which processes are specifically over-represented in regulatory T cells in comparison with effector T cells?
    - Repeat the BiNGO analysis and find out which processes are involving specifically over-represented proteins in regulatory and effector T cells.
2. Some researchers don't trust annotations inferred using automatic annotation. Repeat your analysis filtering those annotations and see how that affects the results.

4.8.1 Final Considerations and Going Beyond BiNGO

Even though altering the layout helps understanding the information you get, the information is still difficult to interpret and you might want to further explore your results beyond getting just a list of terms or a network visualization of the most significantly enriched branches of the ontology. As we said before, it is essential to have a good knowledge of the genetic background from which the proteins in your network come in order to make a correct interpretation of the results. Beyond that, the analysis must be often refined to bring the novelty out of the results.

Fine-tuning the parameters of your BiNGO analysis can help bringing out interesting information, as well as performing specific analysis of certain regions of your network. However, customization of the analysis in BiNGO is limited in comparison with other tools such as ClueGO, where sophisticated options such as the "GO Term Fusion" redundancy reduction tool are available to the user. Although more computation resources-demanding than BiNGO (but still within the capabilities of a standard desktop computer), ClueGO is an excellent alternative for the advanced user when it comes to perform personalized analysis. It is also the tool of choice if what you really want to perform is a differential analysis of the annotation of two different networks/clusters/lists of gene products. The "Compare" option of the ClueGO plugin performs a comparison between the number and percentage of genes that are annotated per term in two different clusters and returns a results table and a color-coded network graph. If you want to learn more about ClueGO and its capabilities, check their excellent documentation in www.ici.upmc.fr/cluego/ClueGODocumentation.pdf.

Beyond that, BiNGO can be nicely complemented with PiNGO [22], its sister tool. With this tool we can easily identify candidate gene products that are significantly associated with a GO term of interest as derived from their network context. It uses the same statistics tools as BiNGO does and the interface is very similar to the one we have described in this tutorial. If you are interested in this type of analysis and want to learn how to use the tool, check a very detailed tutorial provided in their website: www.psb.ugent.be/esb/PiNGO/Tutorial.html.

5 Additional Information

5.1 Installing Plugins in Cytoscape

This set of instructions is specific for the BiNGO plugin, but it can be used for any other plugin you might need to install using the plugins manager in Cytoscape, such as the PSICQUIC client plugin.

1. In Cytoscape, go to "Plugins" → "Manage Plugins."
2. Look for BiNGO using the search box or browsing through the "Functional Enrichment" group of plugins.
3. Press "Install"
4. Check that the plugin was installed, it should be visible in your "Plugins" menu. You might need to re-start Cytoscape if it is not there.

5.2 Further Reading

Below you will find suggestions for further reading.

General review about the basic concepts required to understand protein–protein interactions: De Las Rivas & Fontanillo, 2010 [12].

General review, this one focused on the use of the study of the interactome in relation with human disease: Vidal, Cusick, & Barabási, 2011 [23].

A recent review about differential network biology, the study of the differences between particular biological contexts in contrast with the static interactome: Ideker & Krogan, 2012 [24].

The assessment of confidence values to molecular interactions requires the use of several, complementary approaches. In this study, the performance of different protein interaction detection methods with respect to a golden standard set is evaluated: Braun et al., 2008 [25].

Our group has produced a tutorial in the HUPO discussing the importance of molecular interactions network analysis and applying a similar approach to the one presented here, using BiNGO in combination with the topological cluster analysis plugin clusterMaker. See Koh, Porras, Aranda, Hermjakob, & Orchard, 2012 [11].

Finally, a good example of network analysis using data coming from literature-curated databases can be found in this recent paper in Nature Biotechnology: X. Wang et al., 2012 [26]. They constructed a network with high-quality binary protein–protein interactions where there is information about the interaction interfaces at atomic resolution and integrated disease-related mutation information, finding out an enrichment of disease-causing mutations in interacting interfaces.

5.3 Links to Useful Resources

Useful repositories, databases, and ontologies:

1. The Universal Protein Resource, UniProt: www.uniprot.org
2. The Gene Ontology: http://geneontology.org/
3. The Proteomics IDEntifications database, PRIDE: www.ebi.ac.uk/pride
4. Lots of IMEx-complying interaction databases in the IMEx website: www.imexconsortium.org/about-imex

Summary of useful tools:

1. How do I get interaction data from most of the interaction databases that are out there? Easy answer: use the Proteomics Standard Initiative Common Query Interface (PSICQUIC). You can learn more about it here code.google.com/p/psicquic and here you have a link to its search interface, PSICQUIC View: www.ebi.ac.uk/Tools/webservices/psicquic/view
2. To learn more about Cytoscape or to get access to documentation and tutorials, go to its website: www.cytoscape.org. You can see a list of plugins (also called 'apps') for both Cytoscape 2.8 and 3.0 here: http://apps.cytoscape.org/.
3. More about the BiNGO plugin in their website, with a nice tutorial and useful documentation: www.psb.ugent.be/cbd/papers/BiNGO.
4. PiNGO is BiNGO's sister tool and it can be used to predict candidate gene products, not annotated for a GO term of interest, as inferred from their network interaction neighborhood: www.psb.ugent.be/esb/PiNGO/Home.html.
5. ClueGO, an advanced GO enrichment analysis tool, can be a good alternative to BiNGO for the advanced user. Check their extensive documentation to be able to use the tool to its full capacity: www.ici.upmc.fr/cluego/cluegoDescription.shtml.
6. In order to find hidden functional circuits in large networks it is often useful to try clusterMaker, a Cytoscape plugin for topological cluster analysis. Lots of documentation and useful tutorials in their website: www.cgl.ucsf.edu/cytoscape/cluster/clusterMaker.html.
7. APID2NET is a Cytoscape plugin for integrated network analysis that brings together different useful tools for interaction retrieval and network annotation and visualization: http://bioinfow.dep.usal.es/apid/apid2net.html.

5.4 Icons List

Figure 5 here you have a list of the Cytoscape icons cited through the tutorial for visual reference.

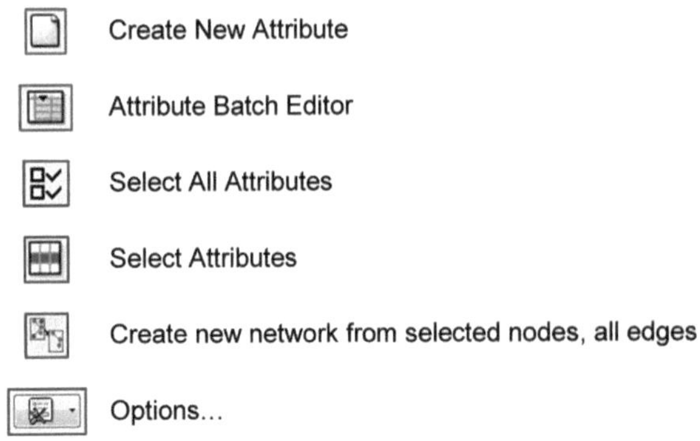

Fig. 5 List of Cytoscape 2.8.3 icons cited through the tutorial

6 Notes

1. There are several ways to get molecular interaction data into Cytoscape apart from the one we present here. For example, from the IntAct web page, the user can generate files in tab-delimited or in Cytoscape-compatible XGMML formats that can be later imported into this software.
2. UniProtKB identifiers are widely used among the different resources we are going to need along the tutorial, so it is highly recommended to use them when dealing with protein datasets. The advantages of using these ACs are that (1) they are stable (they are not changed or updated once assigned); (2) they can reflect isoform information, if provided; and (3) they are recognized by many interaction and annotation databases (in this instance, the two databases we will be using: IntAct and GO). To map this particular list we have used the PICR service (Protein Identifier Cross-Reference Service) that can be accessed in www.ebi.ac.uk/Tools/picr.
3. You can also perform queries using this tool by clicking on the "Search property" tab and selecting "GET_BY_QUERY" in the "Query Mode" option. Then you can search using TaxIDs, gene names, or interaction detection methods and build complex queries with the MIQL syntax reference (check www.ebi.ac.uk/Tools/webservices/psicquic/view and click on the "MIQL syntax reference" link you will find in the far-right upper corner by the search bar).
4. In the version of Cytoscape we use here (2.8.3) you need to have the PSICQUIC client plugin installed to fetch data using PSICQUIC in Cytoscape. Check out how to install plugins from Subheading 7.

5. The PSI-MI-TAB-2.5 format is part of the PSI-MI 2.5 standard and it was originally derived from the tabular format that the BioGrid database used. You can learn more about the fields represented in the format checking their Google Code wiki at http://code.google.com/p/psimi/wiki/PsimiTabFormat.
6. Both the edge and the node attributes in this network are based in the fields defined in the PSI-MITAB format that the IMEx-complying databases use. Go to code.google.com/p/psicquic/wiki/MITAB25Format if you need to know what a particular attribute means.
7. Proteomics data repositories such as PRIDE (www.ebi.ac.uk/pride) store quantitative proteomics data in formats that can be transformed in tab-delimited text files that can be used as attribute tables for Cytoscape.
8. The GO project is an international initiative that aims to provide consistent descriptions of gene products (i.e., proteins). These descriptions are taken from controlled, hierarchically organized vocabularies called "ontologies." GO uses three ontologies covering three biological domains. These are Cellular Component, or the location of the protein within the cell (e.g., cytosol or mitochondrion); Biological Process, or a series of events accomplished by one or more ordered assemblies of molecular functions (e.g., glycolysis or apoptosis); and Molecular Function, which is the activity proteins possess at a molecular level (e.g., catalytic activity or trans-membrane transporter activity). More information can be found in their website, http://geneontology.org/
9. Every GO annotation is associated to a specific reference that describes the work or analysis supporting it. The evidence codes indicate how that annotation is supported by the reference. For example, annotations supported by the study of mutant varieties or knock-down experiments on specific genes are identified with the inferred from mutant phenotype (IMP) code. All the annotations are assigned by curators with the exception of those with the inferred from electronic annotation (IEA) code, which are assigned automatically based in sequence similarity comparisons. See www.geneontology.org/GO.evidence.shtml for more information about evidence codes.
10. The graphical representation of your BiNGO results is just another network that can be modified and analyzed in Cytoscape by making further use of analysis plugins. The "Network Modifications" plugin can be used when you want to roughly see the most diverging differences in the results of two BiNGO analyses.

References

1. Magrane M, U. Consortium (2011) UniProt Knowledgebase: a hub of integrated protein data. Database (Oxford) 2011:bar009
2. Ashburner M, Ball CA, Blake JA et al (2000) Gene ontology: tool for the unification of biology. The Gene Ontology Consortium. Nat Genet 25:25–29
3. Orchard S, Kerrien S, Abbani S et al (2012) Protein interaction data curation: the International Molecular Exchange (IMEx) consortium. Nat Methods 9:345–350
4. Aranda B, Achuthan P, Alam-Faruque Y et al (2009) The IntAct molecular interaction database in 2010. Nucleic Acids Res 38(Database issue):D525–D531
5. Ceol A, Chatr Aryamontri A, Licata L et al (2010) MINT, the molecular interaction database: 2009 update. Nucleic Acids Res 38: D532–D539
6. Salwinski L (2004) The Database of Interacting Proteins: 2004 update. Nucleic Acids Res 32:449D–451D
7. Chaurasia G, Iqbal Y, Hänig C et al (2007) UniHI: an entry gate to the human protein interactome. Nucleic Acids Res 35:D590–D594
8. Goll J, Rajagopala SV, Shiau SC et al (2008) MPIDB: the microbial protein interaction database. Bioinformatics 24:1743–1744
9. Aranda B, Blankenburg H, Kerrien S et al (2011) PSICQUIC and PSISCORE: accessing and scoring molecular interactions. Nat Methods 8:528–529
10. Smoot ME, Ono K, Ruscheinski J et al (2011) Cytoscape 2.8: new features for data integration and network visualization. Bioinformatics 27:431–432
11. Koh G, Porras P, Aranda B et al (2012) Analyzing protein-protein interaction networks. J Proteome Res 11(4):2014–2031
12. De Las Rivas J, Fontanillo C (2010) Protein–protein interactions essentials: key concepts to building and analyzing interactome networks. PLoS Comput Biol 6: e1000807
13. Ashburner M, Ball CA, Blake JA et al (2000) Gene ontology: tool for the unification of biology. The Gene Ontology Consortium. Nat Genet 25:25–29
14. Maere S, Heymans K, Kuiper M (2005) BiNGO: a Cytoscape plugin to assess overrepresentation of gene ontology categories in biological networks. Bioinformatics 21:3448–3449
15. König S, Probst-Kepper M, Reinl T et al (2012) First insight into the kinome of human regulatory T cells. PLoS One 7:e40896
16. Chautard E, Fatoux-Ardore M, Ballut L et al (2011) MatrixDB, the extracellular matrix interaction database. Nucleic Acids Res 39: D235–D240
17. Salwinski L, Miller CS, Smith AJ et al (2004) The Database of Interacting Proteins: 2004 update. Nucleic Acids Res 32:D449–D451
18. Brown KR, Jurisica I (2005) Online predicted human interaction database. Bioinformatics 21:2076–2082
19. Lynn DJ, Chan C, Naseer M et al (2010) Curating the innate immunity interactome. BMC Syst Biol 4:117
20. Bindea G, Mlecnik B, Hackl H et al (2009) ClueGO: a Cytoscape plug-in to decipher functionally grouped gene ontology and pathway annotation networks. Bioinformatics 25:1091–1093
21. Huang DW, Sherman BT, Lempicki RA (2009) Systematic and integrative analysis of large gene lists using DAVID bioinformatics resources. Nat Protoc 4:44–57
22. Smoot M, Ono K, Ideker T et al (2011) PiNGO: a Cytoscape plugin to find candidate genes in biological networks. Bioinformatics 27:1030–1031
23. Vidal M, Cusick ME, Barabási A-L (2011) Interactome networks and human disease. Cell 144:986–998
24. Ideker T, Krogan NJ (2012) Differential network biology. Mol Syst Biol 8:565
25. Braun P, Tasan M, Dreze M et al (2008) An experimentally derived confidence score for binary protein-protein interactions. Nat Methods 6:91–97
26. Wang X, Wei X, Thijssen B et al (2012) Three-dimensional reconstruction of protein networks provides insight into human genetic disease. Nat Biotechnol 30:159–164

Chapter 5

Modeling Signaling Networks with Different Formalisms: A Preview

Aidan MacNamara, David Henriques, and Julio Saez-Rodriguez

Abstract

In the last 30 years, many of the mechanisms behind signal transduction, the process by which the cell takes extracellular signals as an input and converts them to a specific cellular phenotype, have been experimentally determined. With these discoveries, however, has come the realization that the architecture of signal transduction, the signaling network, is incredibly complex. Although the main pathways between receptor and output are well-known, there is a complex net of regulatory features that include crosstalk between different pathways, spatial and temporal effects, and positive and negative feedbacks. Hence, modeling approaches have been used to try and unravel some of these complexities.

We use the mitogen-activated protein kinase cascade to illustrate chemical kinetic and logic approaches to modeling signaling networks. By using a common well-known model, we illustrate here the assumptions and level of detail behind each modeling approach, which serves as an introduction to the more detailed discussions of each in the accompanying chapters in this book.

Key words Cell signaling networks, Network-based modeling, Logical modeling, Stochastic modeling

1 Why Model Cell Signaling Networks?

Signaling networks are complex systems of molecules that relay and interpret signals. They can consist of any number and combination of proteins, nucleotides, fatty acids, and even dissolved gases such as nitric oxide and carbon monoxide that help conserve signal integrity. These systems relay information from the cell surface through to effector proteins that modify the behavior of the cell. Their importance in health can be understood by the severity of disease that occurs when components of these networks are perturbed. To give an example, the epidermal growth factor pathway (EGFR) regulates growth, survival, differentiation, and proliferation in mammalian cells. It does this through a complex system of interactions that converts extracellular signals into a cellular response with high specificity. This specificity can be achieved despite the relatively

Maria Victoria Schneider (ed.), *In Silico Systems Biology*, Methods in Molecular Biology, vol. 1021,
DOI 10.1007/978-1-62703-450-0_5,

low number of intermediate proteins involved in this pathway. Deregulation in the EGFR pathway underscores most cancers; mutation in components leads to uncontrolled up-regulation of the pathway and increased growth and proliferation [1]. As such, it is vital to have a sound mechanistic understanding of both the components ("parts-list") and the interactions of these pathways to better design treatments [2, 3].

2 Data and Why a Parts-List Is Not Enough?

The information in signaling pathways is usually transmitted through phosphorylated proteins—extracellular signals initiate a cascade of such protein modifications from the receptors that receive the stimulus, through to the effector proteins (usually transcription factors) that propagate the signal through to a phenotypic outcome. Hence, the phosphorylated (activated) protein can be seen as a unit of measurement in signaling network systems (not discounting the importance of secondary messengers such as Ca^{2+} and cyclic AMP). Each protein can have many such phosphorylation sites, and the activation/deactivation of each can affect its propensity to transmit a signal in the system. Added to this, the state of these proteins (whether they are active or not) is highly dynamic both in space and time. These details make measurement a challenge and a large variety of experimental techniques have been adapted and developed to achieve these measurements:

2.1 Antibody-Based Measurement

Traditionally, quantitative measurements of proteins in signaling networks have used antibody-based methods. These include protein arrays, reverse-phase protein arrays, and the bead-based xMAP technology from Luminex. A thorough review of these methods can be found in Terfve et al. [4]. Briefly, these methods depend on the quality of antibodies that target phosphorylated proteins in the signaling network. Although such methods can be scaled up to hundreds of samples and target proteins, they are ultimately constrained by the selectivity and affinity of the antibodies for their targets and thus, can be biased towards more well-known pathways such as EGFR.

2.2 Mass Spectrometry-Based Measurement

Mass spectrometry-based approaches have the advantage of being biased-free in that the quantification of proteins in a sample is not limited by prior-knowledge (in terms of antibody design) of the system. However, low-abundance proteins can be difficult to detect and assigning sequences to MS spectra is also challenging [5]. MS-based approaches such as targeted MS/MS have been developed to overcome some of these limitations but fall back into the trap of needing some prior knowledge of the system of interest.

Ultimately, the best approach if time and expense permits is to use a combination of nonspecific and specific experimental approaches to explore and then quantify the system of interest [6].

2.3 Measuring at Single-Cell Resolution

Another dimension to signaling networks has come to the forefront in recent years with the advent of techniques to measure signaling events in single cells. The methods described above rely on the average measure of proteins across many cells. In many cases this gives an acceptable approximation of the dynamics at the single cell level (i.e., where the cell-cycle is not a factor, or variable time delays are minimal). However, it is known that even isogenic populations of cells respond differently to the same cues, producing a heterogeneity of responses. Such heterogeneity can also be functionally important if it is regulated by the cell or is intrinsic to a phenotype [7]. Hence, measurement at the single-cell level is vital to understand properties and dynamics that may be "averaged out" with bulk-cell measurements.

There are a number of techniques that can measure protein levels, localization, and activity at the single cell. These can be broadly divided into flow cytometry and microscopy-based categories.

Flow cytometry is a technology that allows the interrogation of individual cells by suspending them in a stream of fluid and passing them individually through a detection apparatus. The technique allows for quantification of cell-surface proteins in live cells (FACS). However, the cells must be fixed for detection of intracellular proteins. Protein levels can be detected using fluorescent antibodies or, more recently, using mass tags for detection in mass spectrometry [8].

Ultimately, signaling processes can display complex dynamics in short periods of time that can affect downstream outcomes [9]. To gain a true understanding of these rapid dynamics requires tracking signal transduction in real time in live cells. Spiller et al. [10] comprehensively introduce the myriad of techniques available for such measurements. Briefly, the main techniques can be encompassed under genetically encoded activity sensors (e.g., FRET probes), or genetically encoded fluorescent fusions. These techniques can measure diverse biological events such as translocation, protein modifications, and protein–protein interactions. The advantages of using such data for modeling are myriad; observing dynamics at the single cell level can lead to novel mechanistic insight [11] and a truer understanding of decision making in signaling pathways [7]. Unfortunately, the payoff for such detailed measurements is countered by a reduction in scope of measurable components. The design and optimization of fluorescent sensors is complex and factors such as phototoxicity and fluorescent overlapping limit multiplexing potential (i.e., the number of

Method	Software referred to in this book	Data Required	Mechanistic Insight	Prior Knowledge	Network Size
Kinetic / stochastic models	Copasi R	Focused	Greater	Extensive	Smaller
Logic Models	CellNOptR GINsim BoolNet CellNetAnalyzer				
Bayesian Networks	Not covered				
Statistical Models	Not covered	Broad	Less	Little	Larger

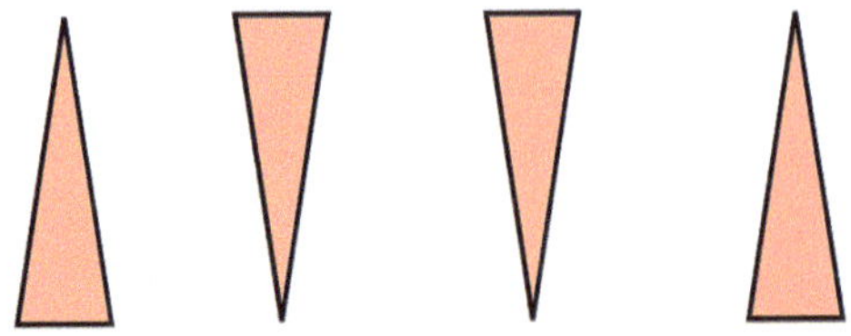

Fig. 1 Modeling approaches can be ranked according to the type of data required, the mechanistic insight that results from such a model, the prior knowledge in the form of protein interactions or biochemical detail that is needed, and the size of the network that can be used with each approach. The software list is not comprehensive but simply what is used or referenced in this book. Copasi [44] was used in the chapter to generate the deterministic and stochastic time courses of the Kholodenko model and R [45] was used to generate plots. CellNOptR [40] trains logic models to data and GINsim [46], BoolNet [47], and CellNetAnalyzer [48] can be used to build, simulate, and analyze logic models (see also Chap. 6)

measurements in parallel). In such cases, normalization methods such as computational multiplexing [12] can be used to compare measurements between different cells.

3 So What Can Be Done with This Data?

Once a "parts-list" is obtained, the question becomes how to understand this potentially multidimensional, complex, and noisy data. The computational or statistical approach to take depends on a number of factors, as summarized in Fig. 1.

Statistical models can be subdivided into predictive and non-predictive categories [4]. Non-predictive methods include clustering and principal component analysis (PCA), where the goal is to classify the data based on multivariate input. Predictive methods include the family of regression-based methods. Such models are nonmechanistic in that there is no attempt to understand the underlying biochemistry; these models are simply concerned with linking a set of input variables (e.g., phosphorylation measurements) with an output or outputs (e.g., cell phenotypes).

4 Introducing Networks

Data-driven nonnetwork approaches such as those described above can be informative in linking signaling pathway inputs and measurements to outputs [13]. However, to gain a functional, deeper understanding of the underlying biology, information about the interactions between members of the "parts-list" of measured species must be added. This information can be described as a network. Networks consist of nodes and edges. Typically, the nodes represent the components of the network (e.g., a phosphorylated protein) and the edges describe the interactions between the components (e.g., protein C is phosphorylated when protein A and B are phosphorylated). Signaling networks can represent interaction detail at varying resolutions—from the coarse knowledge of protein interaction networks (where only the presence/absence of an interaction is known), through to detailed, biochemically accurate representations (typical of more well-known signaling pathways such as EGFR).

Even in the absence of data, such networks contain topological properties that can be analyzed to understand emergent properties [14]. In the presence of data, these networks can be utilized by mapping data types onto the nodes [15]. Another powerful network-based approach, and what we will focus on here, is to transform such representations into models, which can be then used to explain the corresponding data. Such models then offer a framework for improved experimental design, validation, and prediction.

5 The Advantages of Network-Based Modeling

With a modeling approach to signaling networks based on mechanistic understanding, it becomes possible to predict the effects of, say, perturbations and mutations at the cellular level. There are many examples of nonobvious emergent properties that cannot be explored through a knowledge of system components alone. For example, in signaling networks, factors such as protein dynamics [9], localization [16], and stochasticity [17] contribute to cell-specific responses that can only be explored through dynamic, spatial, and stochastic modeling. respectively [3].

As can be seen from Fig. 1, the choice of modeling approach should be made based on factors such as network size, data resolution, and the prior knowledge of the system. We will use this chapter and a small example to introduce a number of approaches sitting on this continuum of approaches—stochastic and deterministic chemical kinetic modeling and logic modeling. This is a brief overview that serves as an introduction to more detailed discussion of these formalisms in other chapters in the book (Parts II, III, and V). There are a number of useful approaches that are outside the scope of this book, such as rule-based, biophysical and spatial modeling. We point the reader to a number of excellent reviews that detail such formalisms (*see* refs. 18—rule-based 19- spatial modeling in signaling networks).

6 The Model: The MAPK Cascade as an Example

In order to illustrate some of the concepts presented in the introduction, as well as the different approaches to modeling introduced in the other chapters, we will use the example of the mitogen-activated protein kinase (MAPK) signaling cascade. This series of three kinases plays a vital role in cell signaling where it regulates cell proliferation, survival, motility, and differentiation. It can be viewed as a module that is downstream of a large number of cell-surface receptors (e.g., tyrosine kinase receptors such as EGFR, tumor necrosis factor receptor) but which also interplays with PI3K and other signaling pathways through feedback mechanisms and crosstalk [20–22]. The main constituents of this signaling pathway are three kinases—Raf, MEK, and ERK. The activation of cell-surface receptors such as EGFR activates the GTPase Ras, which in turn activates the first kinase in the cascade, Raf. This three-kinase cascade has been subject to intense research over the last few decades since its discovery in 1980 [23] for a number of reasons: (1) it is a highly conserved pathway, which allows for experimental data from different model organisms [24], (2) despite its simplicity, its regulation is complex [20, 25] and is still being unraveled, and (3) it is still a mystery how such a system can regulate such a diverse set of cellular processes [26].

Accordingly, the MAPK cascade has been extensively modeled in order to understand emergent properties that explain its regulation and control of downstream signaling events. We will use the example of Kholodenko [27] (*see* Figs. 2 and 3 below) to explore

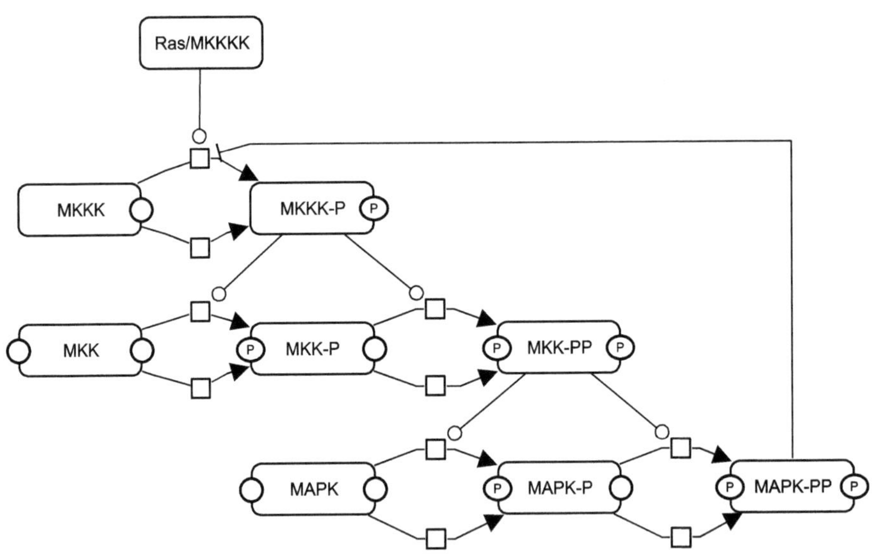

Fig. 2 The kinetic scheme of the MAPK cascade model from [27]. The model is represented as a process diagram and illustrates the dual-phosphorylation mechanism required for activation at each level of the cascade

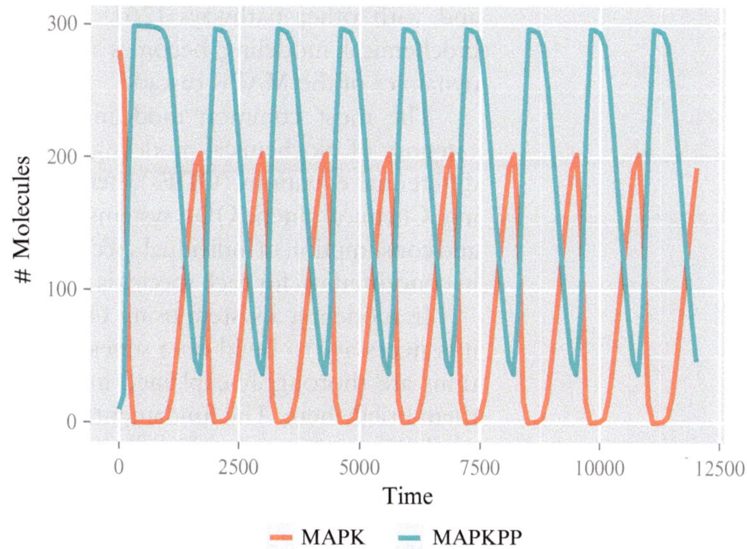

Fig. 3 The oscillatory behavior produced by the Kholodenko model: a product of ultra-sensitivity at each stage of the cascade (i.e., small changes in input can produce large changes in output) and negative feedback

the assumptions, limitations, and differences between the different modeling approaches.

The Kholodenko model describes the MAPK cascade as a series of phosphorylation events, each catalyzed by the preceding kinase. Activation of each kinase requires 1 or 2 phosphorylations, which occur in distinct reactions. It is known that these phosphorylation reactions at each level are non-processive, i.e., the kinase dissociates itself from the substrate after a single phosphorylation, rather than phosphorylating all sites at the same time [24]. The model can be adapted to different formalisms, which can help illustrate the assumptions of logic and chemical kinetic (both deterministic and stochastic) modeling. We refer the reader to more in-depth discussion of these approaches in Chaps. 6, 8 and 9.

7 "Biochemical" Modeling of Signal Transduction

Biochemical (or physicochemical) approaches encompass formalisms that describe biomolecular interactions in terms of equations based on chemical (and sometimes physical) theory [2]. Such approaches require a sound knowledge of the underlying biology. As stated above, we now have biochemically detailed knowledge of the phosphorylation reactions of the MAPK cascade. However, these reactions do not occur in isolation but include interactions with scaffolding proteins and multiple crosstalk and feedback interactions (including transcriptional control) within the cascade and with other pathways [26]. Given this level of knowledge, biochemical modeling becomes a natural platform to study the dynamics of the MAPK cascade.

The most common modeling formalism that falls into the category of biochemical modeling is a system of coupled ordinary differential equations (ODEs), henceforth chemical kinetic modeling. Chemical kinetic ODE systems represent the rates of production and consumption of individual species in the model, hence the change in concentration for each species is represented by a single ODE.

In modeling a system using ODEs based on chemical kinetics, it is necessary to build on a series of assumptions. These assumptions are thoroughly explained in Chap. 8 but we will introduce them briefly here. The fundamental assumption in chemical kinetic modeling is mass-action kinetics—the rate of a chemical reaction is proportional to the concentration of the reactants. The difficulty with mass action kinetics is that it is only valid for elementary reactions (i.e., a reaction with no stable intermediate). Unfortunately, such reactions are often unknown in biology where it may be the case that only the start and end-point of a series of reactions can be measured or deduced. This is often the case for enzyme biology, which brings us back to the MAPK cascade (a series of phosphorylations catalyzed by kinases, a type of enzyme). Using the Kholodenko model already introduced, together with some previous and updated models [28, 29], we can demonstrate how a variety of assumptions are used and how they lead to different behaviors in the system.

7.1 Chemical Kinetic Modeling of the MAPK Cascade

Initial modeling efforts used mass-action kinetics to simulate the two-step non-processive mechanism of phosphorylation that occurs throughout the MAPK cascade (*see* Fig. 1). With this approach, Huang et al. [28] demonstrated that ultra-sensitivity could be achieved within the constraints of this biochemical system (ultra-sensitivity is the phenomenon where a small change in input (substrate) leads to a large change in output (enzyme/kinase activity)).

The Kholodenko model extended this model by including negative feedback and replacing the mass-action kinetics of the phosphorylation reactions with Michaelis–Menten kinetics. Michaelis–Menten kinetics are a simplification of mass-action kinetics where an added assumption is that the concentration of the enzyme–substrate complex (e.g., MKK-PP with MAPK from Fig. 2) does not change significantly over the time of the reaction and relative to other concentrations (see Chap. 8 for a more in-depth discussion). Kholodenko demonstrated that, under these conditions, the MAPK cascade could produce oscillations—a signaling pattern that has been proposed to encode output in signaling pathways [9].

The work of Markevich et al. [29] demonstrated the importance of what the reaction assumptions of the MAPK cascade are. This work explored the possibility of bistability (i.e., two stable outputs from the MAPK cascade) in the absence of feedback. It also demonstrated that the use of Michaelis–Menten kinetics in the context of the MAPK cascade can be an over-assumption. The argument is that, at any level of the cascade (*see* Fig. 2), the concentration of the kinase in complex with its substrate cannot be ignored if the total concentration of kinase, phosphorylated and unphosphorylated, is to be constant. Other work has also shown that properties of the MAPK cascade can make it non-compliant with normal enzyme kinetics; sequestration of the substrate by the kinase undermines Michaelis–Menten kinetics [30]. This has the implication that there are retroactive effects [31–33] that render the behavior of the MAPK cascade more complex than a simple chain of signal transducers.

Ultimately, any assumptions used to help make ODE systems tractable have to come from knowledge of the underlying biology. MAPK cascade models have grown to incorporate upstream effects, such as receptor traffic and internalization of receptors [34], transcriptional feedback, [20] and the influence of scaffolds [35], as well as localized and spatial influences [36]. Despite these increases in model size and complexity, the above work has shown that the "devil is in the detail" with regard to reaction kinetics in this system.

7.2 Stochastic Modeling of Signal Transduction

The emergence of single-cell measurement techniques has highlighted the heterogeneity of cell signaling networks, even under similar conditions and in isogenic populations [7]. The stochastic nature of molecular interactions accounts for much of this heterogeneity (notwithstanding extrinsic sources of noise in any measurement technique) and there has been a realization that such effects need to be taken into account in a modeling framework. Chapter 9 provides a concise introduction to the simulation of stochastic kinetic models. For now, we will again use the Kholodenko model to briefly highlight the main differences between deterministic and stochastic chemical kinetic approaches.

The fundamental difference between stochastic and deterministic approaches in chemical kinetic modeling can be described as the following: deterministic simulation uses reaction kinetics to describe the change in concentration of each species over time. Stochastic simulations instead describe the change in the *number* of molecules of each species over time according to the probability of each reaction occurring (reaction propensities).

There are subtle differences between reaction rates used in deterministic modeling and reaction propensities. These are important when describing reactions with low numbers of molecules. For example, in the case where there is a second order rate law or higher (see Chap. 8), the product of the reactants is replaced by a term to

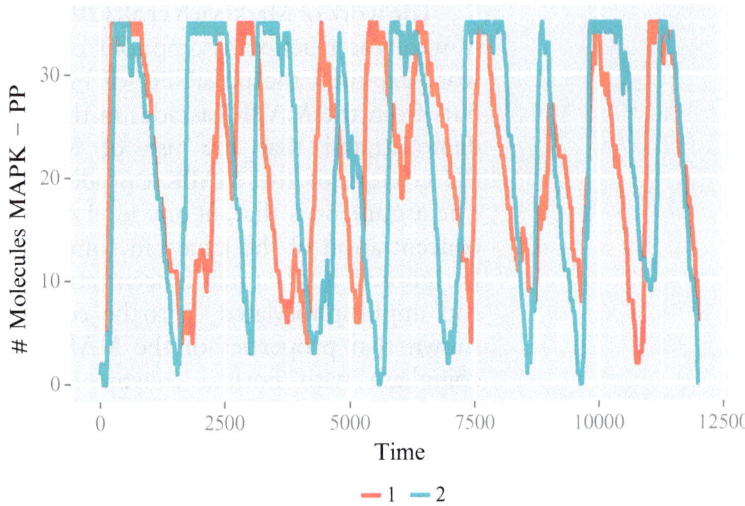

Fig. 4 A time course of the Kholodenko model using a stochastic simulation. The reaction kinetics were converted to propensities internally in Copasi. The volume of the "reaction chamber" was "shrunk" (i.e., the cell volume was set to $6e^{-16}$l) to reduce molecule numbers and more ably demonstrate the variation that can occur. The figure shows the count of MAPK-PP molecules in two simulations over the same time-frame as Fig. 3

insure the reaction only takes place if there are at least two particles present (i.e., x^2 becomes $x \times x{-}1$).

In the case of the Kholodenko model and the Michaelis–Menten approximation of phosphorylation, the assumption of the enzyme-substrate complex being in a steady state may not be valid for stochastic systems [37]. However, we will assume its validity in performing a stochastic simulation of the model to demonstrate the potential importance of stochastic simulations in such oscillating systems.

Figure 4 shows two simulations of the Kholodenko model that demonstrate the stochasticity introduced into the model. When these simulations are averaged (*see* Fig. 5), the oscillations disappear. Such an exercise demonstrates the importance of single cell data (if we take each simulation run as a different cell) to discover underlying mechanism. However, the choice of stochastic versus deterministic chemical kinetics is not straightforward and again comes back to understanding the biology of the system and what one is using the model for [38].

8 Logic Modeling of Signal Transduction

Logic models use logical operators (AND, OR, NOT) to describe the relationships between nodes in a graph. In modeling signaling pathways, these nodes usually represent activated proteins and the

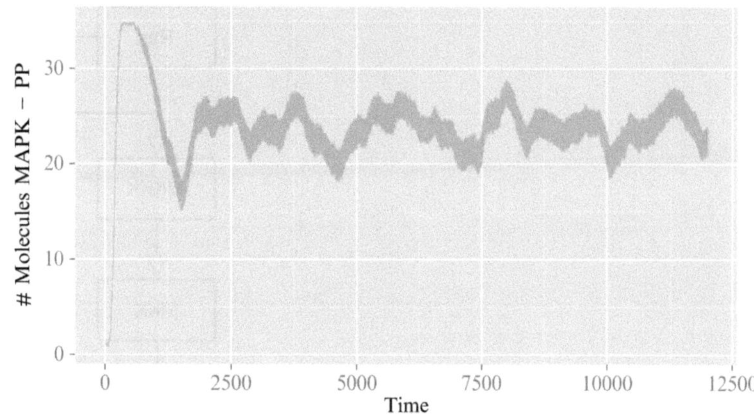

Fig. 5 This figure shows the mean count of MAPK-PP over time for 50 simulations, using the same starting conditions as Fig. 4 for each of the 50 simulations

edges between the nodes describe the dependencies between nodes. Chapter 6 describes how such models are built from prior knowledge resources and also how they are built in comparison to biochemical models.

There are a number of differences in the interpretation of a logic model compared to the equivalent mechanistic model shown in Fig. 2. In Fig. 2, each node represents a physical entity, in this case the individual kinases (together with their phosphorylation status) of the MAPK cascade. In standard (SBGN) notation, this is labeled a process diagram [39]. However, the representation for logic modeling is quite different (*see* Fig. 6 below). Here instead of a physical entity, each node represents an activity (hence the SBGN notation for such a graph: activity flow diagram). Therefore, the node "MKKK" is not the kinase but instead represents "the activation of MKKK." Such nodes are linked by directed activating or inhibitory edges. Logic operators (AND, OR, NOT) then describe how each output node is activated by its inputs.

The differences between chemical kinetic and logic modeling can be illustrated by fitting logic models to the data generated by the Kholodenko model (*see* Fig. 3). CellNOptR (www.cellnopt.org) is a set of R packages (and a user interface in Cytoscape, CytoCopteR) that trains logic models to data [40]. Depending on the type of data used for training and the amount of prior knowledge of the system, the logic simulations of the data can take a number of forms. These range from Boolean (each species (or activity) in the model only has the state ON/OFF) to logic ODEs (where the input–output relationships of the models are defined by ODEs, hence time and the level of activity are continuous).

8.1 A Time-Course Logic Model

We will start with the simplest logic approximation of the data generated from the Kholodenko model that can model oscillations. The *CNORdt* package of CellNOptR fits a scaling factor that

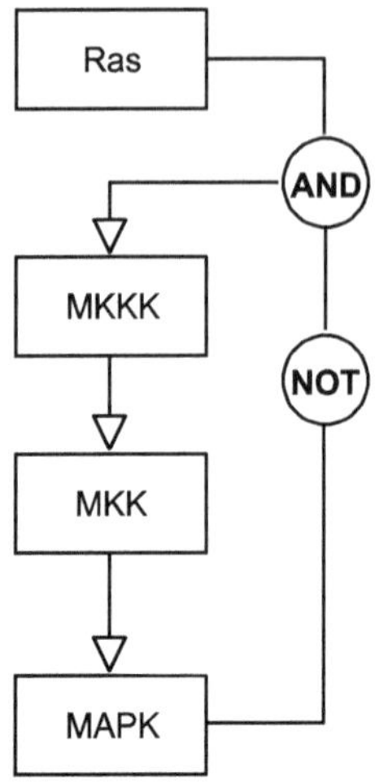

Fig. 6 A logic model representation of Kholodenko [27]. The nodes represent activities, as opposed to physical entities. Each edge describes the relationship between activities in the network. For example, a single arrow denotes an input–output relationship where the output becomes activated when the input is activated. More complex relationships use logic gates [41]. For example, the activation of MKKK here depends on the activation of Ras *and* the inactivation (NOT) of MAPK

allows a Boolean (states can be ON/OFF only) simulation of the model to be compared to time-course data [41]. An important distinction between logic and chemical kinetic simulations is that chemical kinetics describes the *transformation* of entities (in this case proteins), whereas logic models describe the *activation* of an output by its input(s). In the Kholodenko model we can distinguish two categories of reaction where this distinction can be illustrated:

1. Each phosphorylation of a kinase in the cascade is represented by the Michaelis-Menten rate equation. These reaction kinetics produce characteristic curves when the rate of production of "product" (in this case the activated kinase) is plotted against substrate concentration.

   The logic model describes each level of the MAPK as a single variable, without any reaction. This simplification allows us to ignore more complex reaction kinetics as we are only interested in the relationship between input and output. Hence, a logic model that interprets the system correctly in terms of activity relationships can coarsely capture the dynamics of the system (*see* Fig. 7).
2. The feedback from MAPK that negatively affects the activation of MKK is modeled by Kholodenko [27] as noncompetitive inhibition, i.e., binding of the inhibitor to the enzyme is not affected by, or affects, the concentration of bound substrate.

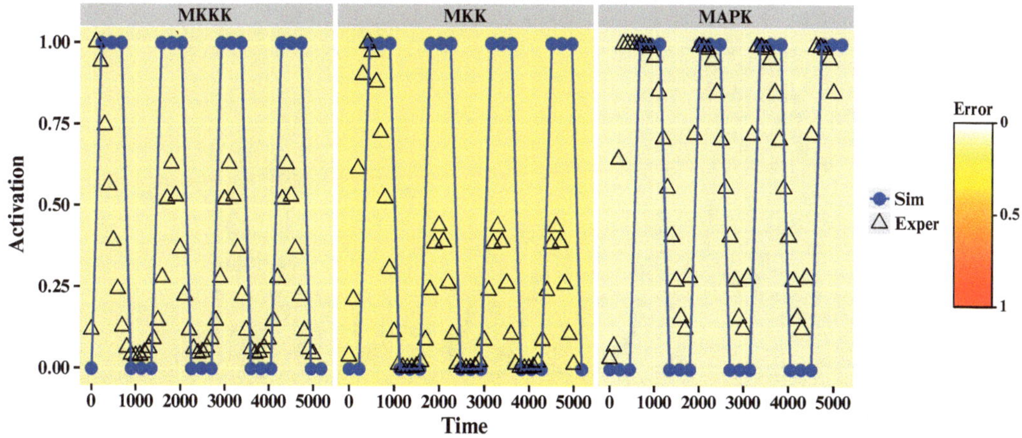

Fig. 7 The fit of the time-course Boolean simulation (*dotted line*) to the in silico generated data of the Kholodenko model (see Fig. 3). The stimulus "RAS" was set to 1 (*black*). The time course has been reduced to 5,000 s for clarity. As can be seen from the background shading, there is an error between the Boolean simulation and the original data as intermediate values cannot be simulated. However, the oscillation frequency closely approximates that of the Kholodenko model

The logic model represents this relationship using an AND gate—the activity of Ras must be equal to 1 and that of MAPK must be 0 for MKK activation to occur. As above, this is a fair approximation to the Kholodenko model when considering simply the ON/OFF (activated/deactivated states) of each kinase in the cascade.

The payoff for the simplicity of approach and parameter-free nature of Boolean modeling is evident from Fig. 7. The state/activity of each node is limited to ON/OFF and hence the intermediate activities of the kinases are poorly fitted. Also, mechanistic insight is limited; the nature of the negative feedback is unknown; and complex catalysis reactions are compressed into a simple input–output relationship. Hence, although such an approach yields limited mechanistic insight, the lack of parameters facilitates training these models to data.

8.2 A Continuous Logic Model

CNORdt fits a Boolean model to time-course data. This allows for a coarse-grained model that is dynamic but can only simulate the states of the model constituents as ON/OFF. *CNORode* [40, 41] fits logic models to data with ODEs describing input–output relationships [42]. Hence, logic ODEs describe the change in activity of each "activity node" over time, as opposed to chemical kinetic ODEs, where the change in concentration of each species is calculated. The logic ODEs used in *CNORode* are based on Hill functions; by varying the parameters n and k of the Hill function, the relationship between input and output can assume a variety of biologically meaningful shapes, such as close to linear or sigmoidal.

Interpreting each relationship in Fig. 6 as a continuous homologue of the Boolean functions used for *CNORdt* generates an

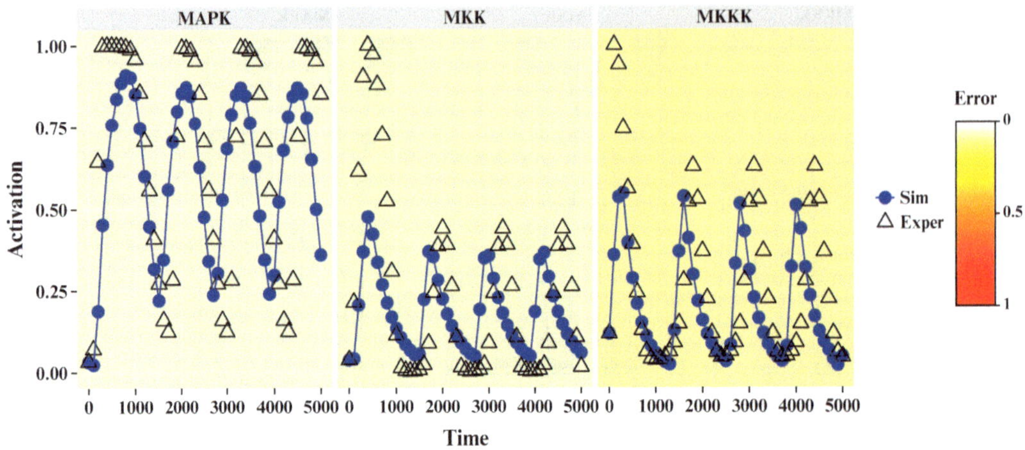

Fig. 8 The fit of the logic ODE model (*dotted line*) to the Kholodenko time-course data. The mean-squared error between the logic model and the data is greatly reduced (*background shading*) compared to the Boolean model (see Fig. 7). The stimulus "RAS" was set to 1 (*black*)

ODE logic model with 3 dynamic states and 11 parameters (the Kholodenko model has 8 dynamic states and 22 parameters). Nine of these parameters (two relating to "Ras" activity were fixed as this node is constitutively active) were optimized to fit the data generated from Kholodenko (*see* Fig. 8). In doing so, we can directly compare a logic-based ODE and a mechanistic biochemical model.

Ostensibly, the fit of the logic ODE model to the data generated by the Kholodenko model is good (*see* Fig. 8). The Hill functions used in the logic ODE model are "plastic" enough to replicate the ultra-sensitivity generated by the cooperative phosphorylation kinetic model of Kholodenko; this ultra-sensitivity being necessary to generate oscillations. However, as a result of this plasticity, the logic ODE Hill functions are essentially a black box and do not reflect the mechanism of cooperative phosphorylation in the manner of chemical kinetics [27].

An additional issue with the increase in the number of parameters when using logic ODEs is parameter identifiability; it is not possible to determine the exact values of the model parameters by fitting them to data. From Fig. 9, we can see that the *n* parameter that controls the steepness of the Hill function curve can vary over a wide range for two of the activations in the model (MKK to MAPK and MKKK), while still producing an optimal fit to the Kholodenko data. The issue of parameter identifiability is not limited to logic ODEs and is an issue with chemical kinetic models of the MAPK cascade. Indeed, it has been shown that the mass action model of the MAPK cascade [28] can return a variety of nonlinear dynamics depending on parameter choice [43]. Such results emphasize the need to couple modeling to experimental work.

As we have seen above, logic models can reproduce complex dynamics with fewer parameters and distinct states compared to

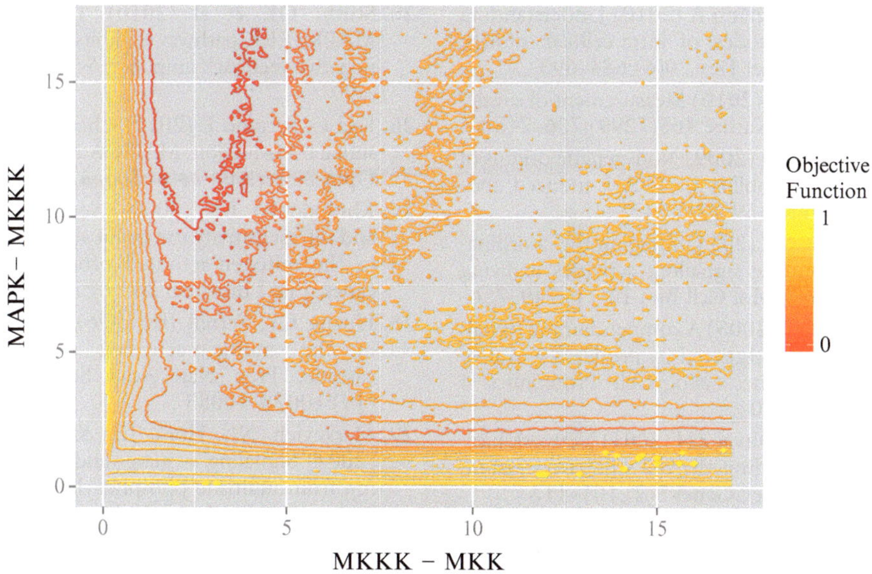

Fig. 9 A contour plot of the objective function (i.e., the difference between the logic ODE model and the Kholodenko data) while varying the n Hill function parameter from MKKK–MKK and MAPK–MKKK

chemical kinetic modeling, with the caveat of less mechanistic detail. The use of the MAPK cascade to illustrate this is slightly disingenuous in that the cascade is biochemically well-studied. In such a case, and if one is considering only this relatively simple system, it makes sense to use biochemical models that can reflect this knowledge. However, less well-studied systems can be explored with logic models, especially where model training to data is being performed (i.e., there is a necessity for fewer degrees of freedom) and in the case of the two formalisms introduced here (*CNORdt* and *CNORode*), the data is in the form of a time-course measurements.

References

1. Hanahan D, Weinberg RA (2011) Hallmarks of cancer: the next generation. Cell 144 (5):646–674
2. Aldridge BB et al (2006) Physicochemical modelling of cell signalling pathways. Nat Cell Biol 8(11):1195–1203
3. Kholodenko B, Yaffe MB, Kolch W (2012) Computational approaches for analyzing information flow in biological networks. Sci Signal 5 (2):re1
4. Terfve C, Saez-Rodriguez J (2012) Modeling signaling networks using high-throughput phospho-proteomics. Adv Exp Med Biol 736:19–57
5. Choudhary C, Mann M (2010) Decoding signalling networks by mass spectrometry-based proteomics. Nat Rev Mol Cell Biol 11 (6):427–439
6. Sabidó E, Selevsek N, Aebersold R (2012) Mass spectrometry-based proteomics for systems biology. Curr Opin Biotechnol 23 (4):591–597
7. Wilkinson DJ (2009) Stochastic modelling for quantitative description of heterogeneous biological systems. Nat Rev Genet 10 (2):122–133
8. Tanner SD, Ornatsky O, Bandura DR (2007) Multiplex bio-assay with inductively coupled plasma mass spectrometry: towards a massively multivariate single-cell technology. Spectrochim Acta Part B At Spectrosc 62 (3):188–195

9. Behar M, Hoffmann A (2010) Understanding the temporal codes of intra-cellular signals. Curr Opin Genet Dev 20(6):684–693
10. Spiller DG et al (2010) Measurement of single-cell dynamics. Nature 465(7299):736–745
11. Spencer SL et al (2009) Non-genetic origins of cell-to-cell variability in TRAIL-induced apoptosis. Nature 459(7245):428–432
12. Welch CM et al (2011) Imaging the coordination of multiple signalling activities in living cells. Nat Rev Mol Cell Biol 12(11):749–756
13. Gaudet S et al (2005) A compendium of signals and responses triggered by prodeath and prosurvival cytokines. Mol Cell Proteomics 4 (10):1569–1590
14. Barabási A-L, Oltvai ZN (2004) Network biology: understanding the cell's functional organization. Nat Rev Genet 5(2):101–113
15. Bossi A, Lehner B (2009) Tissue specificity and the human protein interaction network. Mol Syst Biol 5:260
16. Kiselev VY, Marenduzzo D, Goryachev AB (2011) Lateral dynamics of proteins with polybasic domain on anionic membranes: a dynamic Monte-Carlo study. Biophys J 100 (5):1261–1270
17. Stewart-Ornstein J, Weissman JS, El-Samad H (2012) Cellular noise regulons underlie fluctuations in Saccharomyces cerevisiae. Mol Cell 45(4):483–493
18. Hlavacek WS et al (2006) Rules for modeling signal-transduction systems. Sci STKE 2006 (344):re6
19. Kholodenko BN (2006) Cell-signalling dynamics in time and space. Nat Rev Mol Cell Biol 7(3):165–176
20. Fritsche-Guenther R et al (2011) Strong negative feedback from Erk to Raf confers robustness to MAPK signalling. Mol Syst Biol 7
21. Natarajan M et al (2006) A global analysis of cross-talk in a mammalian cellular signalling network. Nat Cell Biol 8(6):571–580
22. Wang CC, Cirit M, Haugh JM (2009) PI3K-dependent cross-talk interactions converge with Ras as quantifiable inputs integrated by Erk. Mol Syst Biol 5:246
23. Hunter T, Sefton BM (1980) Transforming gene product of Rous sarcoma virus phosphorylates tyrosine. Proc Natl Acad Sci U S A 77 (3):1311–1315
24. Ferrell JE, Bhatt RR (1997) Mechanistic studies of the dual phosphorylation of mitogen-activated protein kinase. J Biol Chem 272 (30):19008–19016
25. Sturm OE et al (2010) The mammalian MAPK/ERK pathway exhibits properties of a negative feedback amplifier. Sci Signal 3(153): ra90
26. Kiel C, Serrano L (2012) Challenges ahead in signal transduction: MAPK as an example. Curr Opin Biotechnol 23(3):305–314
27. Kholodenko BN (2000) Negative feedback and ultrasensitivity can bring about oscillations in the mitogen-activated protein kinase cascades. Eur J Biochem 267(6):1583–1588
28. Huang CY, Ferrell JE (1996) Ultrasensitivity in the mitogen-activated protein kinase cascade. Proc Natl Acad Sci U S A 93 (19):10078–10083
29. Markevich NI, Hoek JB, Kholodenko BN (2004) Signaling switches and bistability arising from multisite phosphorylation in protein kinase cascades. J Cell Biol 164(3):353–359
30. Blüthgen N et al (2006) Effects of sequestration on signal transduction cascades. FEBS J 273(5):895–906
31. Saez-Rodriguez J, Kremling A (2004) Modular analysis of signal transduction networks. IEEE Control Syst Mag 24:35–52
32. Kim Y et al (2011) Substrate-dependent control of MAPK phosphorylation in vivo. Mol Syst Biol 7:467
33. Del Vecchio D, Ninfa AJ, Sontag ED (2008) Modular cell biology: retroactivity and insulation. Mol Syst Biol 4:161
34. Schoeberl B et al (2002) Computational modeling of the dynamics of the MAP kinase cascade activated by surface and internalized EGF receptors. Nat Biotechnol 20(4):370–375
35. Bashor CJ et al (2008) Using engineered scaffold interactions to reshape MAP kinase pathway signaling dynamics. Science 319 (5869):1539–1543
36. Kholodenko BN, Birtwistle MR (2009) Four-dimensional dynamics of MAPK information processing systems. Wiley Interdiscip Rev Syst Biol Med 1(1):28–44
37. Rao CV, Arkin AP (2003) Stochastic chemical kinetics and the quasi-steady-state assumption: application to the Gillespie algorithm. J Chem Phys 118(11):4999
38. Wolkenhauer O et al (2004) Modeling and simulation of intracellular dynamics: choosing an appropriate framework. IEEE Trans Nanobioscience 3(3):200–207
39. Novère NL et al (2009) The systems biology graphical notation. Nat Biotechnol 27 (8):735–741

40. Terfve CDA et al (2012) CellNOptR: a flexible toolkit to train protein signaling networks to data using multiple logic formalisms. BMC Syst Biol 6(1):133
41. MacNamara A et al (2012) State-time spectrum of signal transduction logic models. Phys Biol 9(4):045003
42. Wittmann DM et al (2009) Transforming Boolean models to continuous models: methodology and application to T-cell receptor signaling. BMC Syst Biol 3:98
43. Qiao L et al (2007) Bistability and oscillations in the Huang-Ferrell model of MAPK signaling. PLoS Comput Biol 3(9):1819–1826
44. Hoops S et al (2006) COPASI—a COmplex PAthway SImulator. Bioinformatics 22 (24):3067–3074
45. Team RC (2012) R: a language and environment for statistical computing. Available at: http://www.R-project.org/.
46. Gonzalez AG et al (2006) GINsim: a software suite for the qualitative modelling, simulation and analysis of regulatory networks. Biosystems 84(2):91–100
47. Müssel C, Hopfensitz M, Kestler HA (2010) BoolNet—an R package for generation, reconstruction and analysis of Boolean networks. Bioinformatics 26(10):1378–1380
48. Klamt S, Saez-Rodriguez J, Gilles ED (2007) Structural and functional analysis of cellular networks with Cell NetAnalyzer. BMC Syst Biol 1:2

Chapter 6

From a Biological Hypothesis to the Construction of a Mathematical Model

David Cohen, Inna Kuperstein, Emmanuel Barillot, Andrei Zinovyev, and Laurence Calzone

Abstract

Mathematical models serve to explain complex biological phenomena and provide predictions that can be tested experimentally. They can provide plausible scenarios of a complex biological behavior when intuition is not sufficient anymore. The process from a biological hypothesis to a mathematical model might be challenging for biologists that are not familiar with mathematical modeling.

In this chapter we discuss a possible workflow that describes the steps to be taken starting from a biological hypothesis on a biochemical cellular mechanism to the construction of a mathematical model using the appropriate formalism. An important part of this workflow is formalization of biological knowledge, which can be facilitated by existing tools and standards developed by the systems biology community.

This chapter aims at introducing modeling to experts in molecular biology that would like to convert their hypotheses into mathematical models.

Key words Network diagram, Mathematical model, ODE, Boolean, Formalism

1 Introduction

Mathematical modeling in molecular biology aims at understanding a nonintuitive outcome of an experiment, or predicting how a system would respond to various perturbations (e.g., genetic mutations, environmental stress, or therapeutic molecules).

For most non mathematicians, the construction of a mathematical model that is based on a biological hypothesis can be a puzzling process: what does it do? How can we proceed from a biological hypothesis to a mathematical model? What can we do with it, and what can we not do with it? On the one hand, the mathematical model can be seen as a tool that reproduces what is known about the studied biological processes and proves the coherence of our understanding. On the other hand, it can be challenged, or perturbed to propose predictions that can be validated experimentally. The mathematical model is used for

Maria Victoria Schneider (ed.), *In Silico Systems Biology*, Methods in Molecular Biology, vol. 1021,
DOI 10.1007/978-1-62703-450-0_6,

reasoning and choosing between several scenarios that might explain an observed phenotype in particular conditions.

This chapter is intended to introduce how a mathematical model is constructed to an audience that is not familiar to computational biology or mathematical modeling in particular. With a simple example, we will try to answer the question: how do we proceed from a biological hypothesis to a mathematical model?

2 From Articles to Mathematical Models

The models we refer to are models of biochemical reactions in the cell. The most efficient method for building a mathematical model has been, for us, to follow the workflow presented below. It is neither a unique nor a universal method but, with practice, we identified a series of steps that ultimately tackle the biological hypothesis. Each step in the workflow is formulated as a question.

2.1 What Is the Biological Question That Needs To Be Answered?

Although this seems quite trivial, a well-defined problem enables a "focused" search for data. Once the problem is well defined, it will lead to the next step. However, the vocabulary between a biologist and a mathematician is not always the same. The time taken by the biologist and the mathematician to formulate together the problem, or the hypothesis to test, will facilitate the collaboration.

2.2 What Will Be Included and What Will Not Be Included?

The border and the level of the molecular description should be defined clearly thereby avoiding being unnecessarily broad in data gathering. Here, it needs to be set how many details are required and how detailed should the description of the mechanisms be.

2.3 Where to Find the Information?

Using well-known databases like PubMed (http://www.ncbi.nlm.nih.gov/pubmed/) [1], iHOP (http://www.ihop-net.org/UniPub/iHOP/), Google Scholar (http://scholar.google.com), etc., the scientific articles that contain information about the involved entities, biochemical processes, etc., can be retrieved easily. Databases of molecular pathway (*see* **Note 1**) diagrams such as KEGG, BioCarta and Panther can also be used. It is important to describe accurately the biochemistry behind the processes by gathering what is already known about them. However, the required information is often disseminated throughout hundreds, or even thousands, of experimental articles that need to be gathered and put together to tell a plausible story. Of course, information can also be extracted from *-omics* data using high-throughput technologies that look at gene expression, protein–DNA, RNA–protein, and protein–protein interaction, protein quantities and activities.

2.4 How to Organize the Retrieved Knowledge?

Once the information is collected, it can be stored in different ways: for example, as a database, in the form of a list of known facts, or as a network diagram (*see* **Notes**) recapitulating interactions between the entities of a biological process of interest. Depending on the available information, the level of details and the type of questions that need to be answered, the appropriate diagram will be chosen for a graphical network (*see* **Note 2**) representation.

2.5 How to Translate the Diagram into a Mathematical Model?

The diagram (*see* **Note 3**) is then translated into a mathematical object. According to the type of constructed diagrams, the appropriate formalism must be applied.

2.6 How to Validate the Mathematical Model?

Once the model has been constructed, it has to be able to reproduce what is already known (published data, mutations, or perturbations) or verified with experimental data (in CGH data: loss/gain of a gene to explain a phenotype; in transcriptomics data: differential expression of genes in two different conditions; etc.). Note that the obtained mathematical model may not be the only possible (or the best) model; however, at this point, if it is able to reproduce known facts, it has to be considered as a good and appropriate model that particularly answers the biological hypothesis initially formulated.

2.7 How to Propose Mathematically Derived Predictions and Validate Them?

Once the model has been validated on known data and facts, some not-yet performed perturbations (new mutants or drug treatments) can be simulated *in silico* and predictions on the behavior of the biological system can be formulated. Ideally, experiments will verify if the predictions reflect the in vivo situation.

The whole process can be iterative. Some steps can be omitted according to the formalism chosen.

Our aim, here, is not to explain all steps of this workflow in detail. Rather, we wish to concentrate on the construction of a network diagram and its translation into a mathematical model.

3 Formalisation of Knowledge

Biological knowledge can be retrieved by performing experiments. High-throughput technologies are efficient in generating a vast amount of data. For example, using transcriptome arrays, expression of more than 20,000 genes can be analyzed at the same time, showing possible connections between sets of genes and phenotypes. However, this kind of information about quantities of molecules does not provide the mechanistic details of the pathway functioning and their crosstalk.

To gain insight in the mechanism behind an interaction between (two) molecules, experiments are usually performed concentrating only on the molecules that are putatively involved in the interaction. These types of experiments are well described in many articles and the results of these experiments can provide details about posttranslational modifications, complex formations, degradations, etc. The information retrieved from all these articles together will result ultimately in a network that recapitulates independent experimental settings with different parameters and focus, and, as a result, a generic representation of the biological process or pathway.

4 Representation of Knowledge

As mentioned previously, the knowledge about a particular cellular process can be organized in a database. However, a "wordy" representation of knowledge, though accessible for a computer, cannot be as easily used by humans as a graphical representation of this knowledge. This is the reason why, in addition to creating a database, it is advantageous to summarize the biological knowledge into a map. In geography, a map (*see* **Note 4**) is "a diagrammatic representation of an area of land or sea showing physical features" (*see* **Note 5**). As in geography, in biochemistry, creating maps is a very natural way to represent knowledge about any object (physical or abstract) [2].

4.1 Types of Diagrams

There are different types of diagrams (*see* **Note 3**), each displaying different types of information. We concentrate on four of these types, the last three have been defined and extensively described in Le Novère et al. [3] as the result of a community effort in establishing standards in network visualization (Systems Biology Graphical Notation, SBGN): (1) Interaction diagrams; (2) Activity-flow diagrams; (3) Process-descriptive diagrams; and (4) Entity relationship diagrams. Examples for each of these diagrams are illustrated in Fig. 1.

4.1.1 Interaction Network Diagram

Interaction diagrams indicate relations between two components. These relations can be physical, genetic, correlations, etc. For instance, protein–protein interaction networks inform on which proteins can interact with other proteins within a cell. These diagrams are binary, and nondirectional. Sequences of events cannot be represented, and neither can the mechanical description behind the interactions [4]. Both physical interaction between pair-wise proteins (or complexes) and genetic interaction showing a connection between genotype and phenotype can be visualized [5]. Interaction networks (*see* **Note 2**) can be used for computing differential

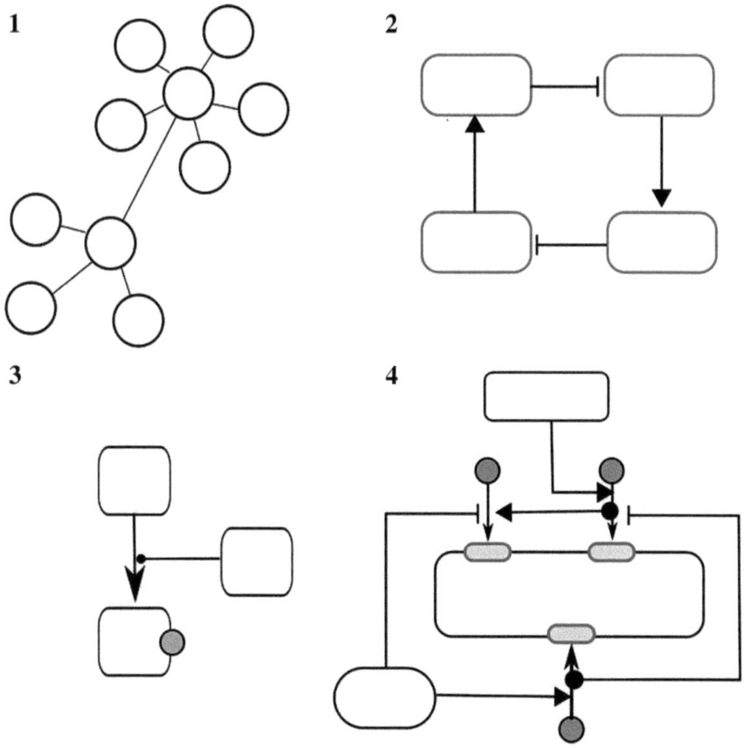

Fig. 1 Four different types of network diagrams. (1) An interaction network diagram. (2) An activity flow diagram. (3) A process descriptive network diagram. (4) An entity relationship diagram. For more details see text (figure provided by N. Le Novère and adapted)

networks, in which edges between nodes are being suppressed or added upon the different conditions [6]. Interaction network diagrams can be compared among different organisms or cell types through the analysis of subgraph structures and the identification of conserved motifs [7].

4.1.2 Activity-Flow Diagrams

In contrast to interaction network diagram, activity-flow diagrams or Regulatory Networks (RN) are directional and can represent sequences of events. The nodes are connected through edges that have either a positive or a negative influence on their neighbors. That way, influences can be shown without the requirement of knowing the detailed mechanism [3]. Although RN are nonmechanistic, the function and/or regulation of a complex biological event such as the role miRNA in ovarian cancer [8], and the transcriptional regulation during epithelial to mesenchymal transition [9] have been represented. Furthermore, RNs integrating different high-throughput sequencing data have been constructed involving pairs of nodes such as transcription factor (TF)-gene, TF-miRNA, and miRNA-mRNA, giving the network a more holistic view of the biological system [10].

4.1.3 Process Description Diagram

The process description diagram shows, in sequential order, the biochemical reactions between entities and is thus directional [3]. Because biochemical reactions can be explicitly displayed, these diagrams show a mechanistic view. Many metabolic and signaling pathways have been depicted using this representation [11–13].

4.1.4 Entity Relationship Diagram

The entity relationship diagram shows the effect of a node onto another node's posttranscriptional modification thereby avoiding multiple representation of the same node in the diagram and showing all different states of an entity [3]. Furthermore, the diagram is directional, the order of events are not sequential, and it provides a mechanistic view [3]. The predecessor of the entity relationship diagram is the molecular interaction map (MIM), which was the first attempt to define symbols and rules for describing molecular interactions [3, 14].

Much effort has been dedicated to the creation of different network diagrams that explicitly details interactions among different entities involved in cellular processes. Many of them are available in the literature or can be retrieved from databases.

4.2 Biological Pathway Databases

As mentioned before, the advantage of representing biological processes in the form of diagrams is not only to bring together components that participate in a process but also to summarize regulatory circuits in one single map. With this philosophy, a significant number of pathway (*see* **Note 1**) databases have been set up for public use to facilitate the retrieval of information (*see* Table 1). Some of these databases concentrate on well-known pathways, e.g., KEGG, Reactome, etc. [15, 16]. Others contain not only generic pathways but also allow visualization of each pathway in the context of different species (*see* Table 1). In addition, a number of pathway databases provide disease-specific pathways like Cancer Cell map (http://cancer.cellmap.org/) or Atlas of Cancer Signalling Networks (http://acsn.curie.fr). The majority of the databases listed in Table 1 have been created manually and curated by qualified biological experts.

To enable data exchange between several databases, common data formats have been introduced (e.g., BioPAX, SBML, PSI-MI, etc., *see* Table 1). Tools have also been developed to extract parts of these databases and to manipulate and merge the pathways of interest (Table 2).

4.3 Drawing

Adequate tools to draw maps in SBGN standard have been made available, among them CellDesigner [17] and SBGN-ED [18]. With CellDesigner, a particular effort has been put on process description diagram representation. SBGN-ED is able to draw all types of diagrams except for interaction networks. For both tools, experimental data can be visualized on the network. To facilitate the usability of the map, it is advised to annotate nodes and edges with

Table 1
Metabolic and signaling pathways databases

Database	Web link	Standard exchange format	Standard visualization (SBGN)	Nodes and edges annotation	Analytical tools	
					Data visualization	Pathways analysis
TRANSPATH	http://www.biobase-international.com	BioPAX, SBML				
ACSN	http://acsn.curie.fr	BioPAX, SBML	●	●		
BioCarta	http://www.biocarta.com	BioPAX		●	●	
BioCyc	http://www.biocyc.org	BioPAX				
Cancer cell map	http://cancer.cellmap.org/cellmap	BioPAX			●	
GenMAPP	http://www.genmapp.org	GPML, BioPAX			●	
Ingenuity	http://www.ingenuity.com	PSI-MI, BioPAX, SBML		●	●	●
KEGG	ftp://ftp.genome.jp/pub/kegg	BioPAX, SBML, KGML	●	●	●	
Pathway Interaction Database	http://pid.nci.nih.gov	PID XML		●		
Panther	http://www.pantherdb.org	BML, BioPAX	●	●	●	
Protein Lounge	http://www.proteinlounge.com/Pathway	SBML, PSI-MI, BioPAX,				
Reactome	http://www.reactome.org	BioPAX, SBML, KGML	●	●	●	●
SPIKE	http://www.cs.tau.ac.il/~spike	SBML, BioPAX				
WikiPathways	http://www.wikipathways.org	GPML, SBML, BioPAX		●		

PSI-MI is a community standard data exchange format used to represent molecular interactions
BioPAX is a standard language that is used to integrate, exchange, visualize, and analyze pathway data
SBML is a machine-readable format used to represent biochemical reaction networks. It can be used to model signal transduction pathways and metabolic networks
GPML is the GenMAPP Pathway Markup Language, a customized XML format compatible with pathway visualization and analysis tools such as Cytoscape, GenMAPP, and PathVisio
KGML is the KEGG Markup Language, an exchange format of the KEGG pathway maps that are manually drawn and updated. KGML provides facilities for computational analysis

Table 2
Tools for map drawing, and visualization/navigation

		Drawing			Visualization/navigation		
Tools	**Web link**	**Drawing editing**	**SBGN format**	**Layout change**	**Web interface**	**Google map navigation**	**Zoom options**
CellDesigner	http://www.celldesigner.org	●	●	●			●
SBGN-ED	http://vanted.ipk-gatersleben.de/addons/sbgn-ed	●	●	●			
CellPublisher	http://cellpublisher.gobics.de		●		●	●	
Pathways projector	http://www.g-language.org/PathwayProjector				●	●	
Cytoscape/BiNoM	http://www.cytoscape.org http://binom.curie.fr	●		●			●
Payao	http://payao.oist.jp:8080/payaologue/index.html		●				
NaviCell	http://navicell.curie.fr		●		●	●	●[a]

[a]Semantic zoom

identifiers when available (Entrez Gene, UNIPROT, etc.), with citations (PubMed IDs), hyperlinks to existing databases, or with any types of notes that could be helpful.

4.3.1 Visualization/Navigation

Due to the size and complexity of molecular networks, dedicated tools for navigation through such networks have been developed such as CellPublisher [19], Pathways projector [20], and NaviCell (http://navicell.curie.fr).

4.4 Network Analysis

CellDesigner and SBGN-ED are good visualization tools but are not able to perform extensive network analysis. Cytoscape is an open source software platform for visualizing and analyzing molecular interaction networks [21], which can be used to import networks that are SBGN compliant [3]. Many plugins for Cytoscape have been developed to facilitate the analysis of pathways. Among them, BiNoM (Biological Network Manager) concentrates on importing, converting, and exporting networks with different standards, such as BioPAX or SBML, on performing path analyses, on extracting specific information from databases, etc. [22, 23]. These analyses can be done prior mathematical modeling to better understand the structure of the studied networks.

5 From Diagrams to Mathematical Modeling

Mathematical models can reproduce what is known about a biological process; can make a link between the physiology and the molecular interactions; can prove that a set of molecular interactions do explain a phenomenon; and can reproduce the system behavior of wild type cells, mutant cells, and in diverse experimental settings. Mathematical models can also provide predictions about not-yet assayed perturbations. However, just building a map is not sufficient to guarantee that the map represents a biologically coherent picture of the biological process. What the dynamical modeling does provide is a formal way to prove that the map, containing the results of many isolated experiments, reflects in vivo observations.

Changing the map into a dynamic mathematical model requires choosing an appropriate type of mathematical formalism. In other words, the content of the map should be translated into formulae using an appropriate mathematical language. This choice will depend on many aspects of the study: the type of data, the type of information that was gathered earlier in the study, the type of questions asked, and the type of answers to be expected.

5.1 Mathematical Models

There are different types of models that are used to describe biological processes at different levels of complexity. For interaction networks, for instance, a statistical model dealing with correlation between the biological entities' quantities is the simplest one [24]. More complex systems depicted in activity flow diagrams, entity relationship, or process description diagrams can be formalized using logical, rule-based, or differential equations [24–27]. We will concentrate here on two of these formalisms: Boolean and chemical kinetics frameworks.

Boolean models are logic-based models that compute the activity of a node depending on the activity of the input nodes [26]. The state or activity of a node can be either on or off and is governed by logical rules linking the effect of the input nodes on the activity of the current node itself. The Boolean gate "OR" is assigned when, for example, either transcription factor A or B binds to a gene and can initiate transcription of the gene. The "AND" gate is used when both transcription factors are required for transcription and "NOT" if transcription factors bind but inhibit the transcription [25]. Once the logical rules are specified, the system can be solved in order to get to the asymptotic solutions (the stable steady states or cyclic attractors), either through synchronous or asynchronous strategy. For the synchronous case, all variables that can be changed according to the logical rules will be updated simultaneously, whereas for the asynchronous case, they will be updated one at a time. Boolean models offer a good starting point and are relatively simple. For the steady-state analysis, they do not require extensive computational power. They can be used for the analysis of large models and do not require a very detailed understanding of the system [24–26]. They are useful when the data are not quantitative and when the molecular details of the process are not known. They can inform on how the cell decides between several fates in response to different inputs [28, 29] or if a proposed mechanism is coherent with observed phenotypes [30, 31].

A *chemical kinetics* model is a set of mathematical equations that describes the rate of change of a set of dynamical variables (nodes in the diagram). In chemical kinetics, variables are simple functions of time [24, 26]. If needed, space can be considered in the form of compartments and pseudo-reactions of transport between these compartments. When taking continuous space into account, which might be essential for the biological question, the *reaction–diffusion equations* formalism needs to be used, but this formalism will not be addressed in this chapter [32]. Also, the effect of stochasticity is not covered here but can be approached by the *stochastic master equation* [33].

We will show two examples of a translation from a diagram to a mathematical model: from a reaction network (process description diagrams) to a chemical kinetics model and from an influence network (activity flow diagram) to a Boolean model. The simple networks that we will construct as toy examples of the methods are centered on the retinoblastoma protein, RB, and its interaction with the transcription factor E2F1. The simplified networks are only illustrative to show the translation from a diagram to a mathematical model.

5.2 Molecular Description of RB Regulation

RB is a tumor suppressor gene which is mutated, or whose regulation is altered, in almost all cancers [34]. Often referred to as the guardian of the cell cycle, the RB protein ensures that the cell remains in G1 phase until proper mitogenic signals are emitted. It is active when hypophosphorylated and inactive when phosphorylated by the cyclin/CDK complexes during the cell cycle. RB sequesters the transcription factors, E2F1, 2, or 3, which are responsible for the transcription of many genes involved in the cell cycle and the apoptosis pathways [35]. In G1 phase, in response to mitogen activation, RB starts to be phosphorylated by the G1 cyclin/CDK complex and Cyclin D1/CDK4(6) and releases its negative control on E2F1 (for simplicity, representing the three activating transcription factors here). E2F1 allows the transition to S phase where Cyclin E/CDK2 and Cyclin A/CDK2 can maintain RB in its phosphorylated state.

Recapitulating the above-mentioned facts and the different choices made for modeling, we can say that:

- RB is active in its unphosphorylated form and inactive in its phosphorylated form. It is phosphorylated by the cyclin/CDK complexes (see chemical equations R1, R2, R3). For simplicity, the cyclin/CDK complexes are considered as parameters. Note that here the cyclins are only able to phosphorylate RB when in complex with E2F1. RB, once phosphorylated and separated from the complex it was forming with E2F1, can be dephosphorylated by a phosphatase (phosphatase not explicitly shown here) (R4). This information can be written down in the form of chemical equations:

$$\text{(R1) E2F1/RB} + \text{CycD1/CDK4} \Rightarrow \text{E2F1/RB-P} + \text{CycD1/CDK4}$$

$$\text{(R2) E2F1/RB} + \text{CycE1/CDK2} \Rightarrow \text{E2F1/RB-P} + \text{CycE1/CDK2}$$

$$\text{(R3) E2F1/RB} + \text{CycA2/CDK2} \Rightarrow \text{E2F1/RB-P} + \text{CycA2/CDK2}$$

$$\text{(R4) RB-P} \Rightarrow \text{RB}$$

- E2F1 is synthesized (R5) and when phosphorylated, it is degraded (R7). CycA2/CDK2 catalyzes the phosphorylation of E2F1. Hence, it appears as both reactant and product in the corresponding reaction (R6). The chemical equations are:

$$\text{(R5)} \Rightarrow \text{E2F1}$$

$$\text{(R6) E2F1} + \text{CycA2/CDK2} \Rightarrow \text{E2F1-P} + \text{CycA2/CDK2}$$

$$\text{(R7) E2F1-P} \Rightarrow ..$$

- E2F1 and the hypophosphorylated form of RB form a complex inhibiting E2F1 function (R8). The complex can dissociate (R9):

$$\text{(R8) E2F1} + \text{RB} \Rightarrow \text{E2F1/RB}$$

$$\text{(R9) E2F1/RB} \Rightarrow \text{E2F1} + \text{RB}$$

- When RB is in its hyperphosphorylated form, the complex dissociates. E2F1 is free and can mediate the synthesis of cell cycle genes:

$$\text{(R10) E2F1/RB-P} \Rightarrow \text{E2F1} + \text{RB-P}$$

The resulting reaction network is depicted in Fig. 2.

5.3 From a Reaction Network to a Chemical Kinetics Model

If time scales and amounts of molecular entities are important for answering the biological hypothesis, then chemical kinetics formalism is more appropriate to use.

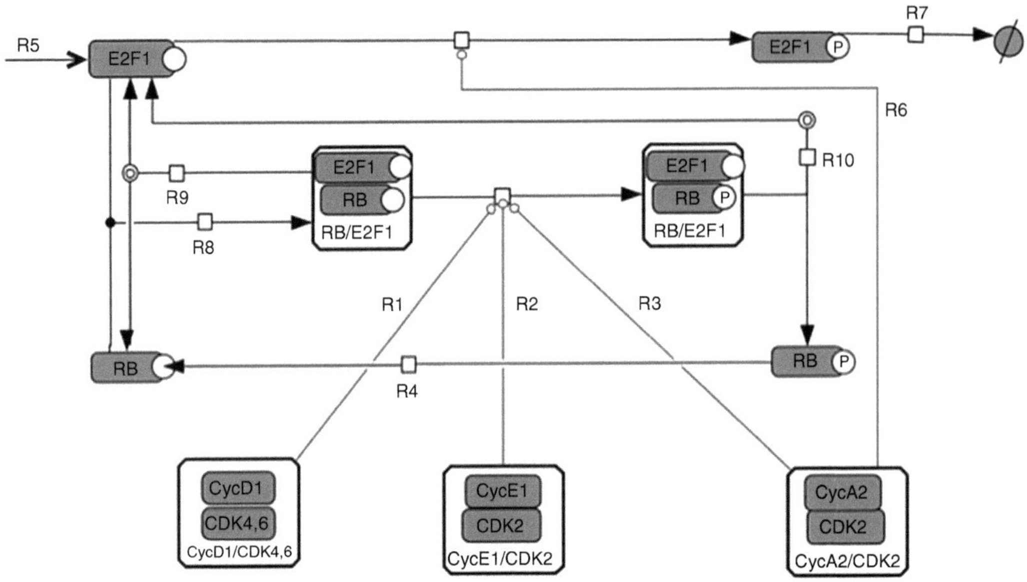

Fig. 2 Molecular reaction network of the RB/E2F interaction. Proteins are represented by simple *gray boxes*, phosphorylated proteins have a "P" in the *white circle* attached to each box. Complexes are surrounded by a *black box* and contain the proteins that compose the complexes

In chemical kinetics, each reaction is assigned a speed referred to as *kinetic reaction rate*. This rate depends on the concentration (or amount) of the reactants through a function called a *kinetic law*. There exist several forms of these kinetic laws, among them, *mass action kinetics* represents the simplest way to define the reaction kinetic rate and corresponds to the multiplication of the reactant concentration to the power of the reactants' stoichiometry. For example, the kinetic rate of the reaction, $A+2B \Rightarrow C$, is written: $k \cdot [A] \cdot [B]^2$, where k is the kinetic reaction rate parameter and the squared brackets are used for the concentration of the chemical species. With mass action kinetics, the speed of reactions can increase infinitely with the increase of the reactants' concentrations, which is not observed in reality. It is often observed that, when the concentration of a chemical entity increases, it does not increase linearly and finally reaches a plateau. More complex kinetic laws allow the description of saturation effect such as *Michaelis-Menten* and *Hill equations*. Michaelis-Menten kinetics was introduced to describe enzyme kinetics and takes the form of: $V_{\text{max}}[S]/(K_{\text{m}} + [S])$ where S is the substrate, V_{max} is the maximum speed of reaction for high concentrations of S, and K_{m} is the value at which *the speed of reaction* has reached half its maximally possible speed V_{max}. Similarly, the Hill equation introduces the idea of cooperativity between two chemical entities: if a ligand S binds to a molecule that has already a ligand attached to it, it may increase its binding affinity to the molecule. It has the form: $[S]^n/(K^n + [S]^n)$, where n is the cooperativity coefficient and K is the value at which the substrate concentration occupies half the amount of the binding site. With n and K tuned, the equation can show a rapid and abrupt activation of the substrate.

Choosing the correct form for the chemical kinetics equation depends on the observed behavior of the species. For instance, RB is known to be phosphorylated at several amino acids residues leading to an abrupt inactivation of the protein when enough of these residues are phosphorylated. Mass action kinetics might not be able to reproduce this behavior, thus we favor a Hill function or use Michaelis-Menten kinetics instead of representing the phosphorylation of RB. In our example, the hyperphosphorylated form of RB will be presented by only one phosphorylation mediated by at least one of the three cyclin/CDK complexes.

The translation of a reaction network into a chemical kinetics model is done knowing that:

- Each node (or chemical species) of the diagram becomes a dynamical variable of the model, representing the amount or concentration of the chemical species
- Each equation represents the rate of change of the variable (d[Species]/dt = rate of the concentration change of the chemical entity as a function of time),
- The right-hand part of the equation describing the rate of change of a chemical species is constructed as follows:
  - When this chemical species participates as a reactant in the reaction i with kinetic rate R_i, then R_i is added to the right-hand side with a negative sign.
  - When this chemical species is a product of the reaction i with kinetic rate R_i, then R_i is added to the right-hand side with a positive sign.

The resulting equations are the following:

$$\frac{d[RB]}{dt} = R9 - R8 + R4$$

$$\frac{d[RBP]}{dt} = R10 - R4$$

$$\frac{d[E2F1]}{} dt = R5 - R6 + R9 - R8 + R10$$

$$\frac{d[E2F1P]}{dt} = R6 - R7$$

$$\frac{d[E2F1/RB]}{dt} = R8 - R9 - (R1 + R2 + R3)$$

$$\frac{d[E2F1/RBP]}{dt} = (R1 + R2 + R3) - R10$$

where

$$R1 = \frac{(k_1 \cdot [\text{CycD1/CDK4}]) \cdot [\text{E2F1/RB}]^n}{K^n + [\text{E2F1/RB}]^n}$$

$$R2 = \frac{(k_2 \cdot [\text{CycE1/CDK2}]) \cdot [\text{E2F1/RB}]^n}{K^n + [\text{E2F1/RB}]^n}$$

$$R3 = \frac{(k_3 \cdot [\text{CycA2/CDK2}]) \cdot [\text{E2F1/RB}]^n}{K^n + [\text{E2F1/RB}]^n}$$

$$R4 = k_{\text{dephosphoRB}} \cdot [\text{RBP}]$$

$$R5 = k_{\text{syn}}$$

$$R6 = k_{\text{phosphoE2F}} \cdot [\text{CycA2/CDK2}] \cdot [\text{E2F1}]$$

$$R7 = k_{\text{deg}} \cdot [\text{E2F1P}]$$

$$R8 = k_{\text{ass}} \cdot [\text{E2F1}] \cdot [\text{RB}]$$

$$R9 = k_{\text{diss}} \cdot [\text{E2F1/RB}]$$

$$R10 = k_{\text{dissP}} \cdot [\text{E2F1/RBP}]$$

In our model, for simplicity, the concentrations of the cyclin/CDK complexes are fixed. They will be considered as parameters.

Verifications can be made to avoid errors in the writing of these equations. For instance, mass should be conserved for the different forms of RB: RB non-phosphorylated, RB phosphorylated, in complex or free. Moreover, some existing software can assist the modeler and automatically generate the differential equations from a description of all individual reactions (BIOCHAM [36], JigCell [37], or CellDesigner [38], MATLAB Systems Biology Toolbox [39]) provided the chosen kinetics (mass action, Michaelis Menten kinetics, or Hill equations). Of course, the translation into a mathematical model is not unique since, for instance, some different choices on the type of kinetics used for each reaction can be made.

The next step would be to determine the parameter values that best reproduce the experimentally observed behavior of the dynamics between RB and E2F1 and the corresponding initial conditions. This task is difficult since measuring these parameters is not an easy procedure and sometimes not feasible.

Once the parameters are set so to reproduce known behaviors, the model can be challenged: some mutants can be simulated. With such a small model, the possibilities are limited, but how would we simulate a deletion of E2F1, for instance? Both the initial conditions of all forms of E2F1 and the synthesis term, k_{syn}, should be set to 0 before running the simulation.

For real applications of differential equation models involving E2F1 and RB and mutant simulations, we advise the reader to explore the articles from [40–42].

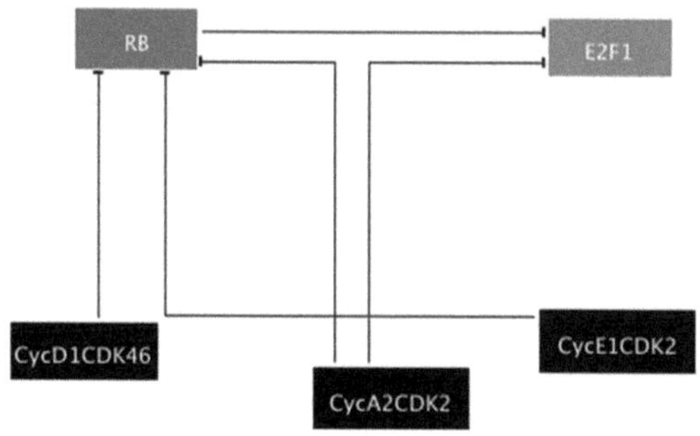

Fig. 3 Activity flow network diagram showing the regulation of RB and E2F1

5.4 From an Influence Network to a Boolean Model

The question answered with logical modeling is more qualitative. It may be appropriate when biochemical mechanisms are not explicitly described but rather represented in ambiguous terms such as "protein A inhibits protein B," or when the kinetic rate parameters are not known. In many cases, the set of biochemical processes can be formalized in terms of influences of one entity on the others, either positive or negative. A node in the influence network corresponds to a variable in a Boolean model. The variable can then have only two values: 1, which is equivalent to present (or active) or 0, which is equivalent to absent (or inactive). That way, RB can be represented by one node: RB, which represents two possible states of RB protein. When the corresponding value is equal to 1, RB is considered to be in its unphosphorylated form and in complex with E2F1. When the value is equal to 0, RB is considered to be phosphorylated by the cyclins and therefore not in complex with E2F1 anymore. In the network diagram below, all edges are negatively signed. As in the reaction network, the cyclin/CDK complexes are fixed and are considered as inputs of the model, i.e., with no incoming edges.

All three cyclin/CDK complexes negatively influence RB since when they are present they phosphorylate and inactivate RB. If one of the three complexes is present, it is sufficient to inactivate RB. E2F1 is active in the absence of RB and of CycA/CDK2 since RB sequesters it and CycA/CDK2 phosphorylates the transcription factor which is then recognized for degradation, respectively (*see* Fig. 3). The translation into Boolean terms is as follows:

RB = **NOT** (CycD1/CDK4 **OR** CycA2/CDK2 **OR** CycE1/CDK2)

E2F1 = **NOT** RB **AND NOT** CycA2/CDK2

RB or E2F1 will be set to 1 if the corresponding right-hand side of the equation is true. Otherwise, they will be set to 0. Since "real" time is not easily represented in this formalism, to show that CycD1/CDK4 can start phosphorylating RB before the other cyclins cannot be represented straightforwardly. Allowing two levels of RB (RB = 1 for CycD1/CDK4 phosphorylation and RB = 2 for CycE1/CDK2 and CycA2/CDK2 phosphorylations) might solve the problem but would increase the complexity of the mathematical model. The steady-state solutions of this system show two types of cases, when RB is on and E2F1 is off corresponding to a G1 arrest and when RB is off and E2F1 is on corresponding to entry in S phase.

The simulation of mutants in Boolean framework is done by setting the variable to 0 in the case of deletion and to 1 in the case of constitutive expression thereby ignoring the initial Boolean rule for the concerned node. For example, E2F1 = **NOT** RB **AND NOT** CycA2/CDK2 will become E2F1 = 0 in the case of deletion.

To simulate Boolean networks, there exist a number of tools that can be used: GINsim [43], for relatively small networks, can provide a thorough analysis of steady states and inform on the different events that lead to these steady states with both synchronous and asynchronous strategies. BoolNet [44] and CellNetAnalyzer [45] use synchronous updating strategies but allow the analysis of bigger networks. Alternative approaches have used hybrid systems including asynchronous strategy with noise and continuous time [46].

More complete and complex models of the RB/E2F pathway using Boolean modeling can be found, for instance, in Fauré et al. [47, 48]

6 Conclusion

In this chapter, a possible workflow has been described that identifies the steps from a biological hypothesis to a mathematical model with the emphasis of constructing a network diagram and its translation into a mathematical model.

The most essential and time-consuming part of this workflow is the formal representation of knowledge on the interplay of the selected biochemical mechanisms involved in the described cellular process. During this step, the biological knowledge that can be dispersed in many publications and expressed in natural language with all its ambiguities is converted into a formal mathematical object visualized as a diagram of particular type and semantics. This diagram, amenable to computer analysis, will serve as a bridge between the written biological results and the formal mathematical methodology.

The constructed diagram can then be translated into a set of mathematical equations though this always requires specifying some additional information such as the kinetic laws or the logical rules. In this chapter, we concentrated our attention on two different modeling methods: chemical kinetics and Boolean models, but according to the question, and although they were not addressed here, other methods might be more appropriate to verify a biological hypothesis. The choice of the methods depends on the type of diagrams, which depend on the type of questions and available information.

7 Notes

In the chapter, the terms diagram, network, pathway, and map are used. We propose here definitions for each of these terms applicable in the context of this chapter

1. Network: is a set of molecular entities and the various types of biochemical interactions they share, such as regulations, influences, molecular transformations, or complex formations.
2. Diagram: is a particular graphical representation of a network, which is typically composed of nodes and edges. These nodes and edges can have different meanings according to the type of diagram.
3. Pathway: is a network of molecular entities belonging to a particular function or process.
4. Map: is a particular representation of a diagram with additional features such as coloring, layout, boundaries and labels.
5. Reference from Oxford dictionary online: http://oxforddictionaries.com/definition/english/map?q=map.

Acknowledgements

We are grateful for receiving funding from the European Union's Seventh Framework Programme (FP7/2007-2013) under grant agreement n°259348. DC, IK, EB, AZ, and LC are members of the team "Computational Systems Biology of Cancer" Equipe labellisée par la Ligue Nationale Contre le Cancer. We would also like to thank Nicolas Le Novère for discussions on different types of diagrams and for providing Fig. 1.

References

1. Hoffmann R, Valencia A (2004) A gene network for navigating the literature. Nat Genet 36:664.
2. Larkin JH, Simon HA (1987) Why a diagram is (sometimes) worth ten thousand words. Cogn Sci 11(1):65–100
3. Le Novère N, Hucka M, Mi H et al (2009) The systems biology graphical notation. Nat Biotechnol 27(8):735–741. doi:10.1038/nbt.1558
4. Wang PI, Marcotte EM (2010) It's the machine that matters: predicting gene function and phenotype from protein networks. J Proteomics 73(11):2277–2289
5. Dixon SJ, Costanzo M, Baryshnikova A et al (2009) Systematic mapping of genetic interaction networks. Annu Rev Genet 43(1):601–625. doi:10.1146/annurev.genet.39.073003.114751

6. Ideker T, Krogan NJ (2012) Differential network biology. Mol Syst Biol 8:565. doi:10.1038/msb.2011.99
7. Przulj N (2011) Protein-protein interactions: making sense of networks via graph-theoretic modeling. Bioessays 33(2):115–123. doi:10.1002/bies.201000044
8. Schmeier S, Schaefer U, Essack M et al (2011) Network analysis of microRNAs and their regulation in human ovarian cancer. BMC Syst Biol 5:183. doi:10.1186/1752-0509-5-183
9. Pratt CH, Vadigepalli R, Chakravarthula P et al (2008) Transcriptional regulatory network analysis during epithelial-mesenchymal transformation of retinal pigment epithelium. Mol Vis 14:1414–1428
10. Cheng C, Yan K-K, Hwang W et al (2011) Construction and analysis of an integrated regulatory network derived from high-throughput sequencing data. PLoS Comput Biol 7(11): e1002190
11. Calzone L, Gelay A, Zinovyev A et al (2008) A comprehensive modular map of molecular interactions in RB/E2F pathway. Mol Syst Biol 4:174. doi:10.1038/msb.2008.7
12. Caron E, Ghosh S, Matsuoka Y et al (2010) A comprehensive map of the mTOR signaling network. Mol Syst Biol 6:453. doi:10.1038/msb.2010.108
13. Patil S, Pincas H, Seto J et al (2010) Signaling network of dendritic cells in response to pathogens: a community-input supported knowledgebase. BMC Syst Biol 4(1):137
14. Kohn KW (1999) Molecular interaction map of the mammalian cell cycle control and DNA repair systems. Mol Biol Cell 10(8):2703–2734
15. Joshi-Tope G, Gillespie M, Vastrik I et al (2005) Reactome: a knowledgebase of biological pathways. Nucleic Acids Res 33(Database issue): D428–D432. doi:10.1093/nar/gki072
16. Kanehisa M (2002) The KEGG database. Novartis Found Symp 247:91–101; discussion 101–103, 119–128, 244–252
17. Kitano H, Funahashi A, Matsuoka Y et al (2005) Using process diagrams for the graphical representation of biological networks. Nat Biotechnol 23(8):961–966. doi:10.1038/nbt1111
18. Czauderna T, Klukas C, Schreiber F (2010) Editing, validating and translating of SBGN maps. Bioinformatics 26(18):2340–2341. doi:10.1093/bioinformatics/btq407
19. Florez LA, Lammers CR, Michna R et al (2010) Cell Publisher: a web platform for the intuitive visualization and sharing of metabolic, signalling and regulatory pathways. Bioinformatics 26(23):2997–2999. doi:10.1093/bioinformatics/btq585
20. Kono N, Arakawa K, Ogawa R et al (2009) Pathway projector: web-based zoomable pathway browser using KEGG atlas and Google Maps API. PLoS One 4(11):e7710. doi:10.1371/journal.pone.0007710
21. Smoot ME, Ono K, Ruscheinski J et al (2011) Cytoscape 2.8: new features for data integration and network visualization. Bioinformatics 27(3):431–432. doi:10.1093/bioinformatics/btq675
22. Zinovyev A, Viara E, Calzone L et al (2008) BiNoM: a Cytoscape plugin for manipulating and analyzing biological networks. Bioinformatics 24(6):876–877. doi:10.1093/bioinformatics/btm553
23. Bonnet E, Calzone L, Rovera D, Stoll G, Barillot E, Zinovyev A (2013) BiNoM 2.0, a Cytoscape plugin for accessing and analyzing pathways using standard systems biology formats. BMC Syst Biol 7:18
24. Bachmann J, Raue A, Schilling M et al (2012) Predictive mathematical models of cancer signalling pathways. J Intern Med 271(2):155–165
25. Ay A, Arnosti DN (2011) Mathematical modeling of gene expression: a guide for the perplexed biologist. Crit Rev Biochem Mol Biol 46 (2):137–151. doi:10.3109/10409238.2011.556597
26. Morris MK, Saez-Rodriguez J, Sorger PK et al (2010) Logic-based models for the analysis of cell signaling networks. Biochemistry 49 (15):3216–3224
27. Karlebach G, Shamir R (2008) Modeling and analysis of regulatory networks. Nat Rev Mol Cell Biol 9:771–780. doi:10.1038/nrm2503
28. Calzone L, Tournier L, Fourquet S et al (2010) Mathematical modelling of cell-fate decision in response to death receptor engagement. PLoS Comput Biol 6(3):e1000702
29. Philippi N, Walter D, Schlatter R et al (2009) Modeling system states in liver cells: survival, apoptosis and their modifications in response to viral infection. BMC Syst Biol 3:97. doi:10.1186/1752-0509-3-97
30. Saez-Rodriguez J, Alexopoulos LG, Zhang M et al (2011) Comparing signaling networks between normal and transformed hepatocytes using discrete logical models. Cancer Res 71(16):5400–5411. doi:10.1158/0008-5472.CAN-10-4453
31. Schlatter R, Schmich K, Avalos Vizcarra I et al (2009) ON/OFF and beyond–a Boolean model of apoptosis. PLoS Comput Biol 5(12): e1000595. doi:10.1371/journal.pcbi.1000595
32. Britton NF (1986) Reaction–diffusion equations and their applications to biology. Academic, London

33. Hegland M, Burden C, Santoso L et al (2007) A solver for the stochastic master equation applied to gene regulatory networks. J Comput Appl Math 205(2):708–724. doi:10.1016/j.cam.2006.02.053
34. Sherr CJ, McCormick F (2002) The RB and p53 pathways in cancer. Cancer Cell 2(2):103–112. doi:10.1016/S1535-6108(02)00102-2
35. Polager S, Ginsberg D (2008) E2F at the crossroads of life and death. Trends Cell Biol 18 (11):528–535. doi:10.1016/j.tcb.2008.08.003
36. Calzone L, Fages F, Soliman S (2006) BIOCHAM: an environment for modeling biological systems and formalizing experimental knowledge. Bioinformatics 22(14):1805–1807. doi:10.1093/bioinformatics/btl172
37. Vass M, Allen N, Shaffer CA, Ramakrishnan N, Watson LT, Tyson JJ (2004) The JigCell model builder and run manager. Bioinformatics 20(18):3680–3681
38. Funahashi A, Matsuoka Y, Jouraku A et al (2008) Cell Designer 3.5: a versatile modeling tool for biochemical networks. Proc IEEE 96 (8):1254–1265
39. Schmidt H, Jirstrand M (2006) Systems Biology Toolbox for MATLAB: a computational platform for research in systems biology. Bioinformatics 22(4):514–515. doi:10.1093/bioinformatics/bti799
40. Aguda BD, Tang Y (1999) The kinetic origins of the restriction point in the mammalian cell cycle. Cell Prolif 32(5):321–335
41. Qu Z, Weiss JN, MacLellan WR (2003) Regulation of the mammalian cell cycle: a model of the G1-to-S transition. Am J Physiol Cell Physiol 284(2):C349–C364. doi:10.1152/ajpcell.00066.2002
42. Novak B, Tyson JJ (2004) A model for restriction point control of the mammalian cell cycle. J Theor Biol 230(4):563–579. doi:10.1016/j.jtbi.2004.04.039
43. Gonzalez AG, Naldi A, Sanchez L et al (2006) GINsim: a software suite for the qualitative modelling, simulation and analysis of regulatory networks. Biosystems 84(2):91–100. doi:10.1016/j.biosystems.2005.10.003
44. Mussel C, Hopfensitz M, Kestler HA (2010) BoolNet–an R package for generation, reconstruction and analysis of Boolean networks. Bioinformatics 26(10):1378–1380. doi:10.1093/bioinformatics/btq124
45. Klamt S, Saez-Rodriguez J, Gilles E (2007) Structural and functional analysis of cellular networks with Cell NetAnalyzer. BMC Syst Biol 1(1):2
46. Stoll G, Viara E, Barillot E et al (2012) Continuous time Boolean modeling for biological signaling: application of Gillespie algorithm. BMC Syst Biol 6:116. doi:10.1186/1752-0509-6-116
47. Faure A, Naldi A, Chaouiya C et al (2006) Dynamical analysis of a generic Boolean model for the control of the mammalian cell cycle. Bioinformatics 22(14):e124–e131. doi:10.1093/bioinformatics/btl210
48. Barillot E, Calzone L, Hupe P, Vert J-P, Zinovyev A (2012) Computational systems biology of cancer. Chapman & Hall, CRC Mathematical & Computational Biology 452 p.

Chapter 7

Practical Use of BiNoM: A Biological Network Manager Software

Eric Bonnet, Laurence Calzone, Daniel Rovera, Gautier Stoll, Emmanuel Barillot, and Andrei Zinovyev

Abstract

The Biological Network Manager (BiNoM) is a software tool for the manipulation and analysis of biological networks. It facilitates the import and conversion of a set of well-established systems biology file formats. It also provides a large set of graph-based algorithms that allow users to analyze and extract relevant subnetworks from large molecular maps. It has been successfully used in several projects related to the analysis of large and complex biological data, or networks from databases. In this tutorial, we present a detailed and practical case study of how to use BiNoM to analyze biological networks.

Key words Biological networks, Graph-based algorithms, Subnetworks, Molecular maps, BiNoM

1 Introduction

The last decade has seen unprecedented advances in the production of high-throughput experimental data in biology, fueled by drastic technological improvements in various ways of measuring biological entities. In return, those large amounts of biological information have stimulated the need of developing standards for an efficient representation and exchange of data. This is especially true for the field of systems biology, which aims at building models and quantitative or qualitative simulations of complex biological systems [1, 2]. To achieve this goal, it is obvious that a good communication and collaboration between modelers and experimentalists having various scientific backgrounds will be facilitated by the standardization of the representation of workflows, data formats, and mathematical models. Several complementary standards have already been created and are increasingly used in a large variety of projects. Most of them are based on an open-source and community-based organization, ensuring an easy access to the detailed specifications of the standard, flexibility, dynamic

Maria Victoria Schneider (ed.), *In Silico Systems Biology*, Methods in Molecular Biology, vol. 1021,
DOI 10.1007/978-1-62703-450-0_7, © Springer Science+Business Media, LLC 2013

evolution, and wide acceptance. Examples of such community standards are the System Biology Markup Language (SBML) [3], a language focused on mathematical modeling, the Biological Pathway Exchange standard (BioPAX) [4], devoted to storing and exchange of pathway information and the Systems Biology Graphical Notation (SBGN), centered on the graphical notation for biological maps [5]. It is worth noting that there are now more than 40 databases and online resources supporting the BioPAX format, while more than 33 databases are using SBML (http://www.pathguide.org). Well-established examples are the Reactome database [6], BioModels [7], and MINT [8]. More and more systems biology software packages are also using standard formats to store and exchange data. For example, CellDesigner is a tool used to edit biological pathways diagrams [9] and is using a compatible SBML dialect to store the all the information related to a given diagram. Cytoscape is a widely spread program used for the visualization, modeling, and analysis of complex molecular and genetic interaction networks [10]. BiNoM was developed as a Cytoscape plugin, with the goal of facilitating the import and export of various systems biology formats, and also proposing a large set of graph-based algorithms for the extraction of relevant subnetworks from large molecular maps [11]. BiNoM was successfully used in several projects related to the analysis of complex biological networks [12]. Here, we present a set of detailed and concrete examples of how to extract relevant information from such maps using BiNoM.

2 Material

The Cytoscape [10] software should be installed on the computer (http://cytoscape.org). The BiNoM plugin can be installed in different ways. The first is to download BiNoM from our web page (http://binom.curie.fr/projects/binom/), (http://binom.curie.fr) and copy the file under the directory "plugins" of the Cytoscape installation folder (administrator privileges might be necessary to perform this operation). The latest version of BiNoM (version 2.0) supports the latest versions of Cytoscape. The previous version (BiNoM 1.0) is also available on our website for older versions of Cytoscape.

Another possibility to install BiNoM is to use the plugin manager of Cytoscape. Starting in version 2.5, the plugin management has been added to allow users to search for, download, install, update, and delete plugins within Cytoscape.

1. Select the function "Plugins > Manage Plugins" from the menu.
2. Navigate in the tree view to the category "Analysis."

3. Select BiNoM v2.0 (or any more recent version available).
4. Click "Install."

All the files used throughout this tutorial can be downloaded from our website (http://bioinfo-out.curie.fr/projects/binom/, http://binom.curie.fr).

3 Methods

3.1 Import and Export

A major function of the BiNoM software is to provide import and export functions for a given number of standard systems biology file formats. It is therefore possible to import data from SBML level 2 files, CellDesigner 3.X and 4.X files, BioPAX level 3, CSML files and also from simple text files formatted to the AIN (Annotated Influence Network) format. The aim of BiNoM import/export functions is not to be a universal converter but rather to favor a number of scenarios where the conversion is making sense (Fig. 1). It is worth mentioning that due to major changes in the specifications, the BioPAX level 3 format is incompatible with the previous level 2 format [4]. The previous version of BiNoM (version 1.0, still available from our website) was managing the BioPAX level 2, but the latest version of BiNoM (2.0) can only deal with BioPAX level 3 files. The current version of Cytoscape does not support a direct import of BioPAX level 3 files yet [10].

Let us take an example. The model of the yeast cell cycle by Novak and colleagues [13] was encoded as a graph using CellDesigner software [9]. We can easily import it in Cytoscape using the BiNoM functions. The file is available on the BiNoM website (file name: M-Phase.xml).

1. Select the function "Plugins > BiNoM 2.0 > BiNoM I/O > Import CellDesigner document from file" from the menu.
2. Select the file from the dialog window.
3. Click OK.

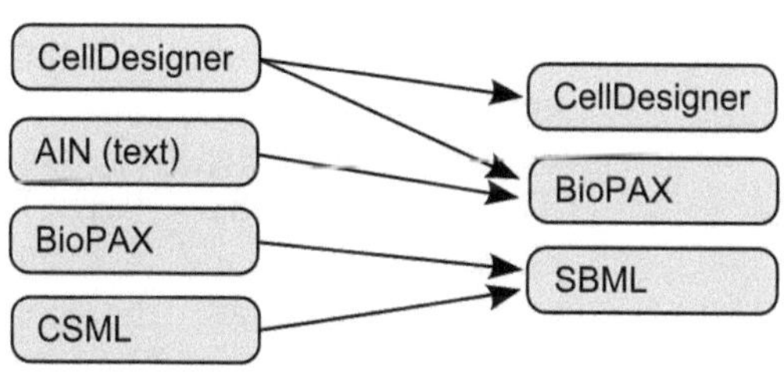

Fig. 1 The BiNoM import/export functions favor a number of scenarios that are illustrated on the figure (*left side*: import file formats, *right side*: export file formats). Note that for the CellDesigner to CellDesigner conversion, it is possible to split the network, change the layout, and change the color and the scale

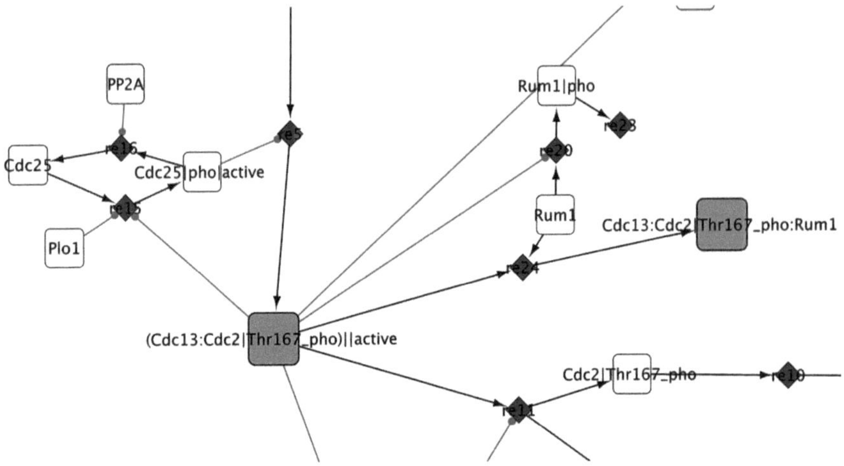

Fig. 2 Zoom on a simple cell cycle network imported from CellDesigner into Cytoscape using BiNoM

4. A new network is created as "M-Phase.xml" with 36 nodes and 42 edges (Fig. 2).

BiNoM uses its own visual mapping to represent the different molecules and their interactions, inspired by the SBGN standard [5], but presents a simplified version of it. For example, simple proteins are represented by white circles, while protein complexes are pictured as gray circles. Similarly, there is a specific mapping for the different relationships between molecules: for example, a catalysis relation will be represented as a red colored edge with a circular end. Figure 3 shows the BiNoM visual styles for BioPAX and CellDesigner. When importing pathway information, BiNoM generates meaningful names for every chemical species, following preestablished rules. Chemical species are defined as physical entities (e.g., a protein) with an optional cellular localization and posttranslational modifications. The name formatting rules are as follows:

(Entity1_name|Modification:Entity2_name|Modification)[_active|_hmN]@compartment

The colon symbol ":"delimitates the different components of a complex, the vertical bar "|" indicates the posttranslational modifications, while the "@" sign indicates the cellular compartment. The optional suffixes "active" or "hm" indicate the state of the chemical species and N-homodimer state, respectively. The parentheses delimitate the components concerned by the N-homodimer state and are useful to eliminate ambiguities (*see* Fig. 2 for examples).

We have recently developed a new import format-denominated AIN. The principle of this format is to encode an influence network, where edges represent either an inhibition or an activation,

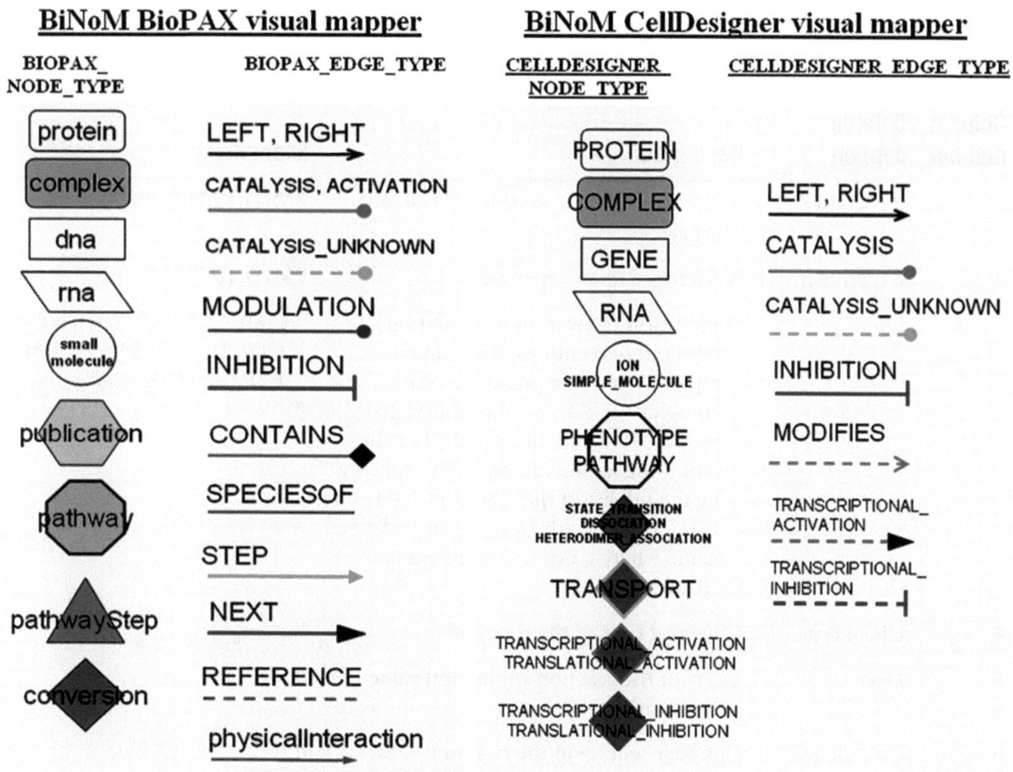

Fig. 3 Comprehensive visual representations followed by the BiNoM software for different entities and their relationships, for both the BioPAX and CellDesigner file formats

into a simple tab-delimited text file (*see* Table 1 for a detailed explanation of the AIN format). Using this format, it is rather straightforward for a biological expert to encode a network from his or her own expertise and/or from published results, using a spreadsheet program such as Excel, and then import it in Cytoscape using BiNoM, rather than using more sophisticated tools such as CellDesigner. All the information contained in the AIN file is automatically converted in the BioPAX format when the file is imported and can be subsequently retrieved with specific BiNoM functions. Let us now import a simple cell cycle model encoded as an AIN file (cell_cycle_AIN.txt, available from the BiNoM website).

1. Select the function "Plugins > BiNoM 2.0 > BiNoM I/O > Import influence network from AIN file" from the menu.
2. Select the file "cell_cycle_AIN.txt" from the dialog window and click OK.
3. Click OK twice, for the windows "Defining families" and "Select constitutive reactions to add."
4. The network is imported as "cell_cycle_AIN" (Fig. 4).

Table 1
Description of the AIN format

<table>
<tr><th>Column number</th><th>Column caption</th><th>Description</th><th>Example(s)</th></tr>
<tr><td>1</td><td>ReviewRef</td><td>A reference (e.g., a PubMed ID) to an article</td><td>PMID:1234</td></tr>
<tr><td>2</td><td>ExperimentRef</td><td>A reference to an experiment</td><td>PMID:10783242</td></tr>
<tr><td>3</td><td>Link</td><td>Connection (activation or inhibition) between two entities. The name can represent a single protein, a protein complex "(C:D)," a phosphorylated protein "(C^p)," or a family. For the latter, the family can be given explicitly by the full list of the members "(C1, C2, C3)," or implicitly by using an undefined name where a dot will represent any character "(C.)"</td><td>"A->B," "A-|B," "((CCNE.):CDK2)->E2F5^p," "(E2F1, E2F2)->CDKN2A"</td></tr>
<tr><td>4</td><td>ChemType</td><td>Chemical type of the reaction</td><td>Binding</td></tr>
<tr><td>5</td><td>Delay</td><td>Delay of the reaction (numerical value and unit)</td><td>0.9 h</td></tr>
<tr><td>6</td><td>Confidence</td><td>Confidence level in the reaction (value between 0 and 1)</td><td>0.8</td></tr>
<tr><td>7</td><td>Tissue</td><td>Tissue where the reaction has been observed</td><td>Fibroblast</td></tr>
<tr><td>8</td><td>Comment</td><td>Comment about the reaction</td><td>"Specific phosphorylated site of E2F5"</td></tr>
</table>

Each line of the table represents a column of the AIN tab-delimited file. Columns are numbered from left to right. Missing values should be indicated by a single dot and text strings should be quoted. The only mandatory column is the Link (column number 3), representing the reaction

3.2 Manipulating Existing Networks

The cell cycle model of Novak et al. (Fig. 2) has 36 nodes and 42 edges in total, making it a rather small network. However, this is not very often the case. On the contrary, most networks publicly available from online databases such as Reactome [6] or large molecular maps built from the literature such as the RB/E2F map [12] have hundreds of nodes and edges, if not more. Such gigantic maps are barely readable and manageable when imported into visualization software such as Cytoscape. One of the main ideas of BiNoM plugin was to provide a set of network tools allowing users to extract meaningful subnetworks from large molecular maps and also to provide means to understand and read these maps [11]. We will see now through a set of examples how to extract such meaningful subnetworks.

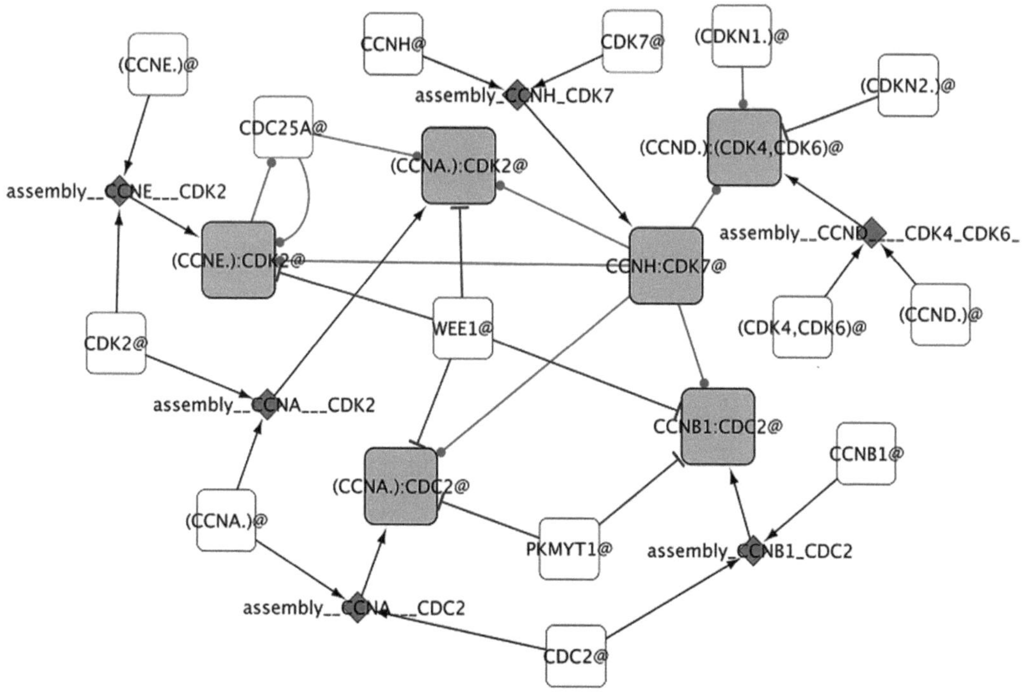

Fig. 4 A cell cycle network imported from an AIN text file (Annotated Influence Network)

As a first exercise, we create a simpler modular view of the M-phase example. Let us first decompose the cell cycle map we have imported in the previous paragraph by pruning the graph. In large networks, this step is important in order to simplify the network: we work on the connected graph rather than the whole graph.

1. Select the network "M-Phase.xml."
2. Choose the function "Plugins > BiNoM 2.0 > BiNoM Analysis > Prune Graph" in the menu.
3. Three networks are created: "M-Phase.xml_in," "M-Phase.xml_scc," and "M-Phase.xml_out."

The function is decomposing any network in three components corresponding to the nodes that are coming in (input), the nodes that go out (output), and the central cyclic part. The central part may sometimes be composed of several strongly connected components. In some situations they can be disconnected, forming several subnetworks. The decomposition of the strongly connected components part can be done in two ways: (1) by cycle decomposition and (2) by material components decomposition. Let us first see how to decompose a network into relevant directed cycles, which usually provides information about the life cycle of a gene or protein of the network.

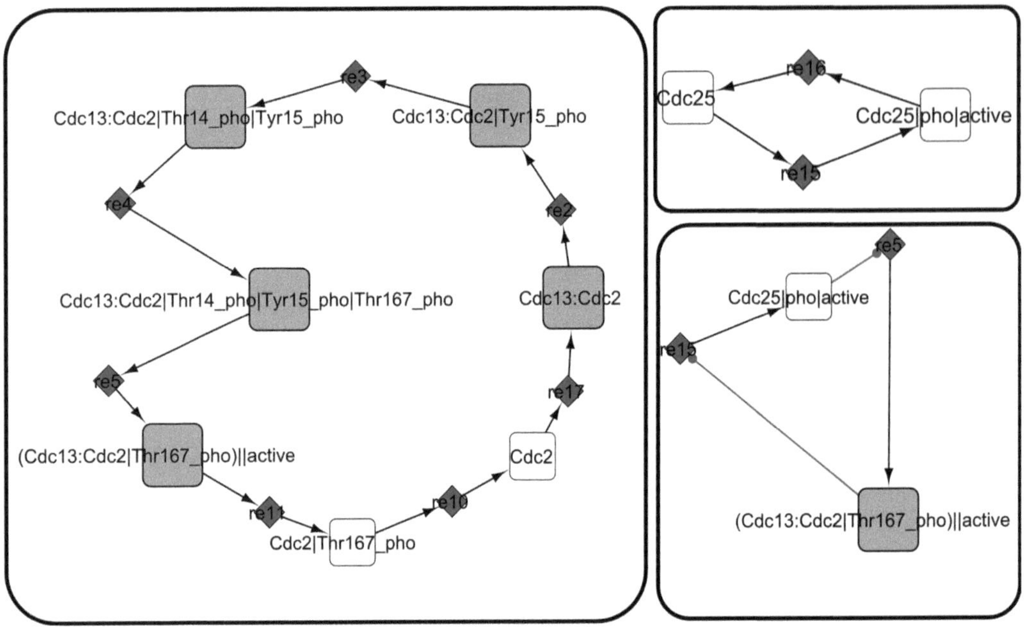

Fig. 5 Subnetworks (cycles) extracted from the M-Phase network using BiNoM functions

1. Select the network "M-Phase.xml_scc" (highlighted in the Cytoscape navigation panel).
2. Select the function "Plugins > BiNoM 2.0 > BiNoM Analysis > Get cycle decomposition" from the menu.
3. Three new networks are created: cycle1, cycle2, and cycle3 (Fig. 5).

Let us now merge in clusters networks that share a certain number of components.

1. Select the function "Plugins > BiNoM 2.0 > BiNoM Analysis > Cluster Networks" from the menu.
2. In the dialog window that appeared, select the networks cycle1, cycle2 and cycle3 (holding down the CTRL key for multiple selection).
3. Set the intersection threshold to 35 % using the sliding bar.
4. Click OK. Two networks are created: "cycle1" and "cycle2/ cycle3."

In fact, only the networks cycle2 and cycle3 were clustered, because they share a component (Cdc25 phosphorylated and active; in a two-component network, they share more than 35 %). Now that the modules are created, we need to include the inputs and outputs that were put aside at the beginning of the analysis.

In order to merge networks, we can use a Cytoscape built-in function.

1. Select the function "Plugins > Advanced Network Merge" from the menu.
2. From the dialog window, select "Union" in the field "Operation."
3. In the list of networks, select "Cycle1," "M_Phase.xml_in," and "M_Phase.xml_out" and then click "Merge."
4. The resulting network is named "Union" and should have 30 nodes and 19 edges.
5. Rename the network to "Union1" by right-clicking on its name and selecting "Edit Network Title."
6. Using the same procedure as above, merge the networks "cycle2/3," "M_Phase.xml_in," and "M_Phase.xml_out." The resulting network should have 22 nodes and 12 edges.
7. Rename it to "Union2/3."

Some edges present in the original file have been lost during all these operations, and they need to be included again. For that, we will now update the networks.

1. Select the function "Plugins > BiNoM 2.0 > BiNoM Utilities > Update connections from other network" from the menu.
2. In the dialog window, select "M-Phase.xml" for the field "From Network" and select the networks "Union1" and "Union2/3" from the list "Networks to Update," and click OK.
3. The networks "Union1" and "Union2/3" are updated to 30 and 20 edges, respectively.

We can now remove unconnected and unnecessary components.

1. Select the network "Union1."
2. Change the layout by using the Cytoscape function "Layout > yFiles > Organic" from the menu (this step allows to visualize the unconnected components more easily).
3. Select all unconnected nodes and remove them.
4. The Wee1 and Rum1 genes should be in a network of their own, so we propose to remove them and all the edges connected to them (they are in fact important proteins that do not share a function with the two modules created).
5. The resulting networks should have 20 nodes and 20 edges for "Union1" (Fig. 6) and 8 nodes and 9 edges for "Union2/3" (Fig. 7).

Note that the analysis requires to make a certain number of choices, based on biological knowledge and related to the final goal of the user. Here, we want to create a modular view of the initial network to highlight the main mechanisms involved in the yeast cell cycle progression. Finally, we can now generate a modular view from the networks Union1 and Union2/3.

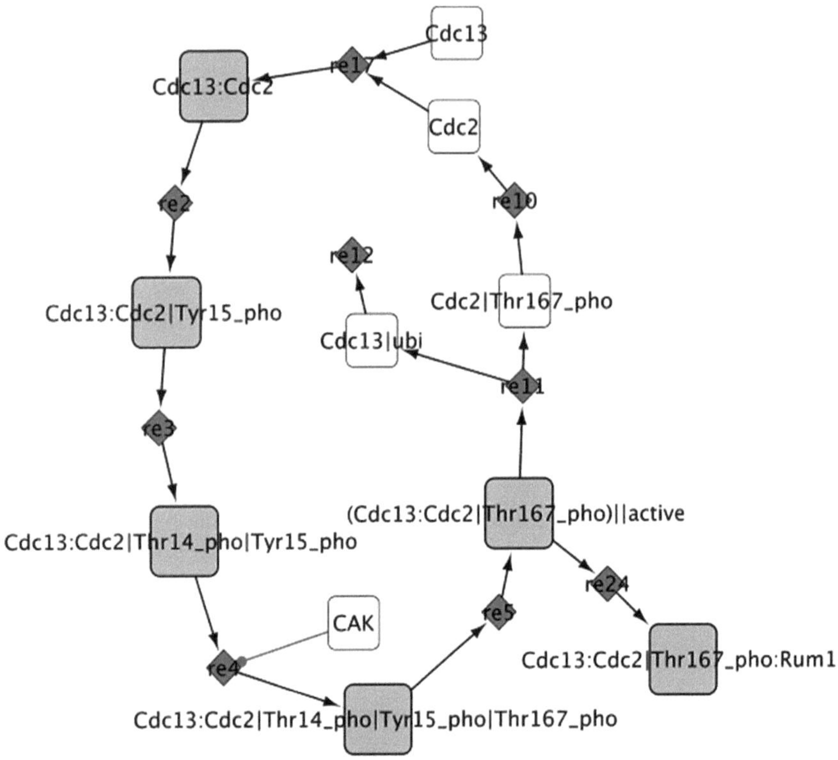

Fig. 6 A subnetwork resulting from the union of the subnetworks "Cycle1", "M_Phase.xml_in" and "M_Phase.xml_out"

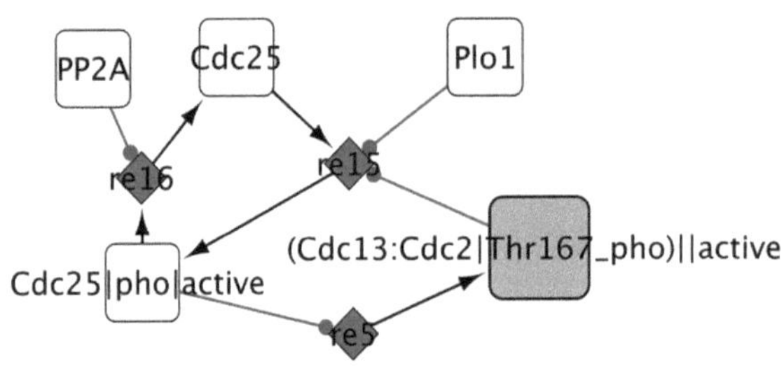

Fig. 7 A subnetwork resulting from the union of the subnetworks "cycle2/3", "M_Phase.xml_in" and "M_Phase.xml_out"

1. Select the function "Plugins > BiNoM 2.0 > BiNoM module manager > Create Network of Modules" from the menu.
2. Select "Union1" and "Union2/3" from the list in the dialog window and click OK.
3. Select the function "Plugins > BiNoM 2.0 > BiNoM module manager > Create connections between modules" from the menu. Select the network "M-Phase.xml" from the list in the dialog window and click OK.

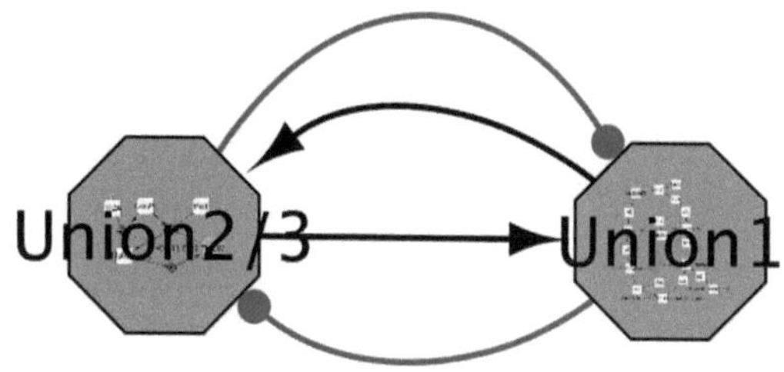

Fig. 8 A modular representation of two subnetworks, Union2/3 and Union1

4. Rename the network to "Module1" by right-clicking on it (Fig. 8).

The resulting map is a modular map of the initial network, in which modules participate in a specific process. For instance, "Union 1" shows all the events that lead to the activation of the maturation promoting factor, a heterodimer composed of the cyclin-dependent kinase Cdc2 and the cyclin B protein Cdc13. Note that in order to navigate from a module to the corresponding subnetwork, you have to perform the following operations:

1. Right-click on the module of interest. A contextual menu appears.
2. In the menu, choose "Nested Network" and then "Go to Nested Network." The corresponding subnetwork is now brought to the front window.

3.3 BiNoM and BioPAX Files

Biological Pathway Exchange (BioPAX) is a standard language that represents biological pathways at the molecular level and facilitates the exchange of pathway data [4]. The current BioPAX specification (level 3, released in July 2010; see http://www.biopax.org), supports representation of metabolic and signaling maps, molecular and genetic interactions, and gene regulation. Furthermore, there are several additional constructs available to store extra details such as database cross-references, chemical structures, sequence feature locations, and links to controlled vocabulary terms encoded in various ontologies (such as the Gene Ontology). BiNoM has a powerful set of functions to manage large BioPAX files, allowing the user to import, export, analyze, and extract the knowledge encoded using this specification [11]. We have recently updated the BiNoM plugin software to provide support for the latest BioPAX specification (level 3; see http://www.biopax.org/specification.php).

3.3.1 Import and Information Extraction from a BioPAX File

In the next example, we will be working with a relatively large molecular map representing the Apoptosis pathway in human, extracted from the Reactome database [6]. The file is available from our website (Apopotosis3.owl). Let us import the file in Cytoscape using BiNoM functions.

1. Select the function "Plugins > BiNoM 2.0 > BiNoM I/O > Import BioPAX 3 Document from file" from the menu.
2. Select the file "Apoptosis3.owl" from the dialog box.
3. A new dialog window appears. The three types of network should be imported. For that, check the boxes "Reaction Network," "Pathway Hierarchy," "Make Root Pathway Node," "Include Pathways," "Include Interactions," and "Interaction map."
4. Click OK. Three new networks are created, corresponding to the reaction network ("Apopotosis3 RN"), Apoptosis Pathway ("Apoptosis3 PS"), and Apoptosis protein–protein interactions ("Apoptosis3 PP").
5. Change the layout of each network for a better readability: choose "Organic" ("Layout > yFiles > Organic") for "Apoptosis3 RN" and "Apoptosis3 PP" and the type "Hierarchic" ("Layout > yFiles > Hierarchic") for "Apoptosis3 PS."

The three networks represent different types of knowledge extracted from the BioPAX file. We call them *network interfaces*, because they allow to access the different parts of the content of a BioPAX file. The Reaction Network (RN) is a graph which contains nodes of two types: "species" and "reactions." Proteins are represented as white rounded squares, complexes as gray rounded squares, and reactions as small gray diamonds (Fig. 9c). The Pathway Hierarchy (PS) contains pathway knowledge with two types of nodes: pathways, pictured as green hexagons, and pathway steps, pictured as pink triangles (Fig. 9b). The last interface contains an interaction map (IM) extracted from the proteins and protein complexes present in the BioPAX file, with edges of type "contains" (Fig. 9a).

The whole network being quite large, it is not always easy to find specific information. In the next example, we propose to query the graph by performing a simple analysis on a BioPAX imported file: the extraction of a path.

1. Select the network "Apoptosis3 RN" by clicking on the name in the navigation panel.
2. Select all nodes and edges by using the function "Select > Select All Nodes and Edges" from the menu.
3. Select the function "Plugins > BiNoM Analysis > Path Analysis" from the menu.
4. A dialog window appears; choose the node "TNF:TNFR1@plasma_membrane" in the "Sources" list and the node "GIG3:RIP:TRADD:TRAP3@cytosol" in the "Targets" list.
5. Take the default search options "Find shortest paths" (you can try the other options as an exercise).

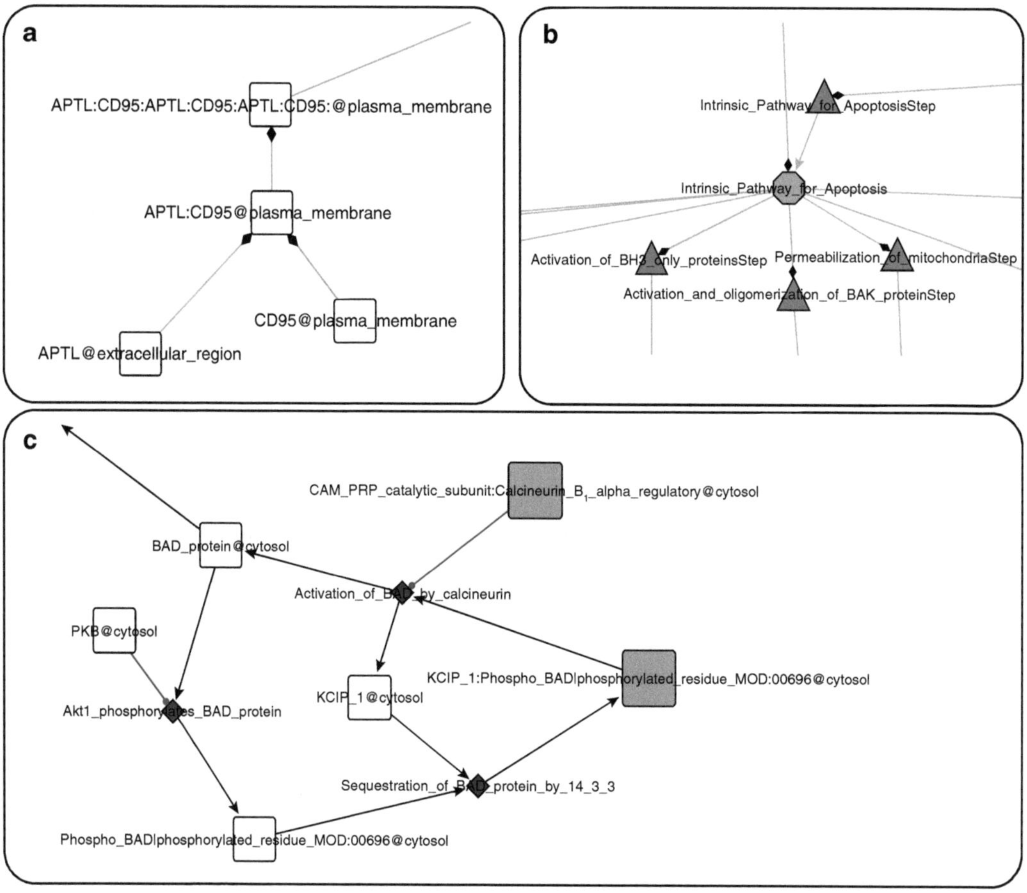

Fig. 9 Three types of networks resulting from the import of a BioPAX file. (**a**) Reaction network (RN). (**b**) Pathway hierarchical structure (PS). (**c**) Protein–protein interaction network (PP)

6. Click OK. The nodes of the shortest path between the two nodes are now highlighted in the network (note that it is not the case for the edges connecting them).
7. Extract the path as a new network by using the function "File > New > Network > From selected nodes, all edges" from the menu.
8. A new subnetwork is created with the name "Apoptosis3 RN—child."

3.3.2 Querying a BioPAX File

The BioPAX format is now used by an increasing number of databases and online repositories such as Reactome (http://www.reactome.org), Cancer Cell Map (http://cancer.cellmap.org), the Pathway Interaction Database (http://pid.nci.nih.gov/), or Pathway Commons (http://www.pathwaycommons.org). The amount of information contained in the files extracted from those databases can be very consequent, making it difficult for the average user to efficiently query and retrieve relevant data. We have included in BiNoM an efficient BioPAX querying system. The BioPAX file is converted to an index, by mapping the BioPAX content on a labeled graph. This index can then be queried by the user for specific elements of interest. The result is returned as a graph directly in Cytoscape and can be further extended to include various elements such as all the complexes in which a protein of interest is involved, the reactions connected to those molecules, and the related publications. For example, let us extract the complexes related to a given protein from the Apoptosis BioPAX file.

1. First, we have to generate the index from the BioPAX file. Select the function "Plugins > BiNoM 2.0 > BiNoM BioPAX 3 Query > Generate Index" from the menu.
2. From the dialog window that appears, select the file "Apoptosis3.owl" for the field "BioPAX File". The second field named "Index File" will be filled automatically with the same file name, just changing the extension to ".xgmml". In this case, it will suggest the name "Apoptosis3.xgmml"; you can change that name if you wish or just accept the proposition. Click OK. The index is generated and saved.
3. Load the index with the function "Plugins > BiNoM 2.0 > BiNoM BioPAX 3 Query > Load Index" from the menu. Select the index file you have just created "Apoptosis3.xgmml" and click OK. The index is now loaded in memory (note that loading the index is an essential step to perform a query; the creation of the index is not enough).
4. Basic statistics related to the index file content can be obtained by the function "Plugins > BiNoM 2.0 > BiNoM BioPAX 3 Query > Load Index" from the menu. A window is displayed containing a table with counts for different elements of the index (proteins, complexes, publications, etc.).
5. Let us now do a basic query. Select the function "Plugins > BiNoM 2.0 > BiNoM BioPAX 3 Query > Select Entities" from the menu.
6. In the text field entitled "Input," type the name "SMAC," and click OK. A new network is created, having a single node named "SMAC@cytosol."
7. For a better visualization, you can set the visual style to "BiNoM BioPAX" on the tab "VizMapper" on the left-hand side of the Cytoscape interface.
8. Now we wish to expand this network by adding all the complexes in which this protein is involved. Select the function "Plugins > BiNoM 2.0 > BiNoM BioPAX 3 Query > Standard Query" from the menu. A window appears, named "BioPAX Standard Query from the index." Check the boxes for the

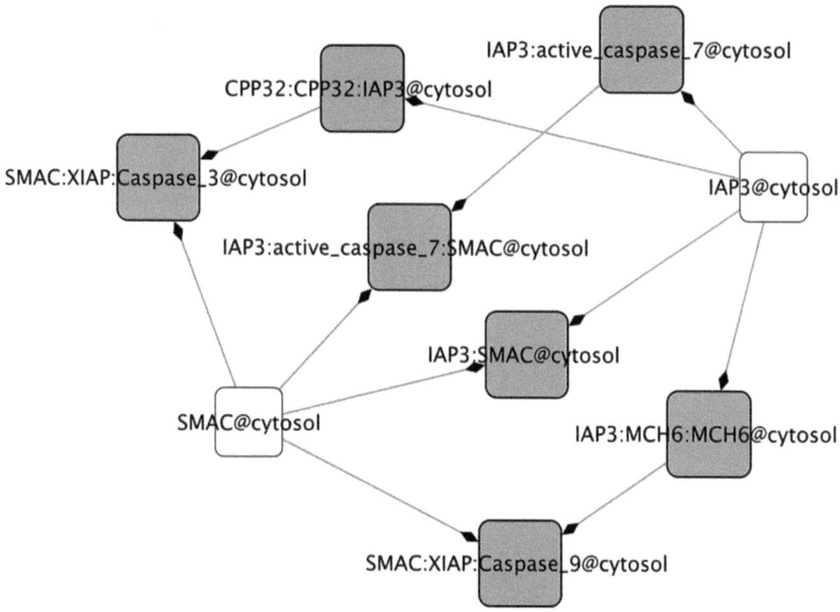

Fig. 10 A network constructed from a BioPAX query, centered on the SMAC protein, and including all protein complexes where this protein is involved. Data extracted from the Apoptosis data of the Reactome database

option "Add complexes" and "expand." Un-select the boxes "Add chemical species," "Add reactions," and "Add publications." Verify that the option "All nodes" is checked for the "Input" section and that "Output in the current network" is checked for the "Output" section (by checking those options, we make sure that all the nodes are by default selected as input and that the result of the query will be added to the current network). Click OK.

9. Several nodes and edges have been added to the current network. For a better visualization, adjust the layout with the function "Layout > yFiles > Organic" from the menu. A green arrow with a diamond ending represents the inclusion of one protein in a complex form. The resulting network should have 9 nodes and 11 edges (Fig. 10).

As we have seen from the standard query interface, it is possible to expand the network by including the chemical species, the reactions connecting all present species that have a common reaction, and the publications related to any of the components of the network (for more information on how to use those options, please consult the BiNoM manual available from our website). Note that the resulting network of interest can be exported as a SBML or BioPAX file as described in the previous paragraphs.

3.4 Other Useful BiNoM Functions

We have seen that BiNoM has several useful functions to extract relevant information from large-scale databases encoded with standards defined by the systems biology community. Very often, the results of those analyses will be one or more subnetworks of interest, possibly grouping a set of molecules involved in a particular biological function (cell death, cell cycle, apoptosis, etc.). An example of such an insightful extraction of a subnetwork is shown in Calzone et al. [12], where a compact modular view of the RB/E2F pathway composed of 16 protein modules and 8 E2F target gene modules (*see* Fig. 3 of this chapter) was extracted from a comprehensive network of hundreds of different molecules and interactions. Once the map is constructed, several options are possible to generate interesting and useful insights. These options include (non-exhaustive list) (1) the creation of a computational predictive model, making possible the analysis of the consequences of deletion or mutation of various elements of the network, and (2) superimposing external and experimental available data related to the function of the network, in order to visually appreciate the effect of different states/perturbations/disease effects. For example, bladder tumor expression data was superimposed on the RB/E2F pathway compact representation mentioned above, for both invasive and noninvasive cases (see http://bioinfo-out.curie.fr/projects/rbpathway/case_study.html). The nodes of the network are then colored according to the averaged expression levels of the different modules, indicating what parts are over- or under-expressed. Clear differences can be seen between the invasive and noninvasive state of the tumor samples, informing of the evolution of tumors at the expression level of genes of the network.

Let us now see an example of how to color a map using BiNoM functions, based on the M-phase network.

1. Import the CellDesigner file M-Phase.xml as described in the Subheading 3.1.
2. Now import values for each gene. They are stored in a simple text file having two columns "NODE_NAME" and "CONCENTRATION". Select the function "File > Import > Attribute from table (Text / MS Excel)". In this case, the values represent expression levels randomly generated, but they could be any type of scoring. Note that for experimental data, proteins with posttranslational modifications will not be colored.
3. Select the input file "M-Phase-Expression.txt." A preview of the file content should appear at the bottom of the dialog box.
4. In the "Advanced" box, click the box "Show text file import options."
5. A new box appears, entitled "Attribute names." Check the box "Transfer first line as attribute name." Now the column titles should read "NODE_NAME" and "CONCENTRATION."

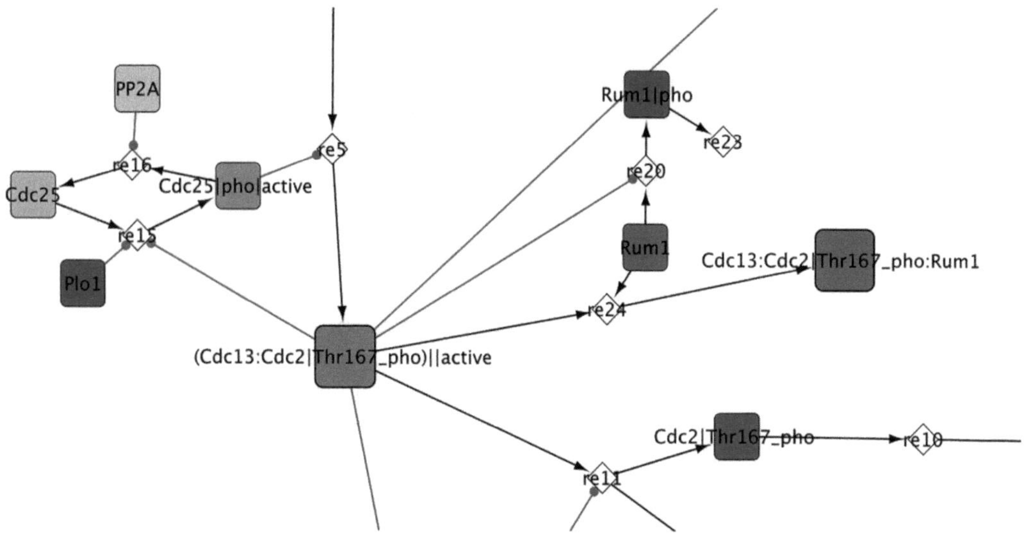

Fig. 11 M-Phase network (zoom) with nodes in shades of gray according to their expression values, ranging from low values (*light gray*) to high values (*dark gray*)

6. Click on "Import." The file is imported, and a new numerical attribute "CONCENTRATION" is created for all the genes.
7. Now click on the "VizMapper" tab, on the left-hand panel of Cytoscape.
8. In the "Visual Mapping Browser" box, click on the "Node Color" small triangle to display the properties.
9. Change the property "Mapping Type" to "Continuous Mapping."
10. Change the value of "Node Color" to "CONCENTRATION".
11. Click on "Graphical View," a new dialog box will appear. Set the minimal and maximal values according to the values of your dataset by clicking on the "Min/Max" button.
12. Set the colors by clicking on the small triangles located above the minimum and maximum values. Click OK.
13. The nodes of the network should be colored according to their expression value, as shown on the Fig. 11.

4 Conclusion

- Model building in systems (and mathematical) biology is a complex multistep process: from the definition of a suitable biological problem, knowledge is first collected and formalized into a network and then translated in mathematical terms.

BiNoM helps with intermediate steps of this process, in the construction of a network of biochemical or regulatory interactions, and in the analysis of the structural properties of this network. For this, BiNoM provides multiple ways: to access pathway databases through their BioPAX representations, manipulate (cut, decompose, reorganize) the network, apply algorithms from graph theory to the network, and map available quantitative data on it.

- The future developments of BiNoM will include functions such as merging several independent networks, finding minimal intervention sets to disrupt or modify the signaling flow from a set of source nodes to a set of target nodes, and the ability to generate a code for web-based representations of biological networks using the Google Map API and semantic zoom.
- BiNoM is not supposed to be a modeling software per se; it does not aim at implementing any engines for numerical simulations, but it has interfaces with external simulators through exporting networks to SBML and GINsim file formats (with use of GINsim Cytoscape plugin [14]). The main application of BiNoM is to facilitate the preparation phase of constructing, annotating, and structuring a biological network for further mathematical modeling and simulation, and this will determine its future development.

5 Notes

- Cycle decomposition can result in a huge number of cycles. It is advised to use it on small to moderate size networks.
- When trying to divide a large network into subnetworks, an alternative to the cycle decomposition described in the Subheading 3.2 is the function "Get Material Components" from the menu "Plugins > BiNoM 2.0 > BiNoM Analysis". This function is using node name semantics to isolate subnetworks in which each protein is involved.

Acknowledgements

EB, LC, DR, GS, EmB, and AZ are members of the team "Computational Systems Biology of Cancer," Equipe labellisée par la Ligue Nationale Contre le Cancer. The research leading to these results has received funding from the European Union Seventh Framework Programme (FP7/2007–2013) ASSET project under agreement number FP7-HEALTH-2010-259348 and from the grant "Projet Incitatif et Collaboratif Computational Systems Biology Approach for Cancer" from Institut Curie.

References

1. Brazma A, Krestyaninova M, Sarkans U (2006) Standards for systems biology. Nat Rev Genet 7(8):593–605. doi:nrg1922[pii] 10.1038/nrg1922
2. Klipp E, Liebermeister W, Helbig A, Kowald A, Schaber J (2007) Systems biology standards–the community speaks. Nat Biotechnol 25(4):390–391. doi:nbt0407-390[pii] 10.1038/nbt0407-390
3. Hucka M, Finney A, Sauro HM, Bolouri H, Doyle JC, Kitano H, Arkin AP, Bornstein BJ, Bray D, Cornish-Bowden A, Cuellar AA, Dronov S, Gilles ED, Ginkel M, Gor V, Goryanin II, Hedley WJ, Hodgman TC, Hofmeyr JH, Hunter PJ, Juty NS, Kasberger JL, Kremling A, Kummer U, Le Novere N, Loew LM, Lucio D, Mendes P, Minch E, Mjolsness ED, Nakayama Y, Nelson MR, Nielsen PF, Sakurada T, Schaff JC, Shapiro BE, Shimizu TS, Spence HD, Stelling J, Takahashi K, Tomita M, Wagner J, Wang J (2003) The systems biology markup language (SBML): a medium for representation and exchange of biochemical network models. Bioinformatics 19(4):524–531
4. Demir E, Cary MP, Paley S, Fukuda K, Lemer C, Vastrik I, Wu G, D'Eustachio P, Schaefer C, Luciano J, Schacherer F, Martinez-Flores I, Hu Z, Jimenez-Jacinto V, Joshi-Tope G, Kandasamy K, Lopez-Fuentes AC, Mi H, Pichler E, Rodchenkov I, Splendiani A, Tkachev S, Zucker J, Gopinath G, Rajasimha H, Ramakrishnan R, Shah I, Syed M, Anwar N, Babur O, Blinov M, Brauner E, Corwin D, Donaldson S, Gibbons F, Goldberg R, Hornbeck P, Luna A, Murray-Rust P, Neumann E, Reubenacker O, Samwald M, van Iersel M, Wimalaratne S, Allen K, Braun B, Whirl-Carrillo M, Cheung KH, Dahlquist K, Finney A, Gillespie M, Glass E, Gong L, Haw R, Honig M, Hubaut O, Kane D, Krupa S, Kutmon M, Leonard J, Marks D, Merberg D, Petri V, Pico A, Ravenscroft D, Ren L, Shah N, Sunshine M, Tang R, Whaley R, Letovksy S, Buetow KH, Rzhetsky A, Schachter V, Sobral BS, Dogrusoz U, McWeeney S, Aladjem M, Birney E, Collado-Vides J, Goto S, Hucka M, Le Novere N, Maltsev N, Pandey A, Thomas P, Wingender E, Karp PD, Sander C, Bader GD (2010) The BioPAX community standard for pathway data sharing. Nat Biotechnol 28(9):935–942. doi:nbt.1666[pii]10.1038/nbt.1666
5. Le Novere N, Hucka M, Mi H, Moodie S, Schreiber F, Sorokin A, Demir E, Wegner K, Aladjem MI, Wimalaratne SM, Bergman FT, Gauges R, Ghazal P, Kawaji H, Li L, Matsuoka Y, Villeger A, Boyd SE, Calzone L, Courtot M, Dogrusoz U, Freeman TC, Funahashi A, Ghosh S, Jouraku A, Kim S, Kolpakov F, Luna A, Sahle S, Schmidt E, Watterson S, Wu G, Goryanin I, Kell DB, Sander C, Sauro H, Snoep JL, Kohn K, Kitano H (2009) The systems biology graphical notation. Nat Biotechnol 27(8):735–741. doi: nbt.1558[pii]10.1038/nbt.1558
6. Joshi-Tope G, Gillespie M, Vastrik I, D'Eustachio P, Schmidt E, de Bono B, Jassal B, Gopinath GR, Wu GR, Matthews L, Lewis S, Birney E, Stein L (2005) Reactome: a knowledgebase of biological pathways. Nucleic Acids Res 33 (Database issue):D428–D432. doi:33/suppl_1/D428[pii]10.1093/nar/gki072
7. Le Novere N, Bornstein B, Broicher A, Courtot M, Donizelli M, Dharuri H, Li L, Sauro H, Schilstra M, Shapiro B, Snoep JL, Hucka M (2006) BioModels database: a free, centralized database of curated, published, quantitative kinetic models of biochemical and cellular systems. Nucleic Acids Res 34(Database issue): D689–D691. doi:34/suppl_1/D689[pii] 10.1093/nar/gkj092
8. Licata L, Briganti L, Peluso D, Perfetto L, Iannuccelli M, Galeota E, Sacco F, Palma A, Nardozza AP, Santonico E, Castagnoli L, Cesareni G (2012) MINT, the molecular interaction database: 2012 update. Nucleic Acids Res 40(Database issue):D857–D861. doi: gkr930[pii]10.1093/nar/gkr930
9. Funahashi A, Morohashi M, Kitano H (2003) Cell Designer: a process diagram editor for gene-regulatory and biochemical networks. Biosilico 1(5):159–162
10. Cline MS, Smoot M, Cerami E, Kuchinsky A, Landys N, Workman C, Christmas R, Avila-Campilo I, Creech M, Gross B, Hanspers K, Isserlin R, Kelley R, Killcoyne S, Lotia S, Maere S, Morris J, Ono K, Pavlovic V, Pico AR, Vailaya A, Wang PL, Adler A, Conklin BR, Hood L, Kuiper M, Sander C, Schmulevich I, Schwikowski B, Warner GJ, Ideker T, Bader GD (2007) Integration of biological networks and gene expression data using Cytoscape. Nat Protoc 2(10):2366–2382. doi:nprot.2007.324[pii] 10.1038/nprot.2007.324

11. Zinovyev A, Viara E, Calzone L, Barillot E (2008) BiNoM: a Cytoscape plugin for manipulating and analyzing biological networks. Bioinformatics 24(6):876–877. doi:btm553[pii] 10.1093/bioinformatics/btm553
12. Calzone L, Gelay A, Zinovyev A, Radvanyi F, Barillot E (2008) A comprehensive modular map of molecular interactions in RB/E2F pathway. Mol Syst Biol 4:173. doi:msb20087 [pii]10.1038/msb.2008.7
13. Novak B, Csikasz-Nagy A, Gyorffy B, Nasmyth K, Tyson JJ (1998) Model scenarios for evolution of the eukaryotic cell cycle. Philos Trans R Soc Lond B Biol Sci 353(1378):2063–2076. doi:10.1098/rstb.1998.0352
14. Gonzalez AG, Naldi A, Sanchez L, Thieffry D, Chaouiya C (2006) GINsim: a software suite for the qualitative modelling, simulation and analysis of regulatory networks. Biosystems 84 (2):91–100. doi:S0303-2647(05)00169-3 [pii]10.1016/j.biosystems.2005.10.003

Chapter 8

Using Chemical Kinetics to Model Biochemical Pathways

Nicolas Le Novère and Lukas Endler

Abstract

Chemical kinetics is the study of the rate of reactions transforming some chemical entities into other chemical entities. Over the twentieth century it has become one of the cornerstones of biochemistry. When in the second half of the century basic knowledge of cellular processes became sufficient to understand quantitatively metabolic networks, chemical kinetics associated with systems theory led to the development of what would become an important branch of systems biology.

In this chapter we introduce basic concepts of chemical and enzyme kinetics, and show how the temporal evolution of a reaction system can be described by ordinary differential equations. Finally we present a method to apply this type of approach to model any regulatory network.

Key words Chemical kinetics, Chemical entities, Quantitative, Metabolic network, Systems biology, Enzyme kinetics, Regulatory network

1 Introduction to Chemical Kinetics

A living cell is built up as a series of compartments of various dimensions. The plasma membrane is an example of a bi-dimensional compartment surrounding the cytosol, which is itself a tridimensional compartment. Microtubules are examples of unidimensional compartments. These compartments can be considered both as containers—we can count the number of instances of a certain type of entity present in, or attached to, a compartment—and as diffusional landscapes—the movements of the entities within the compartment depend on its properties. Within the compartments, the entities can move and react with each other. The object of chemical kinetics is to study the temporal evolution of the positions and quantities of the entities contained in a compartment, sometimes called a reactor. In this chapter, we will not deal with the displacement of the chemical entities within a compartment. We will assume that an entity-pool, that is a set of entities that are indistinguishable as far as the model is concerned, is distributed homogeneously within the compartment. This hypothesis is known as the *well-stirred*

Maria Victoria Schneider (ed.), *In Silico Systems Biology*, Methods in Molecular Biology, vol. 1021,
DOI 10.1007/978-1-62703-450-0_8, © Springer Science+Business Media, LLC 2013

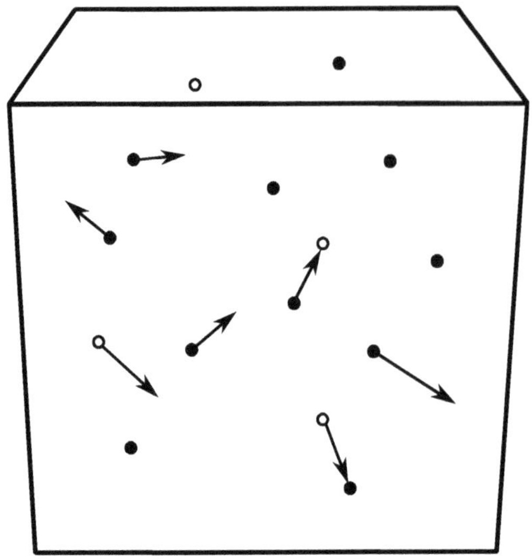

Fig. 1 Representation of a well-stirred container with two types of entities, represented by *empty* and *filled* circles. The *arrows* represent the direction and speed of their movements

approximation (Fig. 1). This approximation is based on the assumption that there is no diffusional anisotropy in the compartment, i.e., the molecules move randomly in any dimension. This is obviously a strong simplification in most of the cases pertaining to biological functions. It has nevertheless proved to be very useful in the past. In addition, the alternative requires to enter the realm of reaction–diffusion modeling, which involves not only more complex methods but also knowledge of distribution and diffusion characteristics of the reacting entities.

1.1 Chemical Reactions

A chemical reaction is the transformation of one set of substances called reactants into another set called products. At a microscopic scale, such a transformation is in general reversible, although there are many cases in which the reverse reaction is of negligible importance compared to the forward one. In all cases, a reversible reaction can be split into forward and reverse reactions. For a given reaction, reactants generally combine in discrete and fixed ratios to form products. These ratios indicate the amount of each substance involved in the reaction. The amounts consumed or produced in one reaction event are called the stoichiometric coefficients or numbers, ν_X, and are positive for products, and negative for reactants. If a substance is neither consumed nor produced by a reaction, its stoichiometric coefficient is 0. Equation 1 depicts a general reaction, in which A and B are reactants combining to form the product P. ν_A would be $-a$, $\nu_B = -b$ and $\nu_P = p$. The list $\{-a, -b, p\}$ is also called the stoichiometry of the reaction.

$$aA + bB \rightarrow pP \tag{1}$$

In many cases in biology only an overall transformation consisting of many sequential reactions is experimentally observable. In the finest grained form these reactions are also known as elementary reactions. An elementary reaction is defined as a minimal irreversible reaction with no stable intermediary products. The lumped stoichiometric coefficients of the overall reaction consist of the sums of the stoichiometric coefficients for each reactant over all elementary reactions.

Chemical kinetics is concerned with the velocity of such transformations, the rates with which substances are consumed and produced. As the rate of change for a reagent depends on its stoichiometric coefficients, it can be different for individual substances. Therefore it is convenient to define the reaction rate, ν, as the rate of change of a substance divided by its stoichiometric coefficient. This effectively represents the number of reaction events taking place per unit of time and unit of compartment size.

$$\nu = \frac{1}{-a}\frac{\mathrm{d}[\mathrm{A}]}{\mathrm{d}t} = \frac{1}{-b}\frac{\mathrm{d}[\mathrm{B}]}{\mathrm{d}t} = \frac{1}{p}\frac{\mathrm{d}[\mathrm{P}]}{\mathrm{d}t}$$

Therefore, we can compute the change of each substance as the product of the reaction rate and its stoichiometric coefficient for this reaction.

$$\frac{\mathrm{d}[\mathrm{A}]}{\mathrm{d}t} = -a \times \nu$$
$$\frac{\mathrm{d}[\mathrm{B}]}{\mathrm{d}t} = -b \times \nu$$
$$\frac{\mathrm{d}[\mathrm{P}]}{\mathrm{d}t} = p \times \nu$$

Reaction rates depend on many factors and can effectively take any form for the purpose of modeling. In the following subsections, we will describe the simple cases where the reaction rates depend solely on the concentrations of the reacting substances.

1.2 Mass-Action Kinetics

For a chemical reaction to take place, the participants have to collide or come into close vicinity of each other. The probability of such collisions depends, among other parameters, on the local density of the reactants, and hence, in well-stirred environments, on their concentrations.[1] This relationship was first described by Guldberg and Waage in the second half of the nineteenth century in a series of articles on the dynamical nature of the chemical equilibrium [1]. They assumed that at equilibrium both the forward and

[1]Under nonideal conditions, as found in biology, activities instead of concentrations should actually be used both for describing rate equations and equilibria. As this is not common practice in biological modeling, we do not distinguish between activities and concentrations in the following. It should be noted, though, that activities can differ significantly from concentrations in cellular environments.

backward reaction forces or velocities were equal, and that these velocities where proportional to the concentrations of the reactants to the power of their stoichiometric coefficients. The relationship of reaction velocities and concentrations is called the "Law of Mass-Action", and rate expressions equivalent to the ones employed in their articles are sometimes referred to as "Mass-Action Kinetics".[2]

The rates of simple unidirectional chemical reactions are usually proportional to the product of the concentrations of the reactants to the power of constant exponents, called *partial reaction orders* or n_X. The sum of all partial orders is called the *order n* of a reaction and the proportionality factor is called the *rate constant k*. As the name indicates, this parameter does not vary in a given system. For example, for the reaction described in Eq. 1 assuming mass-action kinetics the reaction rate appears as follows:

$$v = k \times [A]^{n_A} \times [B]^{n_B}$$

The reaction has an order of $n = n_A + n_B$. In general, the order of elementary reactions is equal to the number of molecules interacting, also known as the *molecularity*. A *unimolecular* reaction $A \rightarrow P$ for example would have an order of one, a *bimolecular* reaction, such as $2A \rightarrow P$ or $A + B \rightarrow P$ would be a second-order reaction etc. However, this equivalence is not always true, and anisotropy or crowding of the reaction environments may affect the motion of molecules, resulting in different, and sometimes nonintegral, reaction orders.

While mass-action kinetics are strictly only valid for elementary reactions, they are widely and successfully applied in various fields of mathematical modeling in biology. Especially for large and vaguely defined reaction networks, as found in signal transduction, mass-action kinetics are commonly employed as a very general initial approach. Most often, the partial orders are taken to be identical to the stoichiometric coefficients. The rate constants can either be calculated from separately measured equilibrium constants and characteristic times, or computationally fitted to reproduce experimental results.

1.2.1 Zeroth Order Reactions

Reactions of order zero have a reaction rate that does not depend on any reactant. Zeroth order reactions can be used for instance to represent constant creations from boundary condition reactants, such as:

$$X \xrightarrow{k} P$$

[2]The term *mass-action* stems from the proportionality of the so-called reaction "force" to the mass of a substance in a fixed volume, which is proportional to the molar concentration of a substance.

where X represent a set of source reactants that are not depleted by the reaction. The reaction rate is then equal to:

$$v = k \times [\mathrm{X}]^0 = k$$

in which k is the rate constant, and has the units of a concentration per time.

1.2.2 First-Order Reactions

In general unimolecular reactions are modeled using first-order mechanisms. In irreversible first-order reactions, the reaction rate linearly depends on the concentration of the reactant. Many decay processes show such kinetics, for example radioactive decay, dissociation of complexes or denaturation of proteins. For a simple reaction:

$$\mathrm{A} \xrightarrow{k} \mathrm{P}$$

the following rate law applies:

$$v = k \times [\mathrm{A}]$$

in which k is the first-order rate constant, and has the units of a reciprocal time, [1/time]. If this is the only reaction affecting the concentration of A in a system, the change of [A] equals the negative reaction rate.Similarly, the change of [P] equals the reaction rate.

$$\frac{\mathrm{d}[\mathrm{A}]}{\mathrm{d}t} = -v = -k[\mathrm{A}]$$
$$\frac{\mathrm{d}[\mathrm{P}]}{\mathrm{d}t} = +v = +k[\mathrm{A}]$$

1.2.3 Second-Order Reactions

Second-order reactions are often used to model bimolecular reactions, either between different types of molecules or between two instances of the same molecules. Examples are complex formation and dimerization reactions. For a simple reaction:

$$\mathrm{A} + \mathrm{B} \xrightarrow{k} \mathrm{P}$$

the following rate law applies:

$$v = k \times [\mathrm{A}] \times [\mathrm{B}]$$

in which k is the second-order rate constant, and has the unit of [1/(time × concentration)]. The change of [P] with time is described by the following differential equation:

$$\frac{\mathrm{d}[\mathrm{P}]}{\mathrm{d}t} = v = k \times [\mathrm{A}] \times [\mathrm{B}]$$

A special case of bimolecular reaction is when two reactant molecules of the same type react to form the product, for example in protein dimerization reactions. For the general reaction:

$$2\mathrm{A} \xrightarrow{k} \mathrm{P}$$

the reaction velocity and the temporal development of [A] and [P] are given by the following equations:

$$v = k \times [\mathrm{A}] \times [\mathrm{A}]$$

$$\frac{\mathrm{d}[\mathrm{A}]}{\mathrm{d}t} = -2v = -2k[\mathrm{A}]^2$$

$$\frac{\mathrm{d}[\mathrm{P}]}{\mathrm{d}t} = v = k[\mathrm{A}]^2$$

Note that this formula is only valid because we assume a very large number of molecules are available to react. If picking one molecule changes significantly the probability to pick a second one, we must replace $[\mathrm{A}]^2$ by $[\mathrm{A}] \times (([\mathrm{A}]V - 1)/V)$, where V the volume of the reactor.

1.3 Representing the Evolution of Multi-Reaction Systems

In the sections above, we only derived expressions describing the temporal evolution of species altered by single reactions. In biological systems, substances are involved in many different processes, leading to complex ordinary differential equation systems, that normally can only be solved numerically and with help of computers. Having carefully designed the elementary processes composing the system, reconstructing the differential equations representing the evolution of the different substances is a systematic and easy procedure. We already saw in Subheading 1.2.2 that the reaction:

$$\mathrm{A} \xrightarrow{k} \mathrm{P}$$

Could be modeled by the system:

$$\frac{\mathrm{d}[\mathrm{A}]}{\mathrm{d}t} = -1v = -1k[\mathrm{A}]$$

$$\frac{\mathrm{d}[\mathrm{P}]}{\mathrm{d}t} = +1v = +1k[\mathrm{A}]$$

If the reaction is reversible, such as:

$$\mathrm{A} \underset{k_\mathrm{r}}{\overset{k_\mathrm{f}}{\rightleftharpoons}} \mathrm{P}$$

then we can consider it as a combination of two irreversible reactions, the rates of which depend on [A] and [P]:

$$v_\mathrm{f} = k_\mathrm{f} \times [\mathrm{A}]$$

$$v_\mathrm{r} = k_r \times [\mathrm{P}]$$

The evolution of both substances therefore depends on the forward and reverse reaction rates. A is consumed by the forward reaction and produced by the reverse reaction. It is the other way around for P.

$$\frac{\mathrm{d}[\mathrm{A}]}{\mathrm{d}t} = -1v_\mathrm{f} + 1v_\mathrm{r} = -1k_\mathrm{f}[\mathrm{A}] + 1k_\mathrm{r}[\mathrm{P}]$$

$$\frac{\mathrm{d}[\mathrm{P}]}{\mathrm{d}t} = +1v_\mathrm{f} - 1v_\mathrm{r} = +1k_\mathrm{f}[\mathrm{A}] - 1k_\mathrm{r}[\mathrm{P}]$$

To understand how to handle non-unity stoichiometric numbers, consider the following dimerization:

$$2\mathrm{A} \underset{k_\mathrm{r}}{\overset{k_\mathrm{f}}{\rightleftharpoons}} \mathrm{P}$$

The forward reaction will be modeled using second-order kinetics, and the rates will therefore be:

$$v_\mathrm{f} = k_\mathrm{f} \times [\mathrm{A}]^2$$
$$v_\mathrm{r} = k_\mathrm{r} \times [\mathrm{P}]$$

As above the evolution of both substances therefore depends on the forward and reverse reaction rates. But this time two molecules of A are consumed by each forward reaction and produced by each reverse reaction. Therefore:

$$\frac{\mathrm{d}[\mathrm{A}]}{\mathrm{d}t} = -2v_\mathrm{f} + 2v_\mathrm{r} = -2k_\mathrm{f}[\mathrm{A}]^2 + 2k_\mathrm{r}[\mathrm{P}]$$
$$\frac{\mathrm{d}[\mathrm{P}]}{\mathrm{d}t} = +1v_\mathrm{f} - 1v_\mathrm{r} = +1k_\mathrm{f}[\mathrm{A}]^2 - 1k_\mathrm{r}[\mathrm{P}]$$

This approach can then be extended, independently of the size of the system considered. An ODE system will contain (at most) one differential equation for each substance. This equation will contain components representing the involvement of the substance in the different reactions of the system. For the substance S_n, involved in a system containing r reactions, the differential equation takes the following form:

$$\frac{\mathrm{d}[\mathrm{S}_n]}{\mathrm{d}t} = \sum_{i=1}^{r} \nu_{ni} v_i$$

ν_{ni} denotes the stoichiometric coefficient of S_n in reaction i, v_i the rate of this reaction. The resulting ODE system can also be represented in matrix notation, by introducing the stoichiometric matrix, **N**, and the reaction rate vector, **v**. The stoichiometric matrix, **N**, contains a row for each of the n species in the system, and a column for each of the r reactions. Its entries, N_{ij}, are the stoichiometric coefficients, of substance i in reaction j. **v** is a column vector with each element v_i indicating the rate of the ith reaction. Using the above, the change of the concentration vector S over time is described by:

$$\frac{\mathrm{d}[\mathrm{S}]}{\mathrm{d}t} = \mathbf{N} \cdot \mathbf{v}$$

2 Numerical Integration of ODE Models

Once a set of differential equations has been determined, to describe the changes of the variables per unit of time, the behaviour of the system can be obtained by fixing initial conditions and

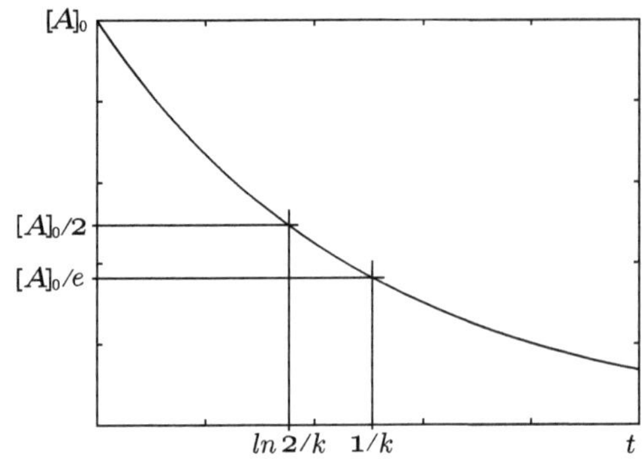

Fig. 2 Decay of a reactant A, that is consumed by a First-order reaction with a constant k from an initial concentration of $[A_0]$. The average lifetime of a given molecule of A, is given by $1/k$. [A] tends toward 0 while [P] tends towards $[A_0] + [P_0]$

solving the equations. In the case of a zeroth order reaction, the solution describing the evolution of P is of course a monotonic increase since a fixed amount of P is created per unit of time:

$$[P](t) = [P_0] + kt$$

The equation describing the evolution of A in a First-order reaction can be easily rearranged and analytically solved, assuming an initial concentration $[A_0]$ at time $t = 0$. Furthermore, since $[P]_t + [A]_t = [P_0] + [A_0]$:

$$[A]_t = [A_0] \times e^{-kt}$$
$$[P]_t = [P_0] + [A_0] \times (1 - e^{-kt})$$

The rate constant in first-order kinetics is directly related to some characteristic times of substances, which are often readily available. For example the average life time of the reactant, τ, and the time it takes for its concentration to half, the half-life $t_{1/2}$, can be derived as (*see* Fig. 2):

$$\tau = \frac{1}{k}$$
$$t_{\frac{1}{2}} = \frac{\ln 2}{k}$$

Integration of the equation describing the evolution of P in a second-order reaction using the initial concentrations $[A_0]$, $[B_0]$ and $[P_0]$ leads to a hyperbolic time dependency:

$$[P](t) = [P_0] + [A_0][B_0]\frac{e^{-kt[B_0]} - e^{-kt[A_0]}}{[A_0]e^{-kt[B_0]} - [B_0]e^{-kt[A_0]}}$$

Contrarily from first-order reactions, the characteristic times in second-order reactions are not independent of the initial conditions, but depend on both the rate constant and the initial

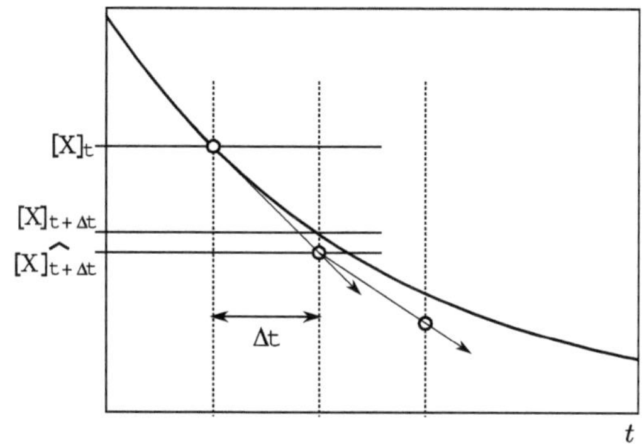

Fig. 3 Graphical representation of the forward Euler method to integrate ordinary differential equations. The *thick curve* represents [X] = *f*(*t*), and the vectors its derivative. Note the progressive error introduced by the coarse time discretization

concentrations of the reactants. In case the two reactants are instances of the same molecular pool (A = B), and assuming the initial concentrations to be $[A_0]$ and $[P_0]$, the resulting time courses for [A] and [P] are described by the following hyperbolic functions:

$$[A](t) = \frac{[A_0]}{2k[A_0]t + 1}$$

$$[P](t) = [P_0] + \frac{[A_0]^2 kt}{2k[A_0]t + 1}$$

However, beside the most elementary systems containing only few well-behaved reactions, we cannot generally solve a system of ordinary differential equations analytically. We have to resort to numerical integration, a method that goes back to the origin of differential calculus, where we approximate the current values of the variables based on the knowledge we have of their values in the (close) past. Many approximations have been developed. The simplest and easiest to grasp (but also the most error prone) is the forward Euler rule. If we discretize the time, one can make the following approximation:

$$\frac{d[X]}{dt} \approx \frac{\Delta[X]}{\Delta t} = \frac{\left([X]_{t+\Delta t} - [X]_t\right)}{\Delta t}$$

We can rearrange the equation above and extract the concentration as follows:

$$[X]_{t+\Delta t} \approx [X]_t + \frac{d[X]}{dt}(t) \times \Delta t$$

We know $d[X]/dt$ as a function of the vector of concentrations, obtained with the method described above, and can therefore compute the difference introduced during one Δt. This procedure is represented in Fig. 3. We can see on the figure that a systematic error is introduced by the time discretization. Such an error becomes larger for more complex dynamics, such as non-monotonic behaviours (e.g., oscillations), or systems with fast and slow components. One can address the error by using tiny time steps but at the expense of computational efficacy. Many methods have been developed over the years to address this problem. A good introduction is given in LeMasson and Maex [2] and a more comprehensive survey of the field by Hairer et al. [3] and Hairer and Wanner [4]. Biological modeling tools such as COPASI [5], JDesigner/Jarnac [6], E-Cell [7] or CellDesigner [8] have their own in-built numerical ODE solver. They also generate the system of ODE to be solved automatically, so the required user input is limited to the list of chemical reactions in some defined format and of the parameters governing those reactions.

3 Modeling Biochemical Networks

Modeling the biochemical pathways does not require much more than what has been presented in Subheading 1. The only complexity we will introduce in the following sections are slightly more complex expressions for the reaction rates.

3.1 Basal Level and Homeostasis

Before modeling the effect of perturbations, such as extracellular signals, it is important to set up the right basal level for the substances that we will consider in the model. This basal level is obtained when the processes producing the substance and the processes consuming it are compensating each other. We then reach a steady state, where input and output are equal. To illustrate this, we will build the simplest system possible that permits to have a steady basal concentration of calcium. The system is made up of a continuous creation of calcium, for instance due to leaky channels in the plasma membrane or in the internal stores, modeled as a zero-order reaction (*see* Subheading 1.2.1). The calcium is then removed from the system for instance by pumps or buffers in excess, modeled as a first-order reaction (*see* Subheading 1.2.2).

$$\emptyset \xrightarrow{k_{in}} Ca^{2+} \xrightarrow{k_{out}} \emptyset$$

The instantaneous changes of calcium concentration then result from the combination of the two reaction rates (Fig. 4).

$$\frac{d[Ca^{2+}]}{dt} = k_{in} - k_{out}[Ca^{2+}]$$

The steady-state level is reached when the changes are null, that is $[Ca^{2+}] = k_{in}/k_{out}$. If the concentration of calcium is higher than this ratio, the second term wins and the concentration decreases. In contrast, if the concentration of calcium is lower than this ratio, the first term wins and the concentration increases. k_{out} can be

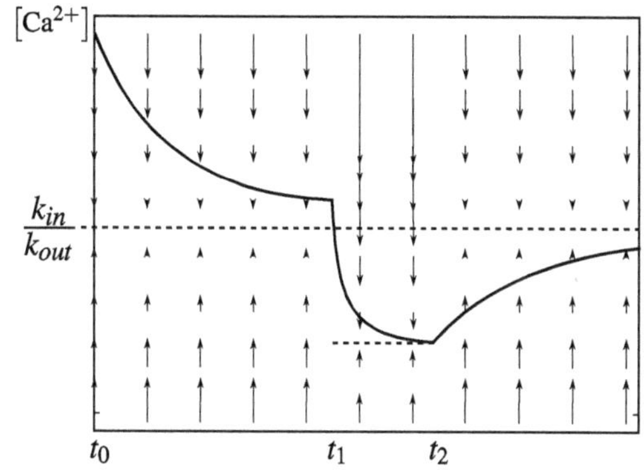

Fig. 4 Evolution of calcium concentration over time. Between t_0 and t_1, the extrusion is stronger than the creation. At t_1, k_{in} strongly decreases, for instance by a block of leak channels, and the concentration is brought to a lower steady-state value. At t_2 the block is removed. The creation becomes stronger than extrusion, and brings back the concentration to the initial steady state. *Vertical arrows* represent the intensity and direction of the reaction's flux for a given concentration of calcium

estimated from the decay observed after stimulation. k_{in} can therefore be computed from the steady state. Changing k_{in} in a discrete manner is a simple way of modeling the opening or closing of calcium channels. Such a homoeostatic control is extremely simple. More complex schemes can be designed, with control loops such as negative feed-backs on the creation steps and positive feed forwards on the extrusion steps.

3.2 Representing Enzymatic Reactions

In order to accelerate chemical reactions and select among different isomers, cells use enzymes, which are protein-based catalysts. They can increase reaction rates to a tremendous degree and often are essential to make reactions occur at a measurable rate. Enzyme catalyzed reactions tend to follow complex sequences of reaction steps, and the exact reaction mechanisms are generally unknown. The single reaction steps can be contracted into an overall description with lumped stoichiometries. However, since the detailed reaction mechanisms are most often unknown, and also parameters for each of these steps are hard to come by, such reactions can rarely be modeled considering each single step and using mass-action kinetics. Depending on how much detail is known, an enzyme catalyzed reaction can be described on different levels. The reaction equations for a simple conversion of a substrate S to a product P catalyzed by an enzyme E, for example, can vary depending on the consideration of intermediate enzyme complexes and reaction reversibility:

$$
\begin{aligned}
&S + E \rightleftharpoons ES \rightleftharpoons EP \rightleftharpoons E + P \\
&S + E \rightleftharpoons ES \rightarrow EP \rightleftharpoons E + P \\
&S + E \rightleftharpoons ES \rightleftharpoons E + P \\
&S + E \rightleftharpoons ES \rightarrow E + P \\
&S + E \rightarrow E + P
\end{aligned} \tag{2}
$$

Knowledge of the mechanism of an enzymatic reaction can be used to derive compact and simplified expressions fitting the overall kinetics. The alternative is to use generic rate laws that are known to loosely fit wide classes of reaction mechanisms, and to choose the ones that seem most appropriate for the reaction in question. The kinetics of the overall reaction are determined by the reaction mechanisms of the elementary steps, but exact derivations can become quite complex and cumbersome to handle. In general it is safer and more convenient to use approximate expressions in biological modeling, even more so as exact mechanisms are rarely known.

Two assumptions are available to simplify complex enzymatic reaction descriptions. The more general one is the quasi steady-state approximation, QSSA. The QSSA considers that some, or all, of the intermediary enzyme-substrate complexes tend to a near constant concentration shortly after the reaction starts. The other widely used assumption, called the rapid equilibrium assumption, is that some steps are much faster than the overall reaction, meaning that the participating enzyme forms are virtually at equilibrium and that their concentrations can be expressed using equilibrium constants. This approach is often used to model fast reactant or modulator binding to the enzyme. The application of these techniques depends very much on how much of the reaction mechanism is known. An excellent introduction into enzyme kinetics is given by Cornish-Bowden [9]. For a more exhaustive treatment with detailed derivations of rate laws for a multitude of mechanisms please refer to the standard work by Segel [10].

At the beginning of the twentieth century, Henri [11] proposed a reaction scheme and an accompanying expression for describing the rate of sucrose hydrolysis catalyzed by invertase. This reaction showed a deviation from normal second-order kinetics and tended to a maximal velocity directly proportional to the enzyme concentration. Making use of the existence of an intermediary substrate-enzyme complex, ES, and assuming that the substrate S and the enzyme E were in a rapid binding equilibrium with the complex, he could derive an expression fitting the experimental observations. A similar approach was taken and expanded in 1913 by Michaelis and Menten [12], who proposed the current form of the reaction rate based on a rapid equilibrium between enzyme and substrate.

$$\mathrm{E} + \mathrm{S} \underset{k_{-1}}{\overset{k_1}{\rightleftharpoons}} \mathrm{ES} \xrightarrow{k_2} \mathrm{E} + \mathrm{P}$$

(k_2 is the catalytic constant, or turnover number, and often called k_{cat}). The QSSA was proposed as a more general derivation by Briggs and Haldane [13]. The substrate binding and dissociation, as well as the product formation step, lead to the following expression for the time dependence of [ES]:

$$\frac{\mathrm{d[ES]}}{\mathrm{d}t} = k_1[\mathrm{E}][\mathrm{S}] - k_{-1}[\mathrm{ES}] - k_2[\mathrm{ES}]$$

At steady state, the concentration of the intermediate complex, [ES], is constant hence $\mathrm{d[ES]}/\mathrm{d}t = 0$. Rearranging this equation and setting $K_M = \frac{k_{-1}+k_2}{k_1}$, we obtain $[\mathrm{E}] = [\mathrm{ES}] \times K_M/[\mathrm{S}]$. Furthermore, because the concentration of enzyme is constant, we have $[\mathrm{E}] = [\mathrm{E_t}] - [\mathrm{ES}]$. Equating both, we obtain:

$$v = \frac{\mathrm{d[P]}}{\mathrm{d}t} = k_2[\mathrm{ES}] = k_2[\mathrm{E_t}]\frac{[\mathrm{S}]}{K_M + [\mathrm{S}]} \quad (3)$$

$k_2 \times [\mathrm{E_t}]$ is sometimes called the maximal velocity v_{max}. This rate expression is often used—and abused—when modeling biochemical processes for which the exact mechanisms are unknown. However, one has to realize that it only holds true if the concentration of the enzyme-substrate complex stays constant, which in turns implies that the concentration of substrate is in large excess. Those conditions are very rarely met in signal transduction systems, resulting in many artifacts.

Plotting the reaction velocity, v, against the substrate concentration, [S], gives a rectangular hyperbolic curve (*see* Fig. 5). The parameter K_M has the unit of a concentration and is of central importance in describing the form of the substrate dependence of the reaction velocity. As can be seen by inserting K_M for [S] in Eq. 3, it denotes the substrate concentration at which the reaction speed is half of the limiting velocity. If $[\mathrm{S}] \ll K_M$, then [S] in the denominator can be disregarded and the reaction becomes linear with regard to S, showing first-order characteristics:

$$[\mathrm{S}] \ll K_M \Rightarrow v \approx \frac{v_{max}}{K_M} \times [\mathrm{S}]$$

On the other extreme, for high substrate concentrations, $[\mathrm{S}] \gg K_M$, the reaction speed becomes virtually independent of [S] and tends toward v_{max}.

$$[\mathrm{S}] \gg K_M \Rightarrow v \approx v_{max} = k_{cat} \times [\mathrm{E_t}]$$

Most enzyme catalyzed reactions show a similar rate behaviour inasmuch as they exhibit first or higher order dependencies on the substrate at lower substrate concentrations and tend to a limiting rate depending only on the enzyme concentration when the reactant concentrations are high.

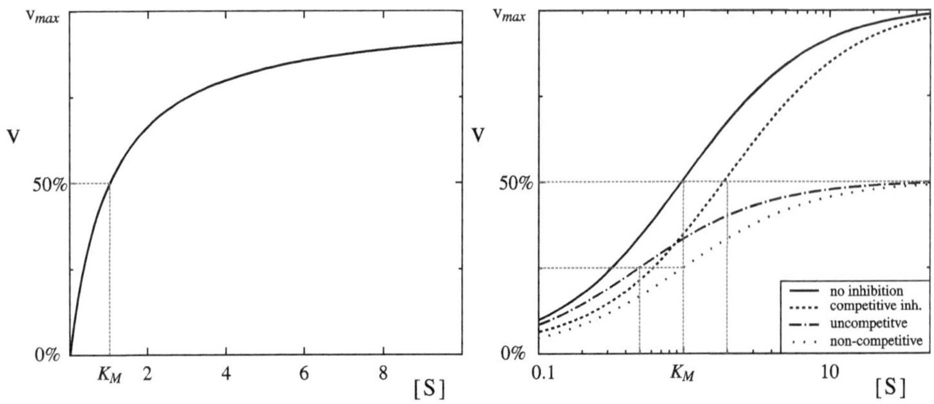

Fig. 5 Dependence of the reaction velocity, v, of the irreversible Michaelis–Menten equation on the concentration of the substrate, S. The *left graph* shows the uninhibited case. On the *right* various forms of inhibition are shown in a semi-logarithmic plot. The *horizontal dotted lines* indicate the apparent half maximal velocities, the *vertical ones* the apparent K_Ms. Competitive inhibition does not alter the maximal velocity, but shifts the K_M to higher values, while non-competitive inhibition simply decreases the apparent V_{max}. The special case of uncompetitive inhibition leads to an apparent increase of substrate affinity of the enzyme, that is a lower K_M, but a reduction of the apparent V_{max}. Mechanistically this is due to the unproductive enzyme-substrate-inhibitor complex ($K_M = 1$; [I] $= 1$; comp., uncomp. and non-comp. inhib.: $K_I = 1$.)

While the original Michaelis–Menten equation was derived to describe the initial velocity of the enzymatic reaction in absence of product, allowing the reverse reaction to be neglected, the QSSA can also be used to derive a reversible Michaelis–Menten equation describing the most extensive reaction scheme in Eq. 2. Using the same procedure as above, the following expression for the reaction velocity in dependence of E_T, S and P can be derived:

$$v = \frac{v_{\mathrm{fwd}} \frac{[\mathrm{S}]}{K_{\mathrm{MS}}} - v_{\mathrm{rev}} \frac{[\mathrm{P}]}{K_{\mathrm{MP}}}}{1 + \frac{[\mathrm{S}]}{K_{\mathrm{MS}}} + \frac{[\mathrm{P}]}{K_{\mathrm{MP}}}} \tag{4}$$

As the net rate of a reversible reaction has to vanish at equilibrium, one of the parameters of Eq. 4 can be expressed using the equilibrium constant by setting the numerator of the expression to zero. The so called Haldane relationship connects kinetic and thermodynamic parameters of an enzymatic reaction. While some mechanisms lead to much more complicated expressions, at least one Haldane relationship exists for every reversible reaction.

$$K_{\mathrm{eq}} = \frac{v_{\mathrm{fwd}} K_{\mathrm{MP}}}{v_{\mathrm{rev}} K_{\mathrm{MS}}} = \frac{k_2 K_{\mathrm{MP}}}{k_{-1} K_{\mathrm{MS}}}$$

3.3 Modeling Simple Transport Processes

Compartmentalization of molecular species and transport across membranes are of great importance in biological systems, and often need to be implicitly accounted for or explicitly included into models.

Transport across membranes can either occur passively by simple diffusion, or be coupled to another reaction to actively move molecules against a chemical potential gradient. In the simplest form of passive diffusion, molecules just directly pass through a membrane or an open channel or pore. As the connected compartments in general have differing volumes, the change of concentration of a substance flowing from one compartment to another is not equal in both compartments. Therefore the rate of translocation is commonly described by the flux, j, of a substance, that is the amount of a substance crossing a unit area per time unit. In case of no other influences on the translocation, but simple diffusion, the flux of a substance S into a cell through a membrane follows a variant of Fick's first law:

$$[S_{out}] \rightleftharpoons [S_{in}]$$
$$js = p_s([S_{out}] - [S_{in}])$$

in which $[S_{out}]$ and $[S_{in}]$ are the concentrations of S on the exterior and inside the cell, respectively. p_S denotes the permeability of the membrane for S. The permeability for direct diffusion is proportional to the diffusion coefficient of S and, for pores or channels, to the number of open channels per area.

To derive an expression of the change of concentration of S, it is important to consider that the flux is given as amount per area and time and not as concentration per time. Therefore the volumes of the exterior and the cell have to be included in the differential expressions of concentrations. The overall rate of translocation, v_t, depends on the surface area, A, of the membrane, and the permeability and area can be contracted to a transport rate constant, $k_S = p_S \times A$. For the change of $[S_{out}]$ and $[S_{in}]$, respectively, the following expressions can be derived:

$$\frac{d[S_{out}]}{dt} = -\frac{v_T}{V_{out}} = -\frac{k_S}{V_{out}}([S_{out}] - [S_{in}])$$
$$\frac{d[S_{in}]}{dt} = \frac{v_T}{V_{in}}$$

with V_{out} and V_{in} being the volumes of the exterior and the cell.

In the case of a molecule that does not simply diffuse through a membrane or pore, but needs to bind a carrier to be translocated from one compartment to the other, the kinetic expressions depend on the exact mechanism of translocation. The simplest case of facilitated, or carrier-mediated, diffusion consists of a carrier with a single binding site, C, which can bind a substance A with equal affinity on each side of the membrane, and flips from one side of the membrane to the other. Using the steady-state approach the following expression can be derived for the translocation rate:

$$v_t = \frac{v_{max}([A_{out}] - [A_{in}])}{K_M + [A_{out}] + [A_{in}] + \frac{K_i[A_{out}][A_{in}]}{K_M}}$$

In this equation ν_{max} is the limiting rate of translocation and depends mostly on the amount of carrier. K_M is the concentration of A on one side at half maximal translocation in case of zero concentration on the other side of the membrane, and K_i, called the interactive constant, depends on the relative mobility of the free and loaded carrier (for details *see* Ref [14]).

4 Modeling Modulation of Dynamical Processes

Reactions in biological systems are not only regulated by the availability of reactants and catalysts, but also by compounds modulating the activity of channels and enzymes, often without any direct involvement in the specific reactions. Examples are neurotransmitters that alter the flow of ions through channels, without direct involvement in the transport process, enzyme allosteric effectors, that will modulate the activity of an enzyme without being involved in the catalytic reaction etc. In this section, we will introduce a generic method to model activation and inhibition of reactions, based on Hill equations.

4.1 Binding of Modulators and Activity

The activities of receptors, channels and enzymes are often regulated by ligands binding to them. One important characteristic of such binding processes is the *fractional occupancy*, $\bar{Y}$ of the bound compound. It is defined as the number of binding sites occupied by a ligand, divided by the total number of binding sites. For a ligand X binding to a single binding site of a protein P, we can express [PX] and $\bar{\Upsilon}$ as follows, using the dissociation constant $K_d = \frac{k_{off}}{k_{on}}$ and the total protein concentration $[P_T] = [P] + [PX]$:

$$\begin{aligned} P + X &\underset{k_{off}}{\overset{k_{on}}{\rightleftharpoons}} PX \\ [PX] &= \frac{[P_T][X]}{K_d + [X]} \\ \bar{\Upsilon} &= \frac{[PX]}{[P_T]} = \frac{[X]}{K_d + [X]} \end{aligned} \quad (5)$$

Equation 5, also known as the Hill–Langmuir equation, is very similar to the Michaelis–Menten equation. Like [S] in Eq. 3, [X] stands for the concentration of *free* ligand, but can be substituted with the total ligand concentration $[X_T] = [X] + [PX]$ in case that $[X_T] \gg [P_T]$. If P is active only when bound to X, one must multiply the reaction rate by $\bar{\Upsilon}$ to describe the actual reaction velocity. On the contrary, if P is active only when not bound to X, one must multiply the reaction rate by $1 - \bar{\Upsilon}$:

$$1 - \bar{\Upsilon} = 1 - \frac{[\mathrm{PX}]}{[\mathrm{P_T}]} = \frac{K_\mathrm{d}}{K_\mathrm{d} + [\mathrm{X}]} \quad (6)$$

A general form of Eq. 5 was developed by Hill [15]. Drawing on observations of oxygen binding to haemoglobin, Hill suggested the following formula for the fractional occupancy $\bar{\Upsilon}$ of a protein with several activator binding sites:

$$\bar{\Upsilon} = \frac{\frac{[\mathrm{X}]^h}{K_H}}{1 + \frac{[\mathrm{X}]^h}{K_H}} = \frac{[\mathrm{X}]^h}{K_H + [\mathrm{X}]^h}$$

where [X] denotes ligand concentration, K_H is an apparent dissociation constant (with the unit of a concentration to the power of h) and h is the *Hill coefficient*, which needs not be an integer. The *Hill coefficient h* indicates the degree of cooperativity, and in general is different from the number of ligand binding sites, n. The Hill equation can show positive and negative cooperativity, for exponent values of $h > 1$ and $0 < h < 1$, respectively. In case of $h = 1$ it shows hyperbolic binding behaviour. With increasing exponents, the ligand binding curve becomes more and more sigmoid, with the limit of a step function with a threshold value of $\sqrt[h]{K_H}$. The constant $K_h = \sqrt[h]{K_H}$ provides the ligand concentration at which half the binding sites are occupied (equivalent to a dissociation constant), or, in purely phenomenological uses, activation or inhibition by the effector is half maximal. Note that negative values of h produce the same decreasing sigmoid function than the above $1 - \bar{\Upsilon}$, so the generalized Hill function can be used for both activators and inhibitors.

4.2 Modeling Regulation of Processes with Hill Functions

The Hill equation can easily be adapted to provide functions to describe interactions with little prior knowledge. Let us assume a gene which expression is regulated in a nonlinear fashion for instance by the binding of a transcription factor A. One can model the gene expression with increasing concentrations of A using a Hill function:

$$v = v_{\max} \times \frac{[\mathrm{A}]^h}{K_\mathrm{A}^h + [\mathrm{A}]^h} \quad (7)$$

Here v is the actual production of mRNA by the gene. $v_{\max}$ indicates the maximal activity of the gene. K_A and h indicate the transcription factor concentration for half maximal activation, and a cooperativity coefficient. If $[\mathrm{A}] = 0$, the correcting factor is close to 0, v is null, i.e., there is no gene expression. If [A] is large, the correction factor is close to 1 and the expression is maximal. Similarly, the effect of a repressor I can be described by:

$$v = v_{\max} \times \frac{K_I^h}{K_I^h + [I]^h} \tag{8}$$

In this equation K_I stands for the transcription factor concentration triggering half maximal inhibition. If $[I] = 0$, the correcting factor is close to 1 the gene expression is maximal, while if [I] is large, the correction factor is close to 0 and so is the gene expression. Note that the same result is obtained with the mathematical expression derived for activation above, with exponents of $-h$. Therefore, one can provide a generic formula that can phenomenologically describe the effects of all independent activators and inhibitors at once.

$$v = v_{\max} \times \prod_{i=1}^{n} \frac{[X_i]^{h_i}}{K_{X_i}^{h_i} + [X_i]^{h_i}} \tag{9}$$

Such a formula can then be used in parameter estimation procedures. For n effectors, one has to estimate $2n$ independent parameters, or only n if the cooperativity is assumed negligible. Note that such a formula is only valid if no significant interactions take place between the effectors.

As an example of Hill equation use, let's study the kinetics of calcium-gated channels. An example containing two different types of activation is given in Borghans et al. [16] for the Ca^{2+} induced Ca^{2+} release (CICR) via the inositol triphosphate (InsP3) receptor. Equation 18 of the paper describes the release of calcium from a calcium sensitive pool. The flux rate is given by:

$$v_{\mathrm{InsP3\,R}} = v_{\max} \frac{[Ca_p]^2}{K_1^2 + [Ca_p]^2} \frac{[Ca_c]^2}{K_2^2 + [Ca_c]^2}$$

In this equation $\nu_{\max}$ denotes the maximal release rate, and $[Ca_p]$ and $[Ca_c]$ the Ca^{2+} concentrations in the pool and the cytoplasm. The release is regulated by the Ca^{2+} concentrations on both sides of the membrane separating the pool and the cytosol, and K_1 and K_2 stand for the threshold concentrations for these activations. Parthimos et al. [17] used an even more complex expression for the CICR from the sarcoplasmic reticulum via the InsP3 receptor. The receptor was modeled to be both activated and inactivated by cytosolic Ca^{2+}, Ca_c, using two Hill functions involving Ca_c. A possible mechanistic explanation for this form would be the existence of independent activation and inhibition sites, with different affinities and degrees of cooperativity for Ca^{2+}. In the flux rate through the InsP3 receptor

$$v_{\mathrm{InsP3R}} = v_{\max} \frac{[Ca_s]^2}{K_1^2 + [Ca_s]^2} \frac{[Ca_c]^4}{K_2^4 + [Ca_c]^4} \frac{K_3^4}{K_3^4 + [Ca_c]^4} \tag{10}$$

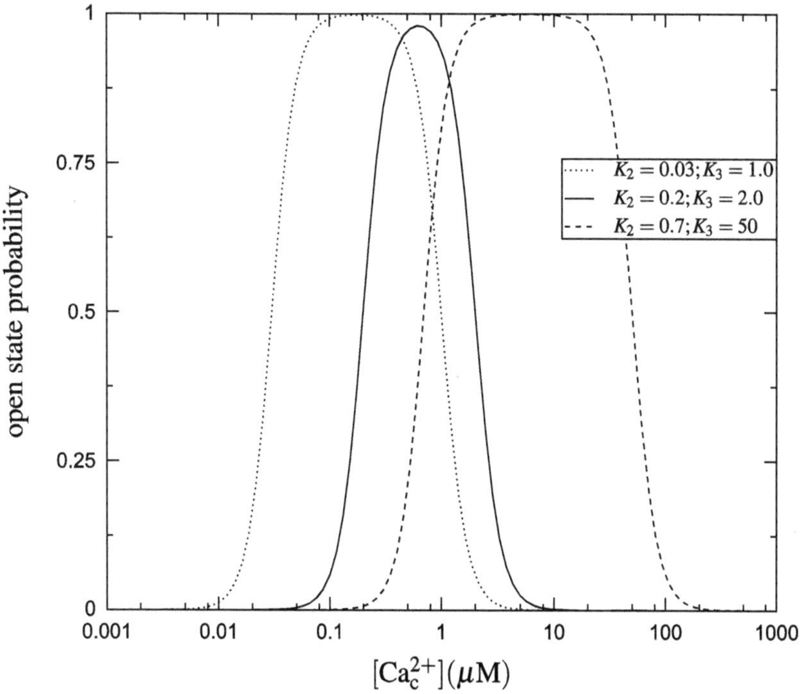

Fig. 6 InsP3 receptor opening probability dependent on cytoplasmic Ca^{2+} after Parthimos et al. [17] as described in Eq. 10. K_2 and K_3 indicate the concentrations of half maximal activation and inhibition, respectively, of the InsP3 receptor. For both activation and inhibition a Hill factor of 4 was assumed

K_2 and K_3 indicate the cytosolic Ca^{2+} concentrations at which activation and inhibition of CICR, respectively, are half maximal. If they are chosen in such a way that $K_2 < K_3$, the flux rate through the receptor reaches a maximum for concentration values between the values of the two constants and vanishes for higher cytosolic Ca^{2+} concentrations (*see* Fig. 6), creating a complex on–off behaviour of the InsP3 receptor in dependence of the Ca^{2+} concentration. In case of nonessential activation or leaky inhibition, a process can still proceed at a basal rate ν_{bas} in absence of the activator or at high concentrations of the inhibitor (Fig. 7). This can be accounted for by using the relative basal rate, $b = \frac{v_{bas}}{v_{max}}$.

$$v = v_{max}(b + (1 - b)\gamma([X]))$$

where $\gamma([X])$ is the function describing the relative activity in dependence of the concentration of the regulating agent X, that is $\bar{\Upsilon}$ or $1 - \bar{\Upsilon}$ mentioned in Eqs. 5 and 6. Note that if there is no reaction in the absence of a modulator, the basal rate is 0, $b = 0$, and the equation is equivalent to Eqs. 7 and 8. One can therefore further generalise Eq. 9 as:

$$v = v_{max} \times \prod_{i=1}^{n} \left(b_i + (1 - b_i) \times \frac{[X_i]^{h_i}}{K_{Xi}^{h_i} + [X_i]^{h_i}} \right)$$

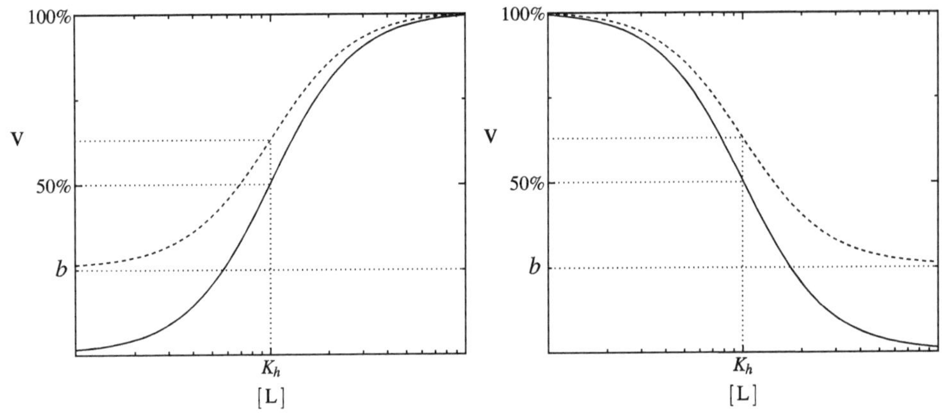

Fig. 7 Activation (*left*) and inhibition (*right*) modeled using Hill functions with a $n_h = 2$. Ligand concentration is shown in units of the concentration of half maximal activation or inhibition, respectively, K_h on a logarithmic scale and the velocity ν in percent of the fully activated or uninhibited velocity, $\nu_{\max}$. The *dashed line* shows cases with a basal rate, ν_{bas}, of 25 % of $\nu_{\max}$ $\left(b = \frac{v_{\text{bas}}}{v_{\max}} = 0.25\right)$

This equation provides an initial framework to model the kinetics of almost any regulatory network in the absence of mechanistic knowledge. Although Hill functions have been frequently used because of their simplicity, other generic frameworks have been proposed to phenomenologically model kinetics of regulatory networks, including logoid function [18], Goldbeter-Koshland switches [19], S-systems [20] etc. Further information can be found in a quite famous review [21]. As mentioned above, it is important to realize that such a framework assumes independence of the modulators. Other more complicated formulae, derived for enzyme regulation, can then be used if the concentration of a modulator affects the effect of another one. The reader should refer to Segel [10] for more details.

5 Further Reading

Biophysical chemistry, James P. Allen. This is a complete and concise presentation of the physical and chemical basis of life [22].

Computational Cell Biology, Christopher P. Fall, Eric S. Marland, John M. Wagner, John J. Tyson. Also known as “the yellow book”, this is an excellent introduction to modeling cellular processes. It contains chapters dedicated to ion channels, transporter, biochemical oscillations, molecular motors and more [23].

Enzyme kinetics, Irwin H. Segel and Fundamentals of Enzyme Kinetics, Athel Cornish-Bowden. Also known as “the black book” and the “the red book”, these are the two reference books if one wants to know how to model an enzymatic reaction, regardless of its complexity.

Solving Ordinary Differential Equations I and II, Ernst Hairer, Syvert P. Norsett, Gerhard Wanner.Extensive coverage of the domain of ordinary differential equations, from Newton and Leibniz to the most advanced techniques implicit solvers.

References

1. Waage P, Guldberg C (1864) Studies concerning affinity. Forhandlinger: Videnskabs-Selskabet i Christiana 35
2. LeMasson G, Maex R (2001) Introduction to equation solving and parameter fitting. In: De Schutter E (ed) Computational neuroscience: realistic modeling for experimentalists. CRC Press, London, pp 1–23
3. Hairer E, Nrsett SP, Wanner G (1993) Solving ordinary differential equations: nonstiff problems. Springer, Berlin
4. Hairer E, Wanner G (1996) Solving ordinary differential equation II: stiff and differential-algebraic problems. Springer, Berlin
5. Hoops S, Sahle S, Gauges R, Lee C, Pahle J, Simus N, Singhal M, Xu L, Mendes P, Kummer U (2006) COPASI–a COmplex PAthway SImulator. Bioinformatics 22(24):3067–3074
6. Sauro H (2000) Jarnac: a system for interactive metabolic analysis. In: Hofmeyr JHS, Rohwer JM, Snoep JL (eds) Animating the cellular map 9th international bio-thermokinetics meeting, Ch. 33. Stellenbosch University Press, pp 221–228
7. Takahashi K, Ishikawa N, Sadamoto Y, Sasamoto H, Ohta S, Shiozawa A, Miyoshi F, Naito Y, Nakayama Y, Tomita M (2003) E-Cell 2: multi-platform E-Cell simulation system. Bioinformatics 19(13):1727–1729
8. Funahashi A, Matsuoka Y, Jouraku A, Morohashi M, Kikuchi N, Kitano H (2008) celldesigner 3.5: a versatile modeling tool for biochemical networks. In: Proceedings of the IEEE, vol 96, pp 1254–1265
9. Cornish-Bowden A (2004) Fundamentals of enzyme kinetics. Portland Press, London
10. Segel IH (ed) (1993) Enzyme kinetics. Wiley, New-York
11. Henri V (1902) Theorie generale de l'action de quelques diastases. Compt Rend Hebd Acad Sci Paris 135:916
12. Michaelis L, Menten M (1913) Die Kinetik der Invertinwirkung. Biochem Z 49:333–369
13. Briggs GE, Haldane JB (1925) A note on the kinetics of enzyme action. Biochem J 19(2): 338–339
14. Kotyk A (1967) Mobility of the free and of the loaded monosaccharide carrier in Saccharomyces cerevisiae. Biochim Biophys Acta 135(1): 112–119
15. Hill AV (1910) The possible effects of the aggregation of the molecules of haemoglobin on its dissociation curves. J Physiol 40:1–7
16. Borghans JM, Dupont G, Goldbeter A (1997) Complex intracellular calcium oscillations. A theoretical exploration of possible mechanisms. Biophys Chem 66(1):25–41
17. Parthimos D, Haddock RE, Hill CE, Griffith TM (2007) Dynamics of a three variable non-linear model of vasomotion: comparison of theory and experiment. Biophys J 93(5): 1534–1556
18. Mestl T, Plahte E, Omholt SW (1995) A mathematical framework for describing and analysing gene regulatory networks. J Theor Biol 176:291–300
19. Goldbeter A, Koshland DE (1981) An amplified sensitivity arising from covalent modification in biological-systems. Proc Natl Acad Sci USA 78:6840–6844
20. Savageau MA, Voit EO (1987) Recasting non-linear differential equations as systems: a canonical nonlinear form. Math Biosci 87:83–115
21. Tyson J, Chen KC, Novak B (2003) Sniffers, buzzers, toggles and blinkers: dynamics of regulatory and signaling pathways in the cell. Curr Opin Cell Biol 15:221–231
22. Allen JP (ed) (2008) Biophysical chemistry. Wiley, Oxford
23. Fall C, Marland E, Wagner J, Tyson J (eds) (2002) Computational cell biology. Springer, New York

Chapter 9

Simulation of Stochastic Kinetic Models

Andrew Golightly and Colin S. Gillespie

Abstract

A growing realization of the importance of stochasticity in cell and molecular processes has stimulated the need for statistical models that incorporate intrinsic (and extrinsic) variability. In this chapter we consider stochastic kinetic models of reaction networks leading to a Markov jump process representation of a system of interest. Traditionally, the stochastic model is characterized by a chemical master equation. While the intractability of such models can preclude a direct analysis, simulation can be straightforward and may present the only practical approach to gaining insight into a system's dynamics. We review exact simulation procedures before considering some efficient approximate alternatives.

Key words Stochastic simulation, Markov jump process, Time discretization

1 Introduction

Computational systems biology is typically concerned with developing dynamic simulation models of biological processes. Such models can be used to test understanding of systems of interest and perform *in silico* experimentation. Statistical methods based on the macroscopic rate equation (MRE), which describes the thermodynamic limit of a system via a set of coupled ordinary differential equations (ODEs), have been widely used [1, 2]. Such an approach may be appropriate when describing average concentrations within a population of cells or when modeling more global physiology, for example, at a tissue or organ level [3]. Single cell experiments and studies of noise in regulatory networks have revealed the importance of stochastic effects in intracellular processes and in turn, this has motivated the need for models that incorporate intrinsic stochasticity [4]. A deterministic modeling approach fails to capture the stochastic (and discrete) nature of chemical kinetics at low concentrations. Reaction events are

The R code used in this chapter can be downloaded from the github repository: https://github.com/csgillespie/In-silico-Systems-Biology.

Maria Victoria Schneider (ed.), *In Silico Systems Biology*, Methods in Molecular Biology, vol. 1021,
DOI 10.1007/978-1-62703-450-0_9, © Springer Science+Business Media, LLC 2013

intrinsically stochastic, driven by Brownian motion. When such events take place, the effect is to change biochemical species numbers by an integer amount. Hence, as reactions take place, species numbers change abruptly and discretely. Such arising random fluctuations are often referred to as *intrinsic noise* [5]. Other sources of noise may be termed *extrinsic* (e.g., due to variations in initial conditions or environmental conditions).

The aim of this chapter is to provide a concise introduction to the simulation of stochastic kinetic models through consideration of some commonly used simulation algorithms. Approximate strategies based on, for example, time discretization or the dispensation of the assumption of discrete states are also explored. The remainder of this chapter is organized as follows. In Subheading 2, we briefly review stochastic chemical kinetics leading to a stochastic kinetic model of a system of interest, formulated as a Markov jump process. We consider exact simulation of the jump process via the Gillespie algorithm [6] in Subheading 3 before examining some recently proposed extensions which aim to increase the computational efficiency of the algorithm. Approximate simulation strategies such as the τ-leap [7], chemical Langevin equation (CLE) [8, 9], and linear noise approximation (LNA) [10] are considered in Subheading 4. The chapter concludes with a discussion in Subheading 6.

2 Stochastic Chemical Kinetics

In this section we represent a biological system of interest with a set of pseudo-biochemical reactions. There are a number of ways in which a system could be represented, from a qualitative diagram to a fully quantitative set of equations. A reaction network provides a flexible representation, allowing the modeler to specify the level of detail deemed appropriate. Once the assumptions about the underlying chemical kinetics have been made, simulation can take place.

2.1 Reaction Networks

To fix notation, consider a biochemical reaction network involving u species $\mathcal{X}_1, \mathcal{X}_2, \ldots, \mathcal{X}_u$ and v reactions R_1, R_2, ..., R_v, written using standard chemical reaction notation as

$$\begin{aligned}
R_1:&\quad p_{11}\mathcal{X}_1 + p_{12}\mathcal{X}_2 + \cdots + p_{1u}\mathcal{X}_u \to q_{11}\mathcal{X}_1 + q_{12}\mathcal{X}_2 + \cdots + q_{1u}\mathcal{X}_u\\
R_2:&\quad p_{21}\mathcal{X}_1 + p_{22}\mathcal{X}_2 + \cdots + p_{2u}\mathcal{X}_u \to q_{21}\mathcal{X}_1 + q_{22}\mathcal{X}_2 + \cdots + q_{2u}\mathcal{X}_u\\
&\qquad\vdots \qquad\qquad\qquad\qquad \vdots \qquad\qquad \vdots\\
R_v:&\quad p_{v1}\mathcal{X}_1 + p_{v2}\mathcal{X}_2 + \cdots + p_{vu}\mathcal{X}_u \to q_{v1}\mathcal{X}_1 + q_{v2}\mathcal{X}_2 + \cdots + q_{vu}\mathcal{X}_u.
\end{aligned}$$

Let $X_{j,t}$ denote the number of molecules of species $\mathcal{X}_j$ at time t, and let X_t be the u-vector $X_t = (X_{1,t}, X_{2,t}, \ldots, X_{u,t})'$.

Further, let $P = (p_{ij})$ be a $v \times u$ matrix of the coefficients p_{ij} with $Q = (q_{ij})$ defined similarly. The $u \times v$ *stoichiometry matrix* S is defined by

$$S = (Q - P)'.$$

The matrices P, Q and S will typically be *sparse*. On the occurrence of a reaction of type i, the system *state* (X_t) is updated by adding the ith column of S. Consequently, if ΔR is a v-vector containing the number of reaction events of each type in a given time interval, then the system state should be updated by ΔX, where

$$\Delta X = S\,\Delta R.$$

The stoichiometry matrix therefore encodes important structural information about the reaction network. In particular, vectors in the left null-space of S correspond to *conservation laws* in the network, that is, any u-vector a satisfying $a'S = 0$ has the property (clear from the above equation) that $a'X_t$ remains constant for all t.

2.2 Markov Jump Process Representation

Let us consider a bimolecular reaction

$$\mathcal{X}_1 + \mathcal{X}_2 \rightarrow \mathcal{X}_3\,.$$

This reaction will occur when a molecule of $\mathcal{X}_1$ collides with a molecule of $\mathcal{X}_2$ while moving around randomly, driven by Brownian motion. Consider a pair of such molecules in a container of fixed volume. Under fairly weak assumptions involving the container and its contents, it is possible to show that the collision hazard (or rate) is constant in a small time interval [8]. Therefore, for the reaction above, the probability of a given pair of molecules reacting in a time interval of length $\mathrm{d}t$ is $c\,\mathrm{d}t$ for some constant c. Suppose now that there are x_1 molecules of $\mathcal{X}_1$ and x_2 molecules of $\mathcal{X}_2$. There are x_1x_2 possible pairs of molecules that could react so the probability of a reaction of this type occurring in a time interval of length $\mathrm{d}t$ is $cx_1x_2\mathrm{d}t$. Note that this probability depends only on the current state of the system—this is the *Markov property.* Moreover, changes to the system state occur at discrete times (say t_1, t_2, ...) and there are finite periods of no change. Taking both of these properties together gives a *Markov jump process*, that is, a continuous time, discrete valued process that satisfies the Markov property.

We now consider the general case. Under the standard assumption of *mass-action stochastic kinetics*, each reaction R_i is assumed to have an associated rate constant, c_i, and a *propensity function*, $h_i(X_t, c_i)$, which define the overall *hazard* of a type i reaction occurring. That is, the system is a Markov jump process, and for an infinitesimal time increment $\mathrm{d}t$, the probability of a type i reaction occurring in the time interval $(t, t + \mathrm{d}t]$ is $h_i(X_t, c_i)\mathrm{d}t$. Under mass-action stochastic kinetics, the hazard function is proportional to a product of binomial coefficients, with

Table 1
Example reactions and their associated hazards

Order	Reactants	Products	Hazard	Description
0	$\emptyset$	$\mathcal{X}_1$	c_1	Influx
1	$\mathcal{X}_1$	$\emptyset$	c_2X_1	Degradation
2	$\mathcal{X}_1+\mathcal{X}_2$	$\mathcal{X}_3$	$c_3X_1X_2$	Catalysation
2	$2\mathcal{X}_1$	$\mathcal{X}_2$	$c_4X_1(X_1-1)/2$	Dimerization
3	$3\mathcal{X}_1$	$\mathcal{X}_3$	$c_5X_1(X_1-1)(X_1-2)/6$	Trimerization

$$h_i(X_t, c_i) = c_i \prod_{j=1}^{u} \binom{X_{j,t}}{p_{ij}}.$$

It should be noted that this hazard function differs slightly from the standard mass action rate laws used in continuous deterministic modeling, but is consistent (up to a constant of proportionality in the rate constant) asymptotically in the high concentration limit.

2.2.1 Reaction Types

Some commonly encountered reaction types are given in Table 1 along with their associated hazards. For notational simplicity we remove the dependence of the current state on time when stating the hazards.

The zeroth order reaction may at first seem a little strange, since it appears that something is created from nothing. However, it can be useful for modeling a constant rate of production of a chemical species. First order reactions can be used to capture the spontaneous change of a molecule such as decay and dissociation. It is often desirable to write third or higher order reactions in terms of a series of reactions of order two or less. For example, the trimerization reaction in Table 1 can be written as

$$2\mathcal{X}_1 \longrightarrow \mathcal{X}_2 \quad \text{and} \quad \mathcal{X}_2 + \mathcal{X}_1 \longrightarrow \mathcal{X}_3.$$

2.3 Chemical Master Equation

The chemical master equation (CME) refers to an ODE satisfied by the transition kernel of the Markov jump process. One such ODE can be derived as follows.

Let $p(x; t)$ denote the transition kernel of the jump process, that is, the probability that there will be at time t $x = (x_1, \ldots, x_u)'$ molecules of each respective species (assuming a well-stirred spatially homogeneous volume Ω, and thermal equilibrium). Once this function is obtained, a fairly complete characterization of the state of the system at time t is apparent. Now write $p(x; t + \Delta t)$ as the sum of the probabilities of the number of ways in which the network can arrive in state x at time $t + \Delta t$. We obtain

$$p(x; t + \Delta t) = \sum_{i=1}^{v} h_i(x - S^i, c_i)P(x - S^i; t)\Delta t + \left\{1 - \sum_{i=1}^{v} h_i(x, c_i)\Delta t\right\} p(x; t), \quad (1)$$

where x is the state of the system at time t and S^i denotes the ith column of the stoichiometry matrix S. Intuitively, the term $h_i(x - S^i, c_i)p(x - S^i; t)\Delta t$ is the probability that the system is one R_i reaction removed from state x at time t and then undergoes such a reaction in $(t, t + \Delta t)$. The second quantity in Eq. 1 is the probability that the system undergoes no reactions in $(t, t + \Delta t)$. We now observe that Eq. 1 leads to the ODE

$$\frac{\mathrm{d}}{\mathrm{d}t} p(x; t) = \sum_{i=1}^{v} \left\{h_i(x - S^i, c_i)p(x - S^i; t) - h_i(x, c_i)p(x; t)\right\}. \quad (2)$$

Equation 2 is most commonly referred to as the CME and is simply Kolmogorov's forward equation for the MJP. Unfortunately, the CME is only tractable for a handful of cases. The exactly solvable cases have been summarized by McQuarrie [11]. Hence, for most systems of interest, an analysis via the CME will not be possible and then stochastic simulation techniques such as those described in the next section will present the only practical approach to gaining insight into a system's dynamics. For further details of the master equation formalism in chemical kinetics, good reviews have been given by van Kampen [10] and Wilkinson [4].

Algorithm 1. Gillespie's Direct Method

1. Set $t = 0$. Initialize the rate constants $c_1, \ldots, c_v$ and the initial molecule numbers $x_1, \ldots, x_u$.
2. Calculate $h_i(x, c_i)$, $i = 1, \ldots, v$ based on the current state, x.
3. Calculate the *combined hazard* $h_0(x, c) = \sum_{i=1}^{v} h_i(x, c_i)$.
4. Simulate the time to the next event, $t' \sim \mathrm{Exp}(h_0(x, c))$ and put $t := t + t'$.
5. Simulate the reaction index, j, as a discrete random quantity with probabilities $h_i(x, c_i)/h_0(x, c)$, $i = 1, \ldots, v$.
6. Update x according to reaction j. That is, put $x := x + S^j$.
7. Output x and t. If $t < T_{\max}$, return to step 2.

3 Exact Simulation Methods

3.1 The Gillespie Algorithm

Let $c = (c_1, c_2, \ldots, c_v)'$ and $h(X_t, c) = (h_1(X_t, c_1), h_2(X_t, c_2), \ldots, h_v(X_t, c_v))'$. Values for c and the initial system state x_0 completely specify the Markov process. Although the Markov jump process is rarely analytically tractable for interesting models, it is straightforward to forward-simulate exact realizations of this Markov process using a discrete event simulation method. This is due to the fact that if the current time and state of the system are t and X_t respectively, then the time to the next event can be shown to have an exponential distribution with rate parameter

$$h_0(X_t, c) = \sum_{i=1}^{v} h_i(X_t, c_i),$$

and the event will be a reaction of type R_i with probability $h_i(X_t, c_i)/h_0(X_t, c)$ independently of the waiting time. Forward simulation of process realizations in this way is typically referred to as *Gillespie's direct method* in the stochastic kinetics literature, after [6]. The procedure is summarized in Algorithm 1.

Note that the assumptions of mass-action kinetics as well as the one-to-one correspondence between reactions and rate constants may both be relaxed. It is also worth mentioning that there is an equivalent alternative algorithm to Gillespie's direct method known as the *first reaction method* [12], although the direct method is typically to be preferred as it is more efficient. In particular, it requires just two random numbers to be simulated per event as opposed to the first reaction method, which requires v. That said, the first reaction method can be turned into a far more efficient method, known as the Gibson–Bruck algorithm [13]. We eschew the method here in favor of further examination of Gillespie's direct method, which we can speed up with a few clever "tricks."

3.2 Speeding Up Gillespie's Direct Method

Not surprisingly, as the number of reactions and species increase, the length of time taken to perform a single iteration of the Gillespie algorithm also increases. We will examine some simple techniques for speeding up the method.

3.2.1 Hazards Update

At each iteration, we update each of the v hazards, $h_i(x, c_i)$, $i = 1, \ldots,$ v–step 2 of Algorithm 2. This requires v computations and is therefore $O(v)$. Naturally, after a single reaction has occurred, a better method is to only update the hazards that have changed. To this end, it is helpful to construct a dependency graph whose nodes represent reactions and a (directed) edge from one node to another indicates that one reaction affects the hazard of another.

3.2.2 Combined Hazard Update

At each iteration, we combine all v hazards to calculate the combined hazard

$$h_0(x, c) = \sum_{i=1}^{v} h_i(x, c_i) \, .$$

This is again $O(v)$. If we have used a dependency graph to determine which reaction hazards have changed after the last reaction occurrence, then we can calculate the combined hazard by subtracting "old" hazard values (before the single reaction occurrence) and adding updated "new" hazard values. This is likely to be less demanding than recalculating h_0 from scratch.

3.2.3 Reaction Selection

In this step we choose a reaction with probability proportional to its hazard, that is, we search for the j satisfying

$$\sum_{i=1}^{j-1} h_i(x, c_i) \; < \; U \times h_0(x, c) \; < \; \sum_{i=1}^{j} h_i(x, c_i)$$

where $U \sim U(0, 1)$. To speed up this step, we can order each h_i in terms of size. One technique is to run a few pre-simulations for a short period of time $t \ll T_{\max}$ [14]. The authors suggest reordering the hazard vector according to the relative occurrences of each reaction in the pre-simulations. Plainly, this method is not ideal as it is not clear how long to run the pre-simulations for, and the pre-simulations will be time consuming. Another method is to move h_i up one place in the hazard vector for each time reaction i is executed[15]. This swapping effectively reduces the search depth for a reaction at the next occurrence of that reaction. Note that the reordering only requires a swap of two memory addresses.

Algorithm 2. Poisson Leap Method

1. Set $t = 0$. Initialize the rate constants and the initial molecule numbers x.
2. Calculate $h_i(x, c_i)$, for $i = 1, \ldots, v$, and simulate the v-dimensional reaction vector r, with ith entry a $Po(h_i(x, c_i)\Delta t)$ random quantity.
3. Update the state according to $x := x + Sr$.
4. Update $t := t + \Delta t$.
5. Output t and x. If $t < T_{\max}$ return to step 2.

4 Approximate Simulation Methods

We have seen how to generate *exact* simulations from a stochastic kinetic model via the Gillespie algorithm and how to make the procedure efficient through the use of a few "tricks." However, if we are prepared to sacrifice the exactness of the simulation method, there is a potential for huge speed-ups.

One method is to divide up the time axis into small discrete chunks over which we approximate the underlying kinetics to allow advancement of the state from the start of one chunk to another in one step. We will work on the assumption that time intervals are small enough to assume constant reaction hazards over the interval.

4.1 Poisson and τ Leap

Consider a Markov process with a constant hazard (say α) of events occurring throughout time, so that the first event follows an exponential Exp(α) distribution. It can then be shown that the number of events, say X, in the interval $(0, t]$ follows a Poisson Po(αt) distribution. A Markov process with constant hazard is known as a (homogeneous) Poisson process.

Given this basic property of the Poisson process, we assume that the number of reactions (of a given type) occurring in a short time interval has a Poisson distribution (independently of other reaction types). We can then simulate Poisson numbers of reaction events and update the system accordingly (Algorithm 2).

The problem with the above method is that of choosing an appropriate time step Δt so that the method is fast but reasonably accurate. Clearly the smaller Δt, the more accurate, and the larger Δt, the faster. Another problem is that although one particular Δt may be good enough for one part of a simulation, it may not be appropriate for another. This motivates the idea of stepping ahead a variable amount of time τ, based on c and the current state of the system, x. This is the idea behind Gillespie's τ-leap algorithm.

The τ-leap method is an adaptation of the Poisson time step method to allow stepping ahead in time by a variable amount τ, where at each time step τ is chosen in an appropriate way in order to try and ensure a sensible trade-off between accuracy and speed. This is achieved by making τ as large (and hence fast) as possible while still satisfying some constraint designed to ensure accuracy. In this context, the accuracy is determined by the extent to which the assumption of constant hazard over the interval is appropriate. Clearly whenever any reaction occurs some of the reaction hazards change, and so an assessment needs to be made of the magnitude of change of the hazards $h_i(x, c_i)$. Essentially, the idea is to choose τ so that the (proportional) change in all of the $h_i(x, c_i)$ is small.

A *preleap* check is typically implemented as follows. We can calculate the expected new state as $x' = x + SE(r)$, where the ith element of $E(r)$ is just $h_i(x, c_i)\tau$. We can then calculate the change in hazard at this "expected" new state and see if this is acceptably small. It is suggested that the magnitude of acceptable change should be a fraction of the cumulative hazard $h_0(x, c)$, i.e.,

$$|h_i(x', c_i) - h_i(x, c_i)| \leq \epsilon h_0(x, c), \quad \forall i.$$

Gillespie provides an approximate method for calculating the largest τ satisfying this property [7]. Note that if the resulting τ is as small (or almost as small) as the expected time leap associated with an exact single reaction update, then it is preferable to do just that. Since the time to the next event is $\text{Exp}(h_0(x, c))$, which has expectation $1/h_0(x, c)$, one should prefer an exact update if the suggested τ is less than (say) $2/h_0(x, c)$. A number of refinements have been made to this basic scheme and are summarized in [16].

4.2 Chemical Langevin Equation

We have considered an approximation to the continuous time, discrete state space Markov jump process by discretizing time. It therefore seems natural to consider a continuous state space approximation, leading to the *CLE*. The CLE can be constructed in a number of more or less formal ways. In particular, it can be derived as a high concentration limit of the Markov jump process, but we will present here an informal intuitive construction, and then provide brief references to more rigorous approaches.

Consider an infinitesimal time interval, $(t, t + \mathrm{d}t]$. Over this time, the reaction hazards will remain constant almost surely. As in the previous section, we can therefore regard the occurrence of reaction events as the occurrence of events of a Poisson process with independent realizations for each reaction type. Therefore, if we write $\mathrm{d}R_t$ for the v-vector of the number of reaction events of each type in the time increment, it is clear that the elements are independent of one another and that the ith element is a $\text{Po}(h_i(X_t, c_i)\mathrm{d}t)$ random quantity. From this we have that $E(\mathrm{d}R_t) = h(X_t, c)\mathrm{d}t$ and $\text{Var}(\mathrm{d}R_t) = \text{diag}\{h(X_t, c)\}\mathrm{d}t$ and so we can write

$$\mathrm{d}R_t = h(X_t, c)\mathrm{d}t + \text{diag}\left\{\sqrt{h(X_t, c)}\right\}\mathrm{d}W_t.$$

Algorithm 3. CLE Method

1. Set $t = 0$. Initialize the rate constants and the initial molecule numbers x.
2. Calculate $h_i(x, c_i)$ and simulate the v-dimensional increment ΔW_t, with ith entry a $N(0, \Delta t)$ random quantity.
3. Update the state according to

$$x := x + S\,h(x, c)\Delta t + S\ \text{diag}\left\{\sqrt{h(X_t, c)}\right\}\Delta W_t$$

4. Update $t := t + \Delta t$
5. Output t and x. If $t < T_{\max}$ return to step 2.

This is the Itô stochastic differential equation (SDE) which has the same infinitesimal mean and variance as the true Markov jump process (where $\mathrm{d}W_t$ is the increment of a v-dimensional Brownian motion). Now since $\mathrm{d}X_t = S\,\mathrm{d}R_t$, we can immediately deduce

$$\mathrm{d}X_t = S\,h(X_t, c)\mathrm{d}t + S\,\mathrm{diag}\left\{\sqrt{h(X_t, c)}\right\}\mathrm{d}W_t \qquad (3)$$

as a SDE for the time evolution of X_t. As written, this SDE is a little unconventional, as the driving Brownian motion is of a different (typically higher) dimension than the state. This is easily remedied by noting that

$$Var(\mathrm{d}X_t) = S\,\mathrm{diag}\{h(X_t, c)\}S'\,\mathrm{d}t,$$

which immediately suggests the alternative form

$$\mathrm{d}X_t = S\,h(X_t, c)\mathrm{d}t + \sqrt{S\,\mathrm{diag}\{h(X_t, c)\}S'}\,\mathrm{d}W_t, \qquad (4)$$

where now X_t and W_t are both u-vectors. Equation 4 is the SDE most commonly referred to as the *CLE* and represents the diffusion process which most closely matches the dynamics of the associated Markov jump process. In particular, while it relaxes the assumption of discrete states, it keeps all of the stochasticity associated with the discreteness of state in its noise term. It also preserves many of the important structural properties of the Markov jump process. For example, Eq. 4 has the same conservation laws as the original stochastic kinetic model.

More formal approaches to the construction of the CLE usually revolve around the Kolmogorov forward equation for the Markov process, given by Eq. 2. A second-order Taylor approximation to this system of differential equations can be constructed and compared to the corresponding forward equation for an SDE model (known in this context as the *Fokker–Planck equation*). Matching the second-order approximation to the Fokker–Planck equation leads to the CLE (Eq. 4), as presented above; see [8, 9] for further details and [17] for a recent discussion.

4.2.1 Numerical Solution

As for ODE models, simulation typically proceeds using an approximate numerical solution, since the SDE in Eq. 4 can rarely be solved analytically. To understand the simplest such scheme, consider an arbitrary d-dimensional diffusion process satisfying

$$\mathrm{d}X_t = \mu(X_t)\mathrm{d}t + \sigma(X_t)\mathrm{d}W_t$$

where $\mu(\cdot)$ is a d-vector known as the *drift* and $\sigma^2(\cdot) = \sigma(\cdot)\sigma(\cdot)'$ is a $d \times d$ matrix known as the *diffusion coefficient*. For small time steps Δt, the increments of the process can be well approximated by using the *Euler–Maruyama* discretization

$$X_{t+\Delta t} - X_t \equiv \Delta X_t = \mu(X_t)\Delta t + \sigma(X_t)\Delta W_t\,,$$

where $\Delta W_t \sim N(0, I\Delta t)$ and I is the $d \times d$ identity matrix. A system at time t can therefore be stepped to $t + \Delta t$ via

$$(X_{t+\Delta t}|X_t = x) \sim N\big(x + \mu(x)\Delta t\,,\ \sigma(x)\sigma(x)'\Delta t\big).$$

Algorithm 3 describes the procedure for numerically integrating the CLE. Note that for simplicity, we use the form of the CLE given in Eq. 3.

As for ODEs, higher order numerical methods (such as the Milstein scheme) can be implemented for SDEs but are less widely used due to the complexity of the implementation [18].

4.3 Linear Noise Approximation

The LNA generally possesses a greater degree of numerical and analytic tractability than the CLE. For example, the LNA solution involves (numerically) integrating a set of ODEs for which standard routines, such as the `lsoda` package [19], exist. Our brief derivation follows the approach of [20] to which we refer the reader for further details.

We begin by replacing the hazard function $h(X_t, c)$ in Eq. 4 with the rescaled form $\Omega f(X_t/\Omega, c)$ where Ω is the volume of the container in which the reactions are taking place. Note that the LNA approximates the CLE increasingly well as Ω and X_t become large, that is, as the system approaches its thermodynamic limit. The CLE then becomes

$$dX_t = \Omega S f(X_t/\Omega, c)dt + \sqrt{\Omega S\,\text{diag}\{f(X_t/\Omega, c)\}S'}dW_t. \quad (5)$$

We then obtain the LNA by writing the solution X_t of the CLE as a deterministic process plus a residual stochastic process [10],

$$X_t = \Omega z_t + \sqrt{\Omega}M_t. \quad (6)$$

Algorithm 4. LNA Method 1

1. Set $t = 0$. Initialize the rate constants and the initial molecule numbers x. Set $z_0 = x/\Omega$, $m_0 = (x - \Omega z_0)/\sqrt{\Omega}$ (i.e., a vector of zeros) and V_0 as the $u \times u$ matrix, with all entries equal to zero.
2. Numerically integrate the system of ODEs satisfied by z_t, m_t, and V_t over $(t, t + \Delta t]$.
3. Update the state by drawing x from a $N(\Omega z_{t+\Delta t} + \sqrt{\Omega} m_{t+\Delta t}, \Omega V_{t+\Delta t})$ distribution.
4. Update $t := t + \Delta t$. Set $m_t = (x - \Omega z_t)/\sqrt{\Omega}$ and V_t as the $u \times u$ matrix, with all entries equal to zero.
5. Output t and x. If $t < T_{max}$ return to step 2.

Substituting into Eq. 5 gives

$$\mathrm{d}z_t + \frac{1}{\sqrt{\Omega}}\mathrm{d}M_t = Sf\left(z_t + M_t/\sqrt{\Omega}, c\right)\mathrm{d}t + \frac{1}{\sqrt{\Omega}} \times \sqrt{S\,\mathrm{diag}\{f(z_t + M_t/\sqrt{\Omega}, c)\}S'}\mathrm{d}W_t. \quad (7)$$

We then Taylor expand the rate function around z_t to give

$$f\left(z_t + M_t/\sqrt{\Omega}, c\right) = f(z_t, c) + \frac{1}{\sqrt{\Omega}}F_t M_t + O(\Omega^{-1}) \quad (8)$$

where F_t is the $v \times u$ Jacobian matrix with (i, j)th element $\partial f_i(z_t, c)/\partial Z_{j,\,t}$, and we suppress the dependence of F_t on z_t and c for simplicity. Substituting Eq. 8 into Eq. 7 and collecting terms of $O(1)$ give the MRE

$$\frac{\mathrm{d}z_t}{\mathrm{d}t} = Sf(z_t, c). \quad (9)$$

Collecting terms of $O(1/\sqrt{\Omega})$ gives the SDE satisfied by the residual process as

$$\mathrm{d}M_t = S\,F_t M_t \mathrm{d}t + \sqrt{S\,\mathrm{diag}\{f(z_t, c)\}S'}\mathrm{d}W_t. \quad (10)$$

Equations 6, 9, and 10 give the LNA of the CLE and therefore of the Markov jump process model.

4.3.1 Solution of the LNA

For fixed or Gaussian initial conditions, that is $M_{t1} \sim N(m_{t0}, V_{t0})$, the SDE in Eq. 10 can be solved explicitly to give

$$(M_t|c) \sim N(m_t, V_t)$$

where m_t is the solution to the deterministic ODE

$$\frac{\mathrm{d}m_t}{\mathrm{d}t} = S\,F_t m_t$$

Algorithm 5. LNA Method 2

1. Set $t = 0$. Initialize the rate constants and the initial molecule numbers x. Set $z_0 = x/\Omega$, and V_0 as the $u \times u$ matrix, with all entries equal to zero.
2. Numerically integrate the system of ODEs satisfied by z_t and V_t over $(t, t + \Delta t]$.
3. Update the state by drawing x from a $N(\Omega z_{t+\Delta t}, \Omega V_{t+\Delta t})$ distribution.
4. Update $t := t + \Delta t$. Set $z_t = x/\Omega$ and V_t as the $u \times u$ matrix, with all entries equal to zero.
5. Output t and x. If $t < T_{\max}$ return to step 2.

and similarly

$$\frac{\mathrm{d}V_t}{\mathrm{d}t} = V_t F_t' S' + S\,\mathrm{diag}\{h(z_t)\}S' + S\,F_t\,V_t\,.$$

Note that we have dropped the dependence of both m_t and V_t on z_t and c to simplify the notation. Hence, the solution of the SDE in Eq. 10 requires the solution of a system of coupled ODEs; in the absence of an analytic solution to these equations, a numerical solver such as that described in [19] can be used. The approximating distribution of X_t can then be found as

$$X_t \sim N\left(\Omega z_t + \sqrt{\Omega} m_t\,,\,\Omega V_t\right).$$

A realization of X_t can then be obtained at discrete times via Algorithm 4. With this approach, the ODE satisfied by z_t is essentially numerically integrated over the entire time horizon of interest. Hence, the accuracy of the LNA applied in this way (relative to the MJP) can become quite poor due to the difference between z_t and the true stochastic solution. An approach advocated by Fearnhead et al. [21] to alleviate this problem is to restart z_t at each simulation time using the value of x_t. Hence, the system of ODEs satisfied by z_t and V_t are (numerically) solved over each interval $[t, t + \Delta t]$ with $z_t = x_t$ and V_t as a $u \times u$ matrix, with all entries equal to zero. Note that m_t is zero for all t and therefore the ODE satisfied by m_t need not be solved. Full details can be found in Algorithm 5. Further discussion of the LNA including details of the LNA solution can be found in [10, 20, 22, 23].

4.4 Hybrid Simulation Strategies

While the CLE and LNA approaches represent a computationally efficient alternative to exact simulation approaches such as the Gillespie algorithm, biochemical reactions describing processes such as gene regulation can involve very low concentrations of reactants [24] and ignoring the inherent discreteness in low copy number data traces is clearly unsatisfactory. The aim of a hybrid simulation strategy is to exploit the computational efficiency of methods such as the CLE and LNA while accurately describing the dynamics of low copy number species, thereby bridging the gap between exact and approximate algorithms. Hybrid simulation strategies for discrete-continuous stochastic kinetic models are reasonably well developed and involve partitioning reactions as fast or slow based on the likely number of occurrences of each reaction over a given time interval and the effect of each reaction on the number of reactants and products. Fast reaction events are then modeled as continuous (using for example the CLE), and the remaining slow reaction events are updated with an exact procedure. A generic hybrid procedure is given in Algorithm 6.

Algorithm 6. Generic hybrid algorithm

1. Set $t = 0$. Initialize the rate constants and the initial molecule numbers x.
2. Classify reactions as fast or slow based on x.
3. Update fast reaction dynamics over $(t, t + \Delta t]$.
4. Based on the fast reaction events over $(t, t + \Delta t]$, determine if a slow reaction has occurred.
5. If no slow reactions have occurred, update x based on the fast reactions only. Set $t := t + \Delta t$ and go to step 7.
6. If (at least) one slow reaction has occurred, identify the time τ and type of the first slow reaction and update the state x to time τ. Set $t := \tau$.
7. Output t and x. If $t < T_{\max}$ return to step 2.

The CLE is used by Salis and Kaznessis [25] to model fast reaction dynamics while modeling slow reaction events with a Markov jump process. Since the slow reaction hazards will necessarily be time-dependent, the time-dependent probability density of the "next reaction" algorithm is used to compute the times of the slow reaction events. Discrete/CLE simulation strategies in the context of a simple gene regulatory system have been considered by Higham et al. [26], while Kiehl et al. [27] and Alfonsi et al. [28] consider discrete/ODE approaches.

5 Example: Lotka–Volterra

As an example, we consider a Lotka–Volterra model of predator and prey interaction consisting of three reactions and two species, developed by Lotka [29] and Volterra [30]. The reaction list is given in Table 2.

Table 2
Reaction list and hazards for the Lotka–Volterra system

Label	Reaction	Hazard	Description
R_1	$\mathcal{X}_1 \xrightarrow{c_1} 2\mathcal{X}_1$	$c_1 X_1$	Prey reproduction
R_2	$\mathcal{X}_1 + \mathcal{X}_2 \xrightarrow{c_2} 2\mathcal{X}_2$	$c_2 X_1 X_2$	Prey death, predator reproduction
R_3	$\mathcal{X}_2 \xrightarrow{c_3} \emptyset$	$c_3 X_2$	Predator death

Although strictly speaking $\mathcal{X}_1$ and $\mathcal{X}_2$ represent animal species, they could equally well be chemical species. In addition, the system is sufficiently complex to explore the autoregulatory behavior that is typical of many biochemical network models.

We aim to investigate the system dynamics through stochastic simulation. We therefore require key ingredients such as the stoichiometry matrix, which is

$$S = \begin{pmatrix} 1 & -1 & 0 \\ 0 & 1 & -1 \end{pmatrix}$$

and the vector of hazards, given by

$$h(X_t, c) = \left(c_1 X_{1,t}, c_2 X_{1,t} X_{2,t}, c_3 X_{2,t}\right)'.$$

The CLE is characterized by the drift and diffusion functions of the SDE in Eq. 4. We obtain

$$S\,h(X_t, c) = \begin{pmatrix} c_1 X_{1,t} - c_2 X_{1,t} X_{2,t} \\ c_2 X_{1,t} X_{2,t} - c_3 X_{2,t} \end{pmatrix}$$

and

$$S\,\mathrm{diag}\{h(X_t, c)\}S' = \begin{pmatrix} c_1 X_{1,t} + c_2 X_{1,t} X_{2,t} & -c_2 X_{1,t} X_{2,t} \\ -c_2 X_{1,t} X_{2,t} & c_2 X_{1,t} X_{2,t} + c_3 X_{2,t} \end{pmatrix}.$$

To compute the LNA, we require $f(z_t, c)$ and the Jacobian matrix F_t. For simplicity, we take a fixed volume of $\Omega = 1$ (and note that for $\Omega \neq 1$, the hazard of R_2 should be $c_2 \Omega^{-1} X_1 X_2$ to scale appropriately with volume). We therefore obtain $f(z_t, c) = h(z_t, c)$ and

$$F_t = \begin{pmatrix} c_1 & 0 \\ c_2 z_{2,t} & c_2 z_{1,t} \\ 0 & c_3 \end{pmatrix}.$$

All simulations used initial conditions of $x_0 = (100, 100)'$ and rate constants $c = (0.5, 0.0025, 0.3)'$ as used in [31].

Figure 1 shows a single stochastic realization of the Lotka–Volterra system generated by Gillespie's direct method. For comparison, the deterministic MRE solution is shown. Note that with the stochastic solution, predator levels will eventually reach zero and the predator population will become extinct. The MRE solution, on the other hand, is a perfectly repeating oscillation, carrying on indefinitely. It should be clear that for this system, the stochastic mean and the deterministic solution do not coincide.

Figure 2 shows the median, inter quartile range, upper and lower 2.5 percentiles for the prey population, using Gillespie's direct method, the CLE and both LNA approaches. The difference between the two LNA approaches is clear. Application of the LNA

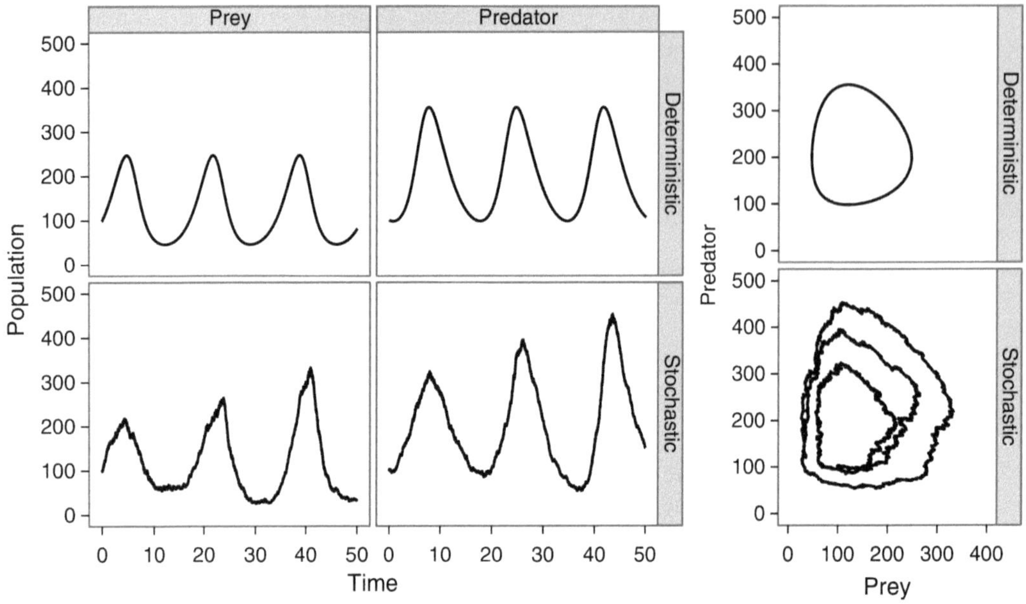

Fig. 1 A single stochastic realization of the Lotka–Volterra system using Gillespie's direct method and the deterministic solution

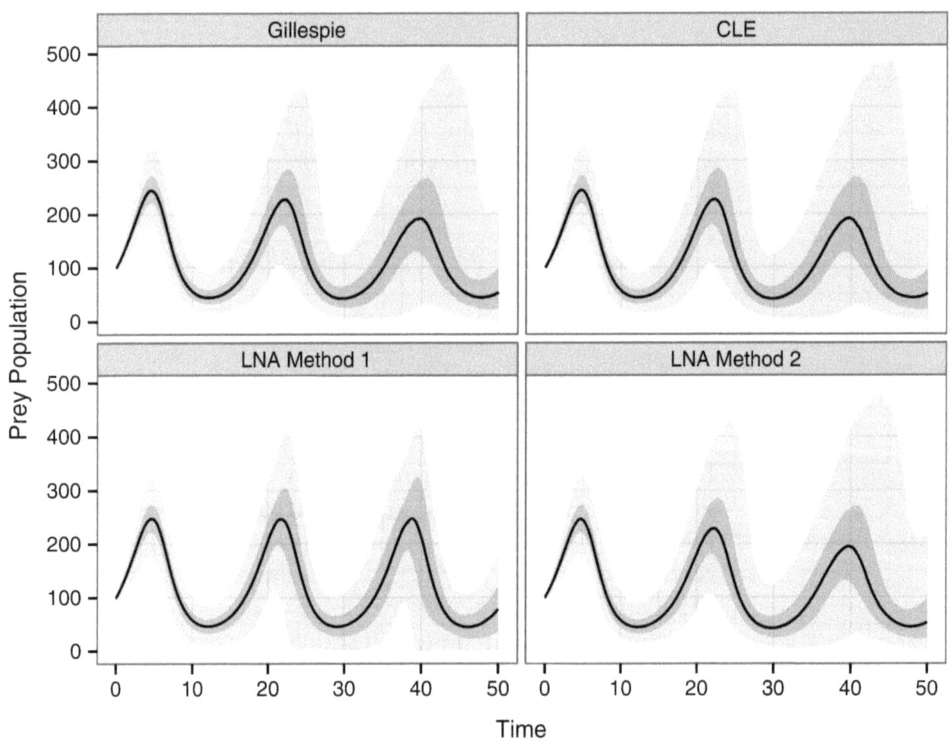

Fig. 2 Median (*solid*), inter-quartile range (*inner shaded region*), upper and lower 2.5 percentiles (*outer shaded region*) based on 10^4 stochastic realizations of the Lotka–Volterra system using (**a**) Gillespie's direct method, (**b**) the CLE with $\Delta t = 0.01$, (**c**) the LNA (Algorithm 4), and (**d**) the LNA (Algorithm 5)

driven by a deterministic solution over the whole time-course of interest leads to a mismatch between the LNA solution and MJP solution. Restarting the deterministic solution at each simulation time, at the simulated value, alleviates this problem.

6 Discussion

Stochastic chemical kinetic theory provides a framework for model building that leads to a Markov jump process model from a simple list of biochemical reactions. Gillespie's direct method provides a straightforward way of simulating such processes on a computer. The algorithm can potentially be computationally intensive and therefore techniques that aim to reduce this cost (such as those considered in Subheading 3.2) can be of benefit. For systems involving many reaction channels and species, the computational cost of an efficient implementation of the Gillespie algorithm may still preclude statistical analysis. The importance of approximate algorithms such as the CLE and LNA is then clear.

The stochastic simulation methods examined here are by no means exhaustive, and indeed there is a vast literature in this area. For example, we can derive moment equations from the CME to obtain fast approximations to the stochastic mean and variance of the system [32–34]. Alternatively, we can take advantage of multi-core processors; models can be partitioned into smaller subsystems and simulated independently [35].

Computing Details

All simulations were performed on a machine with 4 GB of RAM and with an Intel quad-core CPU. The simulation code for the Lotka–Volterra model was written in R [36]. The graphics were created using the ggplot2 R package [37]. The R code used in this chapter can be downloaded from the github repository:

https://github.com/csgillespie/In-silico-Systems-Biology

It was worth noting that using R is useful when developing algorithms that have relatively simple behavior; for larger models, this method does not scale well. Instead, simulators written in compiled languages such as C/C++ or Java are preferred; particularly if they can input/export SBML.

References

1. de Jong H (2002) Modeling and simulation of genetic regulatory systems: a literature review. J Comput Biol 9(1):67–103
2. Finkenstadt B, Heron E, Komorowski M, Edwards K, Tang S, Harper C, Davis J, White M, Millar A, Rand D (2008) Reconstruction of

transcriptional dynamics from gene reporter data using differential equations. Bioinformatics 24(24):2901
3. Calderhead B, Girolami M (2011) Statistical analysis of nonlinear dynamical systems using geometric sampling methods. Interface Focus 1(6):821–835
4. Wilkinson DJ (2009) Stochastic modelling for quantitative description of heterogeneous biological systems. Nat Rev Genet 10:122–133
5. Swain PS, Elowitz MB, Siggia ED (2002) Intrinsic and extrinsic contributions to stochasticity in gene expression. Proc Natl Acad Sci USA 99(20):12795–12800
6. Gillespie DT (1977) Exact stochastic simulation of coupled chemical reactions. J Phys Chem 81:2340–2361
7. Gillespie DT (2001) Approximate accelerated stochastic simulation of chemically reacting systems. J Chem Phys 115(4):1716–1732
8. Gillespie DT (1992) A rigorous derivation of the chemical master equation. Physica A 188:404–425
9. Gillespie DT (2000) The chemical Langevin equation. J Chem Phys 113(1):297–306
10. van Kampen NG (2001) Stochastic processes in hysics and chemistry. North-Holland, Amsterdam
11. McQuarrie DA (1967) Stochastic approach to chemical kinetics. J Appl Prob 4:413–478
12. Gillespie DT (1976) A general method for numerically simulating the stochastic time evolution of coupled chemical reactions. J Comput Phys 22:403–434
13. Gibson MA, Bruck J (2000) Efficient exact stochastic simulation of chemical systems with many species and many channels. J Phys Chem A 104(9):1876–1889
14. Cao Y, Li H, Petzold L (2004) Efficient formulation of the stochastic simulation algorithm for chemically reacting system. J Chem Phys 121(9):4059–4067
15. McCollum JM, Peterson GD, Simpson ML, Samatova NF (2006) The sorting direct method for stochastic simulation of biochemical systems with varying reaction execution behavior. J Comput Biol Chem 30(1):39–49
16. Sandmann W (2009) Streamlined formulation of adaptive explicit-implicit tau-leaping with automatic tau selection. In: Proceedings of the winter simulation conference (WSC) 2009, IEEE, pp 1104–1112
17. Golightly A, Wilkinson DJ (2011) Bayesian parameter inference for stochastic biochemical network models using particle Markov chain Monte Carlo. Interface Focus 1(6):807–820
18. Kloeden PE, Platen E (1992) Numerical solution of stochastic differential equations. Springer, New York
19. Petzold L (1983) Automatic selection of methods for solving stiff and non-stiff systems of ordinary differential equations. SIAM J Sci Stat Comput 4(1):136–148
20. Wilkinson DJ (2012) Stochastic modelling for systems biology, 2nd edn. Chapman & Hall/ CRC Press, Boca Raton
21. Fearnhead P, Giagos V, Sherlock C (2012) Inference for reaction networks using the Linear Noise Approximation. Available from http://arxiv.org/pdf/1205.6920
22. Elf J, Ehrenberg M (2003) Fast evaluation of fluctuations in biochemical networks with a linear noise approximation. Genome Res 13(11):2475–2484
23. Komorowski M, Finkenstadt B, Harper C, Rand D (2009) Bayesian inference of biochemical kinetic parameters using the linear noise approximation. BMC Bioinformatics 10(1):343
24. Guptasarma P (1995) Does replication-induced transcription regulate synthesis of the myriad low copy number proteins of *Escherichia coli*? BioEssays 17:987–997
25. Salis H, Kaznessis Y (2005) Accurate hybrid stochastic simulation of a system of coupled chemical or biochemical reactions. J Chem Phys 122:054103
26. Higham D, Intep S, Mao X, Szpruch L (2011) Hybrid simulation of autoregulation within transcription and translation. BIT Numer Math 51:177–196
27. Kiehl TR, Matteyses RM, Simmons MK (2004) Hybrid simulation of cellular behavior. Bioinformatics 20(3):316–322
28. Alfonsi A, Cances E, Turinici G, Ventura B, Huisinga W (2005) Adaptive simulation of hybrid stochastic and deterministic models for biochemical systems. ESAIM Proc 14:1–13
29. Lotka AJ (1925) Elements of physical biology. Williams and Wilkens, Baltimore
30. Volterra V (1926) Fluctuations in the abundance of a species considered mathematically. Nature 118:558–60
31. Boys RJ, Wilkinson DJ, Kirkwood TBL (2008) Bayesian inference for a discretely observed stochastic kinetic model. Stat Comput 18:125–135
32. Gillespie CS (2009) Moment-closure approximations for mass-action models. IET Syst Biol 3(1):52–58
33. Gómez-Uribe CA, Verghese GC (2007) Mass fluctuation kinetics: capturing stochastic effects in systems of chemical reactions through

coupled mean-variance computations. J Chem Phys 126(2):024109
34. Krishnarajah I, Cook AR, Marion G, Gibson G (2005) Novel moment closure approximations in stochastic epidemics. Bull Math Biol 67 (4):855–873
35. Gillespie CS (2012) Stochastic simulation of chemically reacting systems using multi-core processors. J Chem Phys 136(1):014101
36. R Development Core Team (2012) R: a language and environment for statistical computing. R Foundation for Statistical Computing, Vienna. ISBN 3-900051-07-0
37. Wickham H (2009) ggplot2: elegant graphics for data analysis. Springer, New York

Chapter 10

BioModels Database: A Repository of Mathematical Models of Biological Processes

Vijayalakshmi Chelliah, Camille Laibe, and Nicolas Le Novère

Abstract

BioModels Database is a public online resource that allows storing and sharing of published, peer-reviewed quantitative, dynamic models of biological processes. The model components and behaviour are thoroughly checked to correspond the original publication and manually curated to ensure reliability. Furthermore, the model elements are annotated with terms from controlled vocabularies as well as linked to relevant external data resources. This greatly helps in model interpretation and reuse. Models are stored in SBML format, accepted in SBML and CellML formats, and are available for download in various other common formats such as BioPAX, Octave, SciLab, VCML, XPP and PDF, in addition to SBML. The reaction network diagram of the models is also available in several formats. BioModels Database features a search engine, which provides simple and more advanced searches. Features such as online simulation and creation of smaller models (submodels) from the selected model elements of a larger one are provided. BioModels Database can be accessed both *via* a web interface and programmatically *via* web services. New models are available in BioModels Database at regular releases, about every 4 months.

Key words Controlled vocabularies, model interpretation, SBML, CellML, Reaction network diagram

1 Introduction

Mathematical modelling has been used for decades to help scientists understand the mechanisms and dynamics behind the complex biological processes. As models come from different modelers with different perspectives and are encoded in different formats, their reuse is not always straightforward. Reimplementation of models from scientific publications depends on how well the models are documented. Certain missing details can prevent the models from being reused, which consequently lead to a complete loss of the knowledge provided by the papers. The aim of BioModels Database (*see* **Note 1**) [1, 2] is to overcome these limitations by the following: (1) encoding and simulating the models in standard formats, (2) providing semantic annotation by linking the model elements

Maria Victoria Schneider (ed.), *In Silico Systems Biology*, Methods in Molecular Biology, vol. 1021,
DOI 10.1007/978-1-62703-450-0_10,

to controlled vocabularies and external data resources, and (3) providing a free, centralized, publicly accessible repository for storing, searching, and retrieving curated and annotated computational models. BioModels Database is part of the BioModels.net initiative (*see* **Note 2**) [3]. The following sections elaborate on its significant features, the structure, and use of the resource.

2 General Information About the Resource

BioModels Database offers a wide range of features and serves as an efficient model sharing platform. The increase in both the number and the complexity of the models since its origin in 2005 (*see* Fig. 1) demonstrates its recognition in the field of computational systems biology.

2.1 Models Provenance

Some models are implemented from articles published in peer-reviewed scientific journals by a team of curators. However, an increasing number of models are directly submitted by the modelers themselves. In the past, some models also came from collaboration with other repositories, such as the former SBML model repository (Caltech, USA), JWS Online (*see* **Note 3**) [4] the Database of Quantitative Cellular Signaling (DOQCS) (*see* **Note 4**), and the CellML repository (*see* **Note 5**) [5].

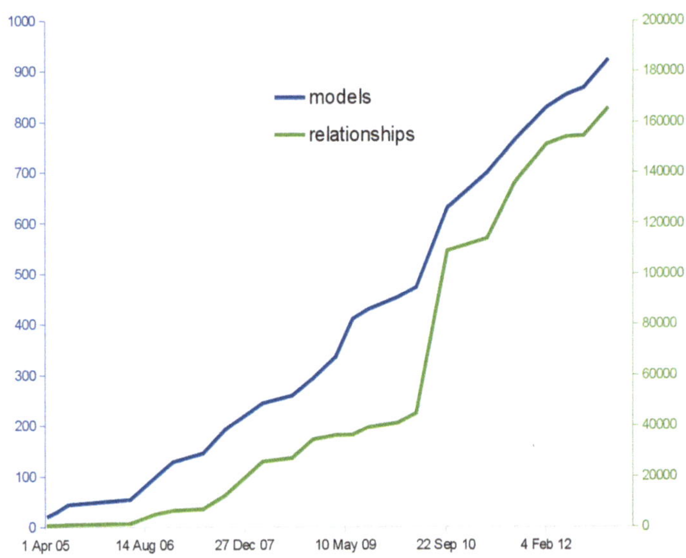

Fig. 1 Growth of BioModels Database: the total number of literature based models (*cyan*) and the total number of relationships (*yellow*) within these models are plotted here. The number of relationships includes SBML “reactions”, “rate rules”, “assignment rules”, “algebraic rules” and “events”. There has been approximately a 20-fold increase in the number of models, since the launch of the resource in 2005, with an average increase in complexity of the models (measured by the number of mathematical relationships) being increased 5 times in the same period.

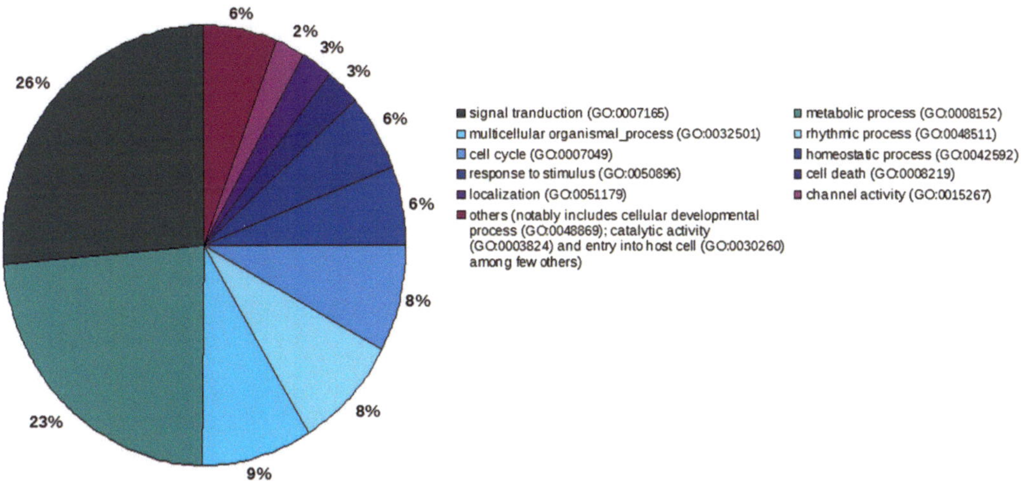

Fig. 2 Type of models: categorization of models in the curated branch of BioModels Database based on the GO terms present in the annotation of the models. This chart was generated by enumerating models in the database, whose annotations refer either to the GO terms listed here or to the children of the GO terms listed here

More than 300 scientific journals recommend submission of models to BioModels Database in their instructions for authors. These include journals from Nature Publishing Group (NPG), the Public Library of Science (PLoS), the Royal Society of Chemistry (RSC), and BioMed Central (BMC). Authors can quote the model identifier (obtained on submitting their model to BioModels Database) in their paper. This allows readers to access the model online and download it once the paper is published.

2.2 Diversity of Models Hosted

BioModels Database covers a wide range of models from several biological categories. It hosts models of simple biochemical reaction systems, larger and complex kinetic models, metabolic network models, and steady-state models. Figure 2 represents the categorization of models in the curated branch using the Gene Ontology (GO) [6] terms present in the model annotation.

2.3 Model Curation Pipeline

Models are not directly visible and retrievable by the public when they are submitted to BioModels Database. From submission to their public release, all models undergo various automated and manually performed curation and annotation procedures to ensure a consistent level of quality and accuracy. This succession of processes is depicted in Fig. 3.

2.4 Model Submission

Submission of models to BioModels Database is free and is open to everyone. Models can be submitted via an online interface and are accepted in two formats, Systems Biology Markup Language

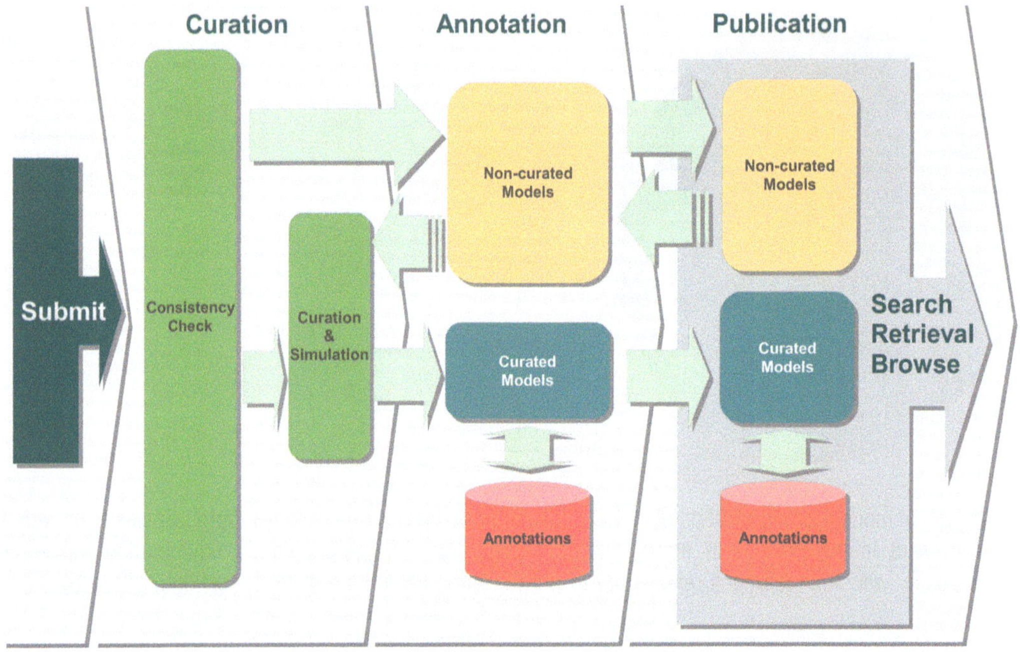

Fig. 3 BioModels Database pipeline: the figure shows the pipeline that handles each model, from its submission to its public release. This illustrates the sequence of steps involved in processing the models and encompasses both public branches of the database (curated and non-curated) as well as the possibility of curating and annotating models already published in the non-curated branch

(SBML) (*see* **Note 6**) [7] and CellML [5]. The reference of the publication describing the model can be provided (either as a PubMed Identifier (*see* **Note 7**) or a Digital Object Identifier (*see* **Note 8**) or an URL) if available at submission time.

While BioModels Database only distributes models that have been described in the peer-reviewed scientific literature, models can be submitted prior to the publication of their associated paper. However, these models are made publicly available only after the publication of the corresponding paper. At the time of submission, each model is assigned a unique and perennial identifier which allows users to access and retrieve it. This identifier can be used by authors as a reference in their publications.

2.5 Curation

The models that are submitted are queued into the curation pipeline and pass through several steps to get published. BioModels Database is composed of two branches: a curated and a non-curated one. Models in both branches are fully SBML compliant. Depending on their curation status, models are moved to one of the two branches.

Models that satisfy the Minimum Information Required in the Annotation of Models (MIRIAM) guidelines [8] progress to the curated branch. These models are thoroughly checked and corrected for accuracy as they must match and reproduce the results published in the reference publication. A representative figure or table reproduced by the model (which is present in the reference publication) together with a description of how it was obtained is available for each model. To run the simulation experiments under the conditions described in the reference publication, a tool that is different from the one that is used by the authors is used. This is to check and demonstrate that the model is far from software-specific behaviors and to avoid hidden dependencies. The tools most commonly used are COPASI [9, 10], the SBML ODE Solver [11], or the facilities provided by the Systems Biology Workbench [12]. This ensures that the encoded form of the model that is provided corresponds to what was described in the paper.

There are several reasons for models to be in the non-curated branch. These are models that either do not satisfy the full requirements for MIRIAM compliance or have not been curated yet due to limited time and resources. Many of these models are pathway maps, network models, or steady-state models (such as Flux Balance Analysis models [13]), without sufficient quantitative results provided for validation. Others are only the subsets of the whole model described in the chapter, as some parts cannot yet be encoded in SBML. And finally, a small number of models could not be made to reproduce the published results due to untraceable errors in the implementation or typos in the publication (often even after contacting the authors of the article).

3 Annotation

All model elements in the curated branch are furthermore thoroughly annotated with cross-references to other database records and ontology terms. The annotations are included in the models using MIRIAM URIs [14]. As model elements are not always named precisely to relate directly to the corresponding biological processes or physical entity, annotations are necessary to enhance interpretability by both users and software tools. To date, model elements in BioModels Database are annotated using around 40 different external resources. Some of the predominantly used external resources for model annotations are Gene Ontology, ChEBI Ontology, Brenda Tissue Ontology, Systems Biology Ontology, Taxonomy, Reactome, KEGG, and UniProt. A collection of resources and their URIs can be obtained from MIRIAM Registry (*see* **Note 9**) [15].

3.1 Model Publication

Following the curation and annotation phases, the final stage in the model processing pipeline is the publication of model. Once the curation and annotations are completed, the model is tagged as ready for publication and becomes available online during the next release of BioModels Database. New releases happen 2–4 times a year.

4 BioModels Database Web Interface

The resource provides several features to allow users to quickly search and locate their models of interest, analyze them, simulate them, extract submodels, or download them in various formats. These facilities are available *via* a web browser or can be accessed from other tools by using the accompanying web services.

4.1 Model Browsing, Searching, and Retrieval

Several features are available through the web interface to facilitate efficient usage of the models. Models can be browsed, independently in the two branches. They can also be located using a pruned Gene Ontology (GO) browser based on the model annotations.

BioModels Database incorporates a powerful search engine that allows users to retrieve models of interest. The search can either be a simple keyword search or a more advanced one. The search engine performs a sequential search by querying metadata, publication information, annotations, SBML file content and supplementary information from external data resources. The system also performs some post-processing of search results in order to deliver the most relevant results.

For example, a search with a taxonomic term traces the whole hierarchy in order to find related models. A search for the term "mammalia" returns models that are not only annotated with *Mammalia* but also with the children and parent taxonomic ranks. Indeed a biological process that happens in *Homo sapiens*, or of *Rattus norvegicus*, occurs in *Mammalia* as well. Similarly, a biological process that happens in *Metazoa* also occurs in *Mammalia* (Fig. 4).

4.2 Model Display

Each model is presented in a tabbed form, providing access to all the information stored by the database about the model. The model elements are hyperlinked between different tabs and the annotations are hyperlinked to their original resource page. The detailed description about the model is separated into six categories, namely, *Model*, *Overview*, *Math*, *Physical entities*, *Parameters*, and *Curation*, which are all accessible via dedicated tabs. This section is highlighted as area "A" in Fig. 5.

- The *Model tab* displays general information about the model such as its reference publication and general annotation to external resources. It also provides access to the submitted (original) version of the model file, as well as information about the encoders, and date and time of the model creation and last modification.

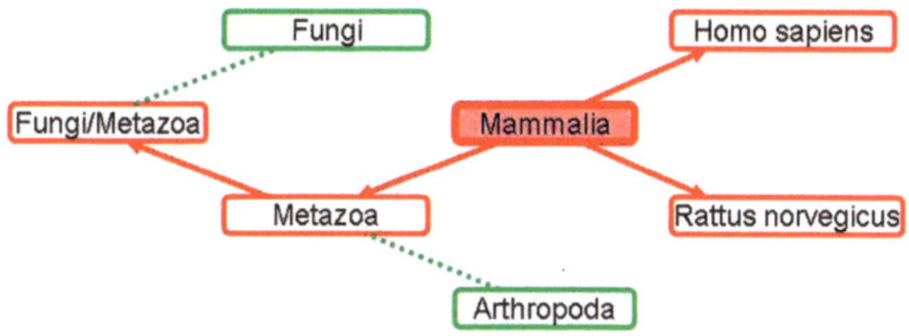

Fig. 4 Taxonomic search: the search engine considers the entire taxonomic tree, when the search is based on a taxonomic term. For example, a search based on the taxonomic term "Mammalia" returns models that are annotated with the term *Mammalia* and related ones, such as *Metazoa*, *Homo sapiens*, or *Rattus norvegicus* (represented as *red* in this figure)

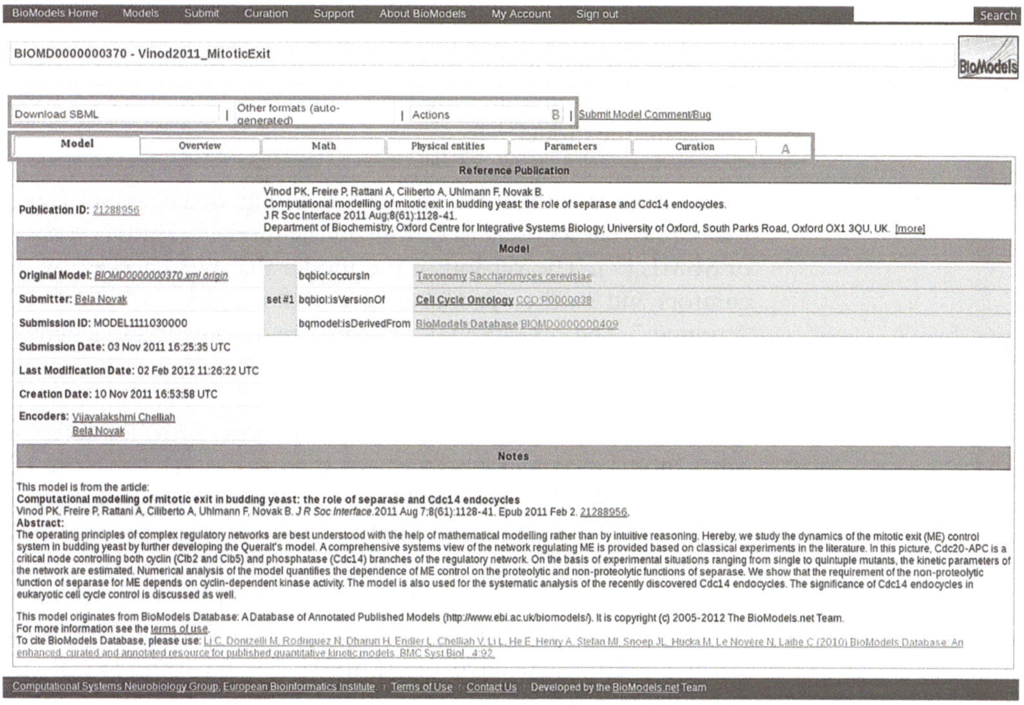

Fig. 5 Model display: this screen image shows the display of an example model in order to illustrate how the model information is displayed on the web interface

- The *Overview tab* provides access to all model components, namely, the mathematical expressions, physical entities, parameters, rules and other elements that constitute the model. Each model element is provided with a hyperlink, which leads to further details displayed on a different *tab*. Smaller models (submodels) can also be created by selecting a few model elements and followed by using the "Create a submodel with selected elements" command. This generates a valid SBML model by extracting all the selected elements and the additional elements required to have a valid model. The resulting model is displayed in the new *submodel tab* and can be downloaded in SBML.

- The *Math tab* lists all the mathematical constructs, which include reactions, rules and events. Each mathematical expression is associated with a rendering of the mathematical equation and hyperlinked annotations.
- The *Physical entity tab* lists all the entity pools and compartments included in the model, along with their initial quantities and annotations.
- The *Parameter tab* lists all the parameter used in the mathematical expressions. Parameters whose values are determined by mathematical expressions are linked to the associated model element of the *Math tab*.
- The *Curation tab* displays a representative curation result (in the form of graphical plots or tables), that matches with that of the reference publication, and which was reproduced by simulating the model.

4.3 Model Exports

This section is highlighted as area "B" in Fig. 5. The menu *Download SBML* allows users to download the model in various versions of SBML [7]. The version of the model that was checked by the curators and used to produce the curation figures is indicated as "curated." The other SBML versions are automatically generated and provided for convenience, as certain tools support only some specific levels or versions of SBML.

The menu *Other formats* (*auto-generated*) provides access to other model representation formats, such as BioPAX (*see* **Note 10**) (level 2 and level 3) [16] and the Virtual Cell Markup Language (VCML) [17]. BioModels Database also provides downloadable configuration files for open tools such as XPPAUT [18], SciLab (*see* **Note 11**), and Octave (m-file) (*see* **Note 12**). In addition, a human-readable summary of the models in the Portable Document Format (PDF), generated using SBML2LaTeX tool [19], is also available from this menu.

The menu *Actions* grants access to the graphical representations of the model's reaction networks, following the Systems Biology Graphical Notation (SBGN) [20] and available in PNG (Portable Network Graphics (*see* **Note 13**)) and SVG (Scalable Vector Graphics (*see* **Note 14**)) formats. The reaction networks are also provided in a dynamic way via an interactive Java applet.

The menu *Actions* also provide access to the online simulation tools. BioModels Database embeds SOSlib [11] to provide a basic online simulation tool. The simulation results are returned both in graphical and textual form. For many models, an additional and more flexible simulation tool is available using JWS Online [4]. Some models have an additional feature called *Model of the Month* (*see* **Note 15**) which is a brief article that discusses the biological background, significance, structure and results of the model.

The *Models of the Month* are also accessible through BMC Systems Biology Gateway (*see* **Note 16**).

4.4 Web Services

BioModels Database features programmatic access *via* web services [21] (*see* **Note 17**). They allow, for example, direct retrieval of complex searches for models and the creation of submodels. The available services are described in a Web Services Description Language (WSDL) (*see* **Note 18**) file that enables software to understand available functions and their usage. The web services use the Simple Object Access Protocol (SOAP) (*see* **Note 19**) to encode requests and responses. The complete list of available methods, as well as a Java library and the associated documentation, is provided on the BioModels Database website.

BioModels Database is developed under the GNU General Public License, and the software is freely available from its SourceForge repository (*see* **Note 20**).

4.5 Conclusions

BioModels Database has significantly grown both in size and number of the models, since its origin in 2005 and serves as an efficient model sharing platform. The submission of models by modelers/authors themselves is increasing rapidly, and this demonstrates the popularity and recognition of BioModels Database in the community. The resource helps modelers to reuse already existing models or model components, to modify them by implementing their own theory, to publish articles describing new models, and to submit those new models to BioModels Database. For example, BIOMD0000000176 and BIOMD0000000177 are derived from BIOMD0000000172 which in turn is derived from BIOMD00000000064. BioModels Database is also used as a source of trusted models for benchmarking model simulation software packages.

BioModels Database has the potential to serve as a comprehensive repository for computational systems biology models, similar to the functionality of GenBank and Protein Data Bank (PDB), the data resources for genes and protein 3D structures.

5 Notes

1. http://www.ebi.ac.uk/biomodels
2. http://biomodels.net
3. http://jjj.biochem.sun.ac.za/
4. http://doqcs.ncbs.res.in/
5. http://models.cellml.org/cellml
6. http://sbml.org/
7. http://www.pubmed.gov/

8. http://www.doi.org
9. http://www.ebi.ac.uk/miriam/
10. http://www.biopax.org/
11. http://www.scilab.org/
12. http://www.gnu.org/software/octave/
13. http://tools.ietf.org/html/rfc2083
14. http://www.w3.org/Graphics/SVG/
15. http://www.ebi.ac.uk/biomodels-main/modelmonth
16. http://www.biomedcentral.com/gateways/systemsbiology
17. http://www.ebi.ac.uk/biomodels-main/webservices
18. http://www.w3.org/TR/wsdl/
19. http://www.w3.org/TR/soap/
20. http://sourceforge.net/projects/biomodels/

References

1. Le Novère N, Bornstein B, Broicher A (2006) BioModels Database: a free, centralized database of curated, published, quantitative kinetic models of biochemical and cellular systems. Nucleic Acids Res 34:D689–D691
2. Li C, Donizelli M, Rodriguez N et al (2010) BioModels Database: an enhanced, curated and annotated resource for published quantitative kinetic models. BMC Syst Biol 4:92
3. Le Novère N (2006) Model storage, exchange and integration. BMC Neurosci 7(Suppl 1):S11
4. Snoep JL, Olivier BG (2002) Java Web Simulation (JWS); a web based database of kinetic models. Mol Biol Rep 29:259–263
5. Lloyd CM, Halstead MDB, Nielsen PF (2004) CellML: its future, present and past. Prog Biophys Mol Biol 85:433–450
6. Ashburner M, Ball CA, Blake JA et al (2000) Gene ontology: tool for the unification of biology. The Gene Ontology Consortium. Nat Genet 25:25–29
7. Hucka M, Finney A, Sauro HM (2003) The systems biology markup language (SBML): a medium for representation and exchange of biochemical network models. Bioinformatics 19:524–531
8. Le Novère N, Finney A, Hucka et al (2005) Minimum information requested in the annotation of biochemical models (MIRIAM). Nat Biotechnol 23:1509–1515
9. Hoops S, Sahle S, Gauges (2006) COPASI–a COmplex PAthway SImulator. Bioinformatics 22:3067–3074
10. Mendes P, Hoops S, Sahle S (2009) Computational modeling of biochemical networks using COPASI. Methods Mol Biol 500(17–59): 2009
11. Machné R, Finney A, Müller S (2006) The SBML ODE Solver Library: a native API for symbolic and fast numerical analysis of reaction networks. Bioinformatics 22(1406–1407): 2006
12. Bergmann FT, Sauro HM (2006) SBW—a modular framework for systems biology. pp. 37–1645.
13. Orth JD, Thiele I, Palsson BØ (2010) What is flux balance analysis? Nat Biotechnol 28:245–248
14. Juty N, Le Novère N, Laibe C (2012) Identifiers.org and MIRIAM Registry: community resources to provide persistent identification. Nucleic Acids Res 40:D580–D586
15. Laibe C, Novère NL (2007) MIRIAM Resources: tools to generate and resolve robust cross-references in systems biology. BMC Syst Biol 1:58
16. Demir E, Cary MP, Paley S et al (2010) The BioPAX community standard for pathway data sharing. Nat Biotechnol 28:935–942
17. Moraru II, Schaff JC, Slepchenko BM (2008) Virtual cell modelling and simulation software environment. IET Syst Biol 2:352–362
18. Ermentrout, B. (2002) Simulating, analyzing, and animating dynamical systems: a guide to XPPAUT for researchers and students

19. Dräger A, Planatscher H, Motsou W et al (2009) SBML2L(A)T(E)X: conversion of SBML files into human-readable reports. Bioinformatics 25:1455–1456
20. Le Novère N, Hucka M, Mi H, Moodie S et al (2009) The systems biology graphical notation. Nat Biotechnol 27:735–741
21. Li C, Courtot M, Le Novère N, Laibe C (2010) BioModels.net web services, a free and integrated toolkit for computational modelling software. Brief Bioinform 11: 270–277

Chapter 11

Supporting SBML as a Model Exchange Format in Software Applications

Sarah M. Keating and Nicolas Le Novère

Abstract

This chapter describes the Systems Biology Markup Language (SBML) from its origins. It describes the rationale behind and importance of having a common language when it comes to representing models. This chapter mentions the development of SBML and outlines the structure of an SBML model. It provides a section on libSBML, a useful application programming interface (API) library for reading, writing, manipulating and validating content expressed in the SBML format. Finally the chapter also provides a description of the SBML Toolbox which provides a means of facilitating the import and export of SBML from both MATLAB and Octave (http://www.gnu.org/software/octave/) environments.

Key words SBML, Systems biology standards, Reproducibility, Exchange format, Language

1 Introduction

Systems Biology projects take into account the interactions between a very large number of physical entities and the analysis of many parameters. As seen in the previous chapters, the quantitative relationships between entities and their interactions are often described using mathematical models and there exists a variety of software applications that can be used for different types of analysis. Early modellers and software developers in systems biology quickly realised that if their efforts were to be of benefit to the wider community it must be possible to share and reuse the models. The best way to facilitate this, and to enable concurrent use of multiple software packages with different capabilities, was to agree a common format for describing the models.

There are many ways to describe models in a standardised manner. One can use natural languages, graphical languages, sets of equations, logical relationships between elements, etc. The need for a language to exchange models became manifest at the end of the last century, with efforts starting in the field of metabolic networks [1] and physiology modelling [2]. A similar need was expressed

Maria Victoria Schneider (ed.), *In Silico Systems Biology*, Methods in Molecular Biology, vol. 1021,
DOI 10.1007/978-1-62703-450-0_11, © Springer Science+Business Media, LLC 2013

during the first Workshop on Software Platforms for Systems Biology held at the California Institute of Technology in early 2000.

The result was the Systems Biology Markup Language (SBML) [3]; a machine-readable model definition language based on XML, the eXtensible Markup Language [4].

2 SBML

An SBML document contains all the information pertaining to the structure of a model, including the list of symbols, variables and constants, the list of mathematical relationships linking them, and all the numbers needed to instantiate simulations. SBML was originally viewed as being aimed towards models of molecular pathways [5]. However, its versatility means that SBML can be, and today is being, used in a variety of modelling contexts. For instance, BioModels Database [6] contains SBML representations of models including cell signalling [7], metabolism [8], gene regulation [9] but also nonmolecular representations of cellular processes [10], models of neurons [11], treatment of tumours [12] or even of zombie invasions [13]. In general, SBML enables the encoding of any mathematical model based on pools of entities and processes that modify them. This versatility is currently expanding towards rule-based modelling, reaction–diffusion, logical modelling etc. Since its creation in 2000, SBML has continued to evolve as an international community effort, and has grown in terms of the levels of acceptance to the point where, at the time of this writing, it is used by over 200 software packages worldwide and required as a format for model encoding by many journals.

2.1 SBML Development

SBML has been, and continues to be, developed in stages, with specifications released at the end of each development stage. This approach, which effectively freezes SBML development at incremental points, allows users to work with stable standards and gain experience with the standard before further development. Future development can then benefit from the practical experiences of users and developers.

Major editions of SBML are termed *levels* and represent substantial changes to the composition and structure of the language. The latest level being developed is Level 3 [14] representing a major evolution of the language through Level 2 [15] from the introduction of Level 1 in the year 2001 [3, 16]. SBML Level 3 is being developed as a modular language, with a central core comprising a self-sufficient model definition language, and extension packages layered on top of this core to provide additional, optional sets of features.

The separate levels of SBML are intended to coexist. All of the constructs of Level 1, i.e., the elements and attributes of the SBML representation can be mapped to Level 2; likewise, the majority of the constructs from Level 2 can be mapped to Level 3 Core. In addition, a subset of Level 3 constructs can be mapped to Level 2 and a subset of Level 2 constructs can be mapped to Level 1. However, the levels remain distinct; a valid SBML Level 1 document is not a valid SBML Level 2 document and so on.

Minor revisions of SBML are termed *versions*, and constitute changes within a Level to correct, adjust and refine language features.

2.2 Structure of SBML Level 3 Core

SBML is a structured language with a strict syntax and very precise semantics. A serious understanding of the language can only be achieved through the SBML specification document [14]. In this section we will present a basic overview of the common constructs of SBML. The later sections look at code designed to intuitively work with the SBML constructs and attributes and will provide more insight into the SBML language itself.

This text restricts itself to SBML Level 3 Core and does not go into any detail relating to the L3 packages that are being developed to extend the core both in terms of SBML development and also the development of the software discussed in the later sections of the chapter.

2.2.1 The SBML Element

An SBML document is essentially an XML document containing an **sbml** element which declares the *namespace*, *level* and *version* of SBML. The **sbml** element MUST contain a **model** element which itself consists of lists of one or more components. This SBML snippet illustrates an **sbml** element containing a **model** element.

```
<?xml version="1.0" encoding="UTF-8"?>
<sbml
xmlns="http://www.sbml.org/sbml/level3/version1/
core"
      level="3" version="1">
   <model/>
</sbml>
```

2.2.2 The Model Element

Some components in SBML (see Table 1) represent items that have a numerical value that may be constant or may vary throughout a simulation. The constructs that represent possible variables are compartment, species and parameter. In all these cases, the *id* attribute of the component is used throughout the model to represent the numerical value of that component at the point in time specified by any simulation/analysis that is being undertaken.

It is also possible to introduce variable stoichiometry into reactions using the id of the speciesReference listed as a product or reactant within a reaction.

Other components represent mathematical constructs that define some level of interaction between the components that can be varied. These constructs include the reaction, rules and event components.

The remaining constructs: initialAssignment, functionDefinition, constraint and unitDefinition provide methods of adding further information or mathematical detail to a model.

Table 1
Components of an SBML Level 3 core model element

Component description	
Compartment	A container of finite size for well-stirred substances
Species	A pool of undistinguishable entities
Parameter	A quantity of whatever type is appropriate
Reaction	A statement describing some transformation, transport or binding process that can change one or more species
Rule	A mathematical expression that is added to the model equations
Event	A set of mathematical formulas evaluated at specified moments in the time evolution of the system
Initial assignment	A mathematical formula to assign the initial value of a component
Function definition	A named mathematical function that can be used in place of repeated expressions
Constraint	A mathematical formula for stating the assumptions under which the model is designed to operate
Unit definition	A name for a unit used in the expression of quantities in a model

2.2.3 Elements Providing Variables

In Subheading 2.2.2 it was noted that the compartment, species and parameter components of the **model** element represent items that have a numerical value that may be varied in the simulation of a model. The SBML specification assigns attribute values to these components that allow the user to specify initial values, units and whether the particular instance of a component can be varied by other constructs within the model. Some attributes are required and others have default values in some SBML Levels. Full details are available in the SBML specifications [14].

As a quick illustration consider a model that contains two species and a parameter. Since species must be located within a compartment, it will also contain a compartment. The **compartment** is a constant, 3D container of volume 2.3 L. Within this compartment are two **species**. There is also a fixed **parameter** with value 3000 per second. It is possible to define this unit in SBML but that is left as an exercise for the reader.

The **species** component in SBML does not represent a single molecule but rather a pool, that is an ensemble of indistinguishable entities, represented by its concentration or amount in a **compartment**. The environment is well stirred and thus no concentration gradients need to be considered. The first species has an initial amount of 4.6 mol and the second an initial amount of 1 mol. These are reacting and therefore the amounts will vary throughout a simulation.

An SBML model specifying these components is shown below.

```
<model>
   <listOfCompartments>
      <compartment id="cell" spatialDimensions="3"
            size="2.3" units="litre"
          constant="true"/>
   </listOfCompartments>
   <listOfSpecies>
      <species id="s1" compartment="cell"
            initialAmount="4.6"
            substanceUnits="mole"
          hasOnlySubstanceUnits="false"
            boundaryCondition="false"
          constant="false"/>
      <species id="s2" compartment="cell"
            initialAmount="1"
            substanceUnits="mole"
          hasOnlySubstanceUnits="false"
            boundaryCondition="false"
          constant="false"/>
      <species id="s2" compartment="cell"
            initialAmount="1"
            substanceUnits="mole"
          hasOnlySubstanceUnits="false"
            boundaryCondition="false"
          constant="false"/>
      </listOfSpecies>
      <listOfParameters>
          <parameter id="p" value="3000"
    constant="true"/>
   <listOfParameters>
</model>
```

The species specified above have initial amounts specified in moles. However, the *hasOnlySubstanceUnits* attribute has a value of false, indicating that whenever the *id* of the species appears in the model it refers to concentration.

Thus for any analysis, it may be necessary to convert between amount and concentration using

$$\text{Concentration} = \frac{\text{Amount}}{\text{Size}}$$

where size refers to the *size* of the **compartment** in which the **species** is located.

It is possible to create models in SBML without the need to consider units and thus units have largely been ignored within this text. However, in the situation where a model uses species that have been located within a compartment whose size is not unity the issue of concentration and amount **must** be considered.

2.2.4 Elements Providing Relationships

There are several elements in SBML that allow the user to define relationships between variables within the model.

We will use a **reaction** to illustrate this, but there are other constructs (listed in Table 1) that can also be used.

A **reaction** in SBML represents any kind of process that can change the quantity of one of more **species**. It may be a mass action reaction, or involve transport, catalysis or any process that changes the species involved (note that transport changes species because they are located in compartments). It is necessary to define the participating reactants and/or products. This is done using a **speciesReference** component that identifies the species from the model's **listOfSpecies** and assigns a *stoichiometry* value to that species role within the reaction. Species that merely influence a reaction, such as a catalyst, are listed as objects of type **modifier-SpeciesReference**. This construct is similar to **speciesReference** without the *stoichiometry* attribute. Attributes for a **reaction** object allow the modeller to specify whether the reaction is *reversible* or *fast*. The mathematics describing the velocity of the reaction can be encoded in the **kineticLaw** component. SBML uses a subset of the MathML 2.0 standard [17] to encode mathematical formula directly within SBML components. Note that an SBML **kinetic-Law** represents the extent of the reaction per unit of time, and not the rate of the reaction. In other words, the result is not a concentration per time, but a quantity per time. This is why the rate is multiplied by the volume in the **kineticLaw**.

The SBML snippet shows the description of the reaction

$$s1 \xrightarrow{k} s2$$

with a rate of

$$p \times s1$$

where *s*1 and *s*2 are two species residing in compartment *cell* and *p* is a parameter.

```
<model>
...
<listOfReactions>
<reaction reversible="false"
   fast="false">
<listOfReactants>
      <speciesReference species="s1"
              stoichiometry="1"/>
   </listOfReactants>
   <listOfProducts>
      <speciesReference species="s2"
          stoichiometry="1"/>
</listOfProducts>

<kineticLaw>
   <math xmlns="http://www.w3.org/1998/Math/MathML">
      <apply>
```

```
              <times/>
              <ci> p </ci>
              <ci> s1 </ci>
              <ci> cell </ci>
          </apply>
       </math>
      </kineticLaw>
     </reaction>
    </listOfReactions>
  ...
</model>
```

2.2.5 The Mathematical Model

An SBML document contains all the information pertaining to the structure of a model. However, it does not directly contain the system of mathematical equations that describes the behaviour of the model. It is therefore necessary for software using SBML to reconstruct the mathematics needed to perform the required analysis. In some cases, such as assignments, the correct equations can be extracted fairly directly from the SBML constructs.

The SBML snippet here shows two rules, an **assignmentRule** and a **rateRule** from which the equations can be easily extracted.

$$y = 2x + 1$$

$$\frac{dg}{dt} = g - 1$$

```
<model>
  ...
  <listOfRules>
    <assignmentRule variable="y">
      <math xmlns="http://www.w3.org/1998/Math/MathML">
        <apply>
          <plus/>
          <apply>
            <times/>
            <cn> 2 /cn>
            <ci> x </ci>
          </apply>
          <cn> 1 </cn>
        </apply>
      </math>
    </assignmentRule>
    <rateRule variable="g">
      <math xmlns="http://www.w3.org/1998/Math/MathML">
        <apply>
          <minus/>
          <ci> g </ci>
          <cn> 1 </cn>
        </apply>
      </math>
    </rateRule>
  </listOfRules>
  ...
</model>
```

In other cases, the process necessitates more complex procedures for instance extracting the equations from a set of reactions. As mentioned in Section/refsec:reactions one of the specificities of SBML is the representation of biological networks as a set of processes (SBML **Reactions** that are converting pools of entities (SBML **Species**) into others). Each process (SBML **Reaction**) is associated with a kineticLaw and lists the entities affected (SBML **SpeciesReferences**) characterised by their stoichiometry for this process. SBML does not contain the ODEs describing the evolution of the pools. Software must reconstruct them from the list of all reactions and associated stoichiometries. This approach, which is in direct contrast to CellML [2] where the focus is on describing the mathematical model, makes the maintenance of models easier. Adding or removing a reaction does not require carefully exploring all the equations. It also permits the use of a given computational model within different simulation frameworks, for example both deterministic or stochastic simulations can be constructed from the same set of SBML **Reactions**.

3 LibSBML

LibSBML [18] is an application programming interface (API) library for reading, writing, manipulating and validating content expressed in the SBML format. It is written in ISO C and C++, provides language bindings for .NET, Java, Python, Perl, Ruby, MATLAB and Octave, and includes many features that facilitate the adoption and use of both SBML and the library. LibSBML is freely available as source code and binaries for all major operating systems under the LGPL open source terms.

Developers can embed libSBML in their applications, saving themselves the work of implementing their own SBML parsing, manipulation and validation software.

3.1 SBML and the libSBML API

LibSBML uses objects (classes) that correspond to SBML components with member variables that represent the attribute values. The API is constructed to provide an intuitive way of relating SBML and the code needed to create or manipulate it. LibSBML has extensive documentation available both online and as a separate documentation archive available with each libSBML release. The C++ documentation is the most extensive but documentation specifically tailored for many of the other language bindings is available.

Here we provide an outline of key features of the library used in conjunction with some of the constructs of SBML that should provide a user with a basic starting point. The code examples use a combination of sample code and notes to provide further detail and should be considered as part of the text. Code snippets use python but the API is identical for all the major languages supported. It should be noted that the bindings for MATLAB and Octave do differ from the general libSBML API and are considered separately in Subheading 4.

3.2 Reading and Writing SBML

LibSBML enables reading from and writing to either files or strings. Once read in libSBML stores the SBML in an **SBMLDocument** object which can later be written out. Thus any of functions shown could be used to read and write SBML from a file or string.

```
>>> import libsbml
# read a document
>>> doc = libsbml.readSBMLFromFile(filename)
>>> doc = libsbml.readSBMLFromString(string)
# helper function that takes either a string
# or filename as argument
>>> doc = libsbml.readSBML(filename)
>>> doc = libsbml.readSBML(string)
# write a document
>>> libsbml.writeSBMLToFile(doc, filename)
>>> True
>>> libsbml.writeSBMLToString(doc)
>>> '<?xml version="1.0" encoding="UTF-8"?>\n
<sbml
xmlns="http://www.sbml.org/sbml/level3/version1/
core"
   level="3" version="1">\n
  <model/>\n
</sbml>\n'
```

The doc object produced is an instance of the **SBMLDocument** class which represents the **sbml** element.

The **SBMLDocument** contains one instance of a **Model** object which in turn contains the ListOfXYZ classes representing the SBML elements contained within the model element (see Fig. 1). The API allows the user to retrieve sub elements from a parent. Assuming that the SBML snippet of Subheading 2.2.1 has been read in, each of these ListOfXYZ objects is an empty list since the **model** element is empty.

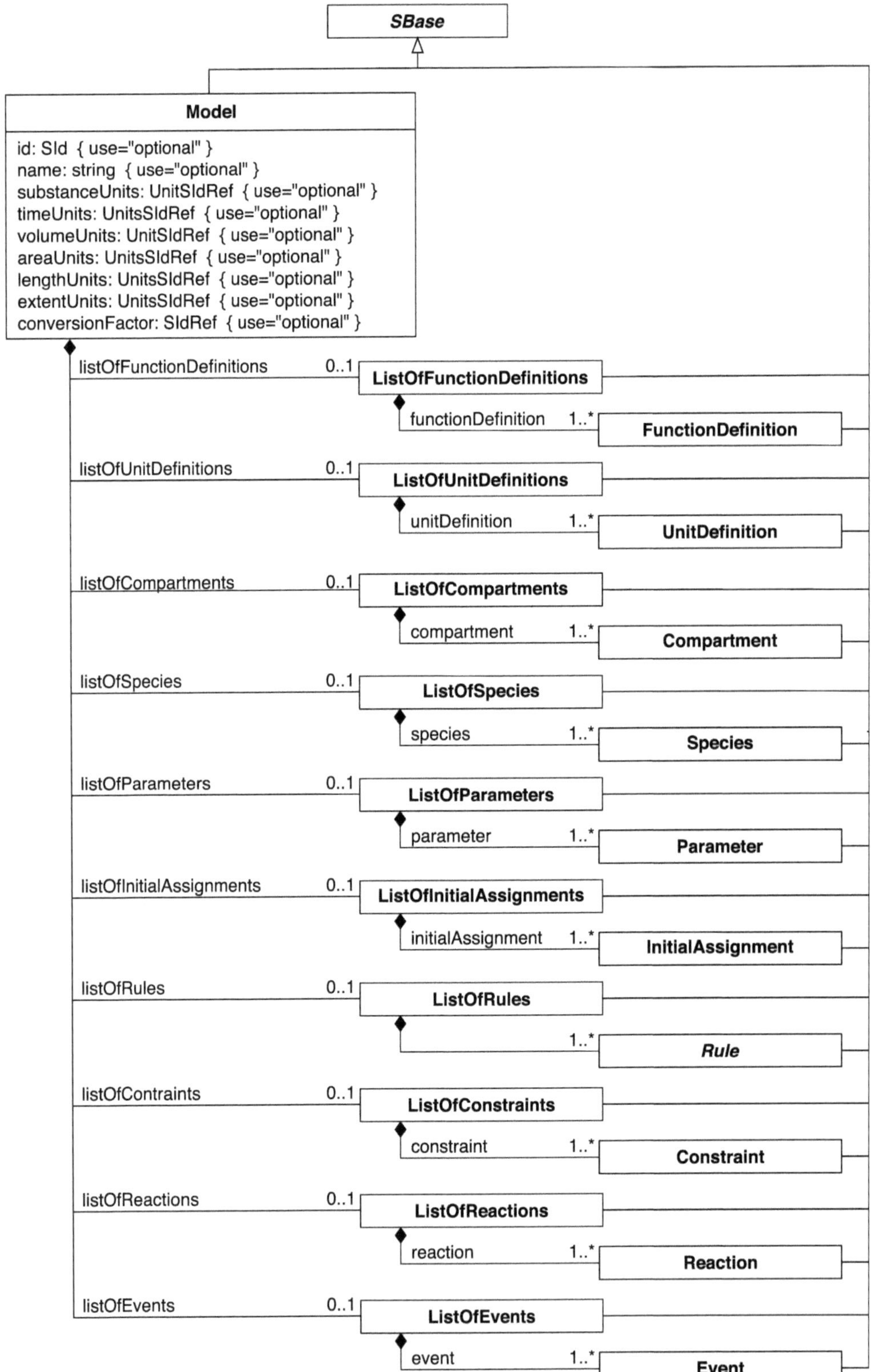

Fig. 1 This figure illustrates the structure of the **model** element. The various components are located within the relevant "ListOfXYZ" components

```
>>> import libsbml
>>> reader = libsbml.SBMLReader()
>>> doc = reader.readSBMLFromFile(filename)

>>> model = doc.getModel()

# there are no compartments in the list
>>> model.getNumCompartments()
0

# the listOfSpecies will exist but be empty
>>> model.getListOfSpecies()
<libsbml.ListOfSpecies;
  proxy of <Swig Object of type 'ListOfSpecies *'
  at 0x02D48F98> >
>>> model.getNumSpecies()
0

# individual objects (if they exist) can be retrieved
# by index or identifier
>>> p1 = model.getParameter(1)
>>> p2 = model.getParameter("p")
```

3.3 Creating and Manipulating SBML

The libSBML API allows easy creation of objects and sub objects representing SBML elements and the sub elements contained within them.

```
>>> import libsbml
# create an SBML Level 3 Version 1 document
>>> sbmlns = libsbml.SBMLNamespaces(3,1)
>>> doc = libsbml.SBMLDocument(sbmlns)

#create the model as a sub element of the document
>>> model = doc.createModel()
#create a compartment as a sub element of the model
>>> compartment = model.createCompartment()
# note the compartment is created and
# added to the listOfCompartments within the model
>>> model.getNumCompartments()
1
```

3.3.1 SBMLNamespaces

LibSBML allows you to work with all levels and versions of SBML. In order to facilitate this there is an **SBMLNamespaces** object that records the level, version and appropriate namespaces for each object. A sub object inherits the **SBMLNamespaces** from the containing document object. It is however possible to create objects prior to adding them to a document, in this case libSBML does check that there is consistency between **SBMLNamespaces** before adding objects. It is considered best practice to create a document and then create the sub elements directly from the document as in the example above.

```
<model>
  ...
  <listOfCompartments>
    <compartment id="cell" spatialDimensions="3"
                 size="2.3" units="litre" constant="true"/>
  </listOfCompartments>
  ...
</model>
```

SBML snippet 1: An SBML compartment

3.3.2 SBML Component Example: Compartment

The **compartment** component has attributes that specify its *spatialDimensions*, its *size* and the corresponding *units*, plus a *constant* attribute that determines whether the size can change or not during a simulation. The SBML snippet 1 represents a constant, 3D compartment with volume 2.3 L.

The libSBML API provides functions to set and retrieve each attribute, functions to check whether an attribute has been set and, in some cases, functions to unset an attribute. For all components and attributes these functions follow a standard form.

```
>>> import libsbml

# create an SBML Level 3 Version 1 document
>>> sbmlns = libsbml.SBMLNamespaces(3,1)
>>> doc = libsbml.SBMLDocument(sbmlns)

#create the model as a sub element of the document
>>> model = doc.createModel()

#create a compartment as a sub element of the model
>>> compartment = model.createCompartment()
# set the attributes on the compartment
# note a return value of 0 indicates success
>>> compartment.setId("cell")
0
>>> compartment.setSize(2.3)
0
>>> compartment.setSpatialDimensions(3)
0
>>> compartment.setUnits("litre")
0
>>> compartment.setConstant(True)
0

# get the attribute values
>>> compartment.getId()
```

```
'cell'
>>> compartment.getSpatialDimensions()
3

# examine the status of the attribute
>>> compartment.isSetSize()
True
>>> compartment.getSize()
2.3

#unset an attribute value
>>> compartment.unsetSize()
0
>>> compartment.isSetSize()
False
>>> compartment.getSize()
Nan
```

3.4 Validation

In addition to a strict syntax for the structure of the language the SBML specifications list a number of Validation Rules that indicate further requirements for fully valid SBML. The rules also include optional conditions for applying particular non-required types of consistency and those things recommended as best practices.

LibSBML provides a rich API for selecting and applying the various validation rules. It is possible to disable validators that deal with concepts that are not of interest to the user. For example, unit consistency is not a requirement of SBML. Software developers have differing views on whether they want units to be consistent and thus the ability to deselect unit validation is particularly useful. The following illustrates a case where turning unit validation off removes a warning from the list of validation errors reported.

```
>>> import libsbml

>>> reader = libsbml.SBMLReader()
>>> doc = reader.readSBMLFromFile(filename)

# validating the document returns the number of errors
>>> doc.validateSBML()
1
# querying the error log displays the error
>>> doc.getErrorLog().toString()
```

```
"line 9: (99505 [Warning]) In situations where a mathematical expression contains literal numbers or parameters whose units have not been declared,it is not possible to verify accurately the consistency of the units in the expression. The units of the <assignmentRule> <math> expression 'k * 2' cannot be fully checked. Unit consistency reported as either no errors or further unit errors related to this object may not be accurate.\n\n"

# turn the unit consistency checks off
>>> doc.setConsistencyChecks(libsbml.LIBSBML_CAT_UNITS_CONSISTENCY, False)

# validate the document again with unit checking off
>>> doc.validateSBML()
0
```

3.5 Working with Multiple SBML Levels/Versions

It has been noted in Subheading 2.1 that the levels and versions of SBML are intended to coexist. LibSBML provides a uniform API that seamlessly covers all SBML Levels and Versions, making it significantly easier for software developers to support the different definitions in their applications. The functions used to set attribute values will return error codes to indicate that the particular attribute or indeed the given value is not appropriate for the level and version of SBML being used. The code below provides examples of some of the error codes that may be returned.

```
>>> import libsbml
# create a L3 compartment and set spatialDimensions to
   4.5
# return code 0 indicates success
>>> compartmentL3 = libsbml.Compartment(3,1)
>>> compartmentL3.setSpatialDimensions(4.5)
0 # libsbml.LIBSBML_OPERATION_SUCCESS

# try the same with an L2 compartment
# return code -4 indicates that the value of 4.5 is not
allowed
>>> compartmentL2 = libsbml.Compartment(2, 4)
>>> compartmentL2.setSpatialDimensions(4.5)
-4 # libsbml.LIBSBML_INVALID_ATTRIBUTE_VALUE
```

```
# with L1
# return code -2 indicates that the spatialDimensions
# attribute does not exist in L1
>>> compartmentL1 = libsbml.Compartment(1,2)
>>> compartmentL1.setSpatialDimensions(4.5)
-2 # libsbml.LIBSBML_UNEXPECTED_ATTRIBUTE
```

In some cases developers prefer to target their own analysis software to one particular Level/Version of SBML.

LibSBML provides a method of converting models between levels and versions thus facilitating this approach.

It should be noted that it is not always possible to convert constructs from higher levels to lower levels. Events cannot be described in the Level 1 format. However there are some attributes, for example *sboTerm*, that provide semantic information that may not be crucial to the mathematical understanding of the model and thus converting to a lower level may remove the attribute, whilst leaving the mathematical model intact. Other constructs, such as initialAssignment and functionDefinition, can be converted. LibSBML will only perform a conversion if the set of mathematical equations that would be derived from both the source and target model are identical. The functions report success or failure; and querying the error log of the document will return any warnings about the conversion that has been done or errors that indicate why it could not proceed.

```
>>> import libsbml

>>> reader = libsbml.SBMLReader()
>>> doc = reader.readSBMLFromFile(filename)

# by default conversion will not happen if the source
model is invalid
# or the resulting model would be invalid
>>> doc.setLevelAndVersion(1,2)
False
>>> doc.getErrorLog().toString()
'line 4: (91003 [Warning]) Conversion of a model with
<constraint>s to
SBML Level 1 may result in loss of information.\n\n
line 1: (91014 [Warning]) SBML Level 2 Version 4 removed
the requirement
that all units be consistent. This model contains units
that produce
inconsistencies and thus conversion to Level 1 would
produce an invalid
model.\n\n'
```

```
# this behaviour can be overridden by changing the strict flag to false
>>> strict = False
>>> doc.setLevelAndVersion(1,2,strict)
True
```

```
<species id="Ca_calmodulin" metaid="cacam" sboTerm="SBO:0000297"
         compartment="C" hasOnlySubstanceUnits="false"
         boundaryCondition="false" constant="false">
  <annotation>
    <rdf:RDF  xmlns:rdf="http://www.w3.org/1999/02/22-rdf-syntax-ns#"
              xmlns:bqbiol="http://biomodels.net/biology-qualifiers/">
      <rdf:Description rdf:about="#cacam">
        <bqbiol:hasPart>
          <rdf:Bag>
            <rdf:li rdf:resource="http://identifiers.org/uniprot/P62158"/>
            <rdf:li rdf:resource="http://identifiers.org/obo.chebi/CHEBI%3A29108"/>
          </rdf:Bag>
        </bqbiol:hasPart>
      </rdf:Description>
    </rdf:RDF>
  </annotation>
</species>
```

SBML snippet 2: An SBML species that corresponds to a "protein complex" (SBO:0000297). It is made up of two parts, the protein calmodulin (described by the UniProt entry P62158) and the divalent calcium cation (ChEBI term CHEBI:29108)

3.6 Working with MIRIAM RDF Annotation

In addition to the model semantics, that is the variables and their mathematical relationships, SBML provides a mechanism for adding attribution information and a layer of biological semantics on top of each component of the model. This is discussed in detail in another chapter.

LibSBML provides an API for interacting with **CVTerms** and **ModelHistory** elements. The following code would produce the SBML Snippet 2.

```
>>> import libsbml

# create the species and set the metaid attribute
>>> s = libsbml.Species(3,1)
>>> s.setMetaId("cacam")

# create a new cvTerm that is the biological qualifier "hasPart"
>>> cv = libsbml.CVTerm(libsbml.BIOLOGICAL_QUALIFIER)
>>> cv.setBiologicalQualifierType("hasPart")
```

```
0

# add resources to the cvterm
>>>    cv.addResource("http://identifiers.org/uni-
prot/P62158")
0

>>>    cv.addResource("http://identifiers.org/obo.
chebi/CHEBI\%3A29108")
0

# add the cvterm to the species
>>> s.addCVTerm(cv)
0
```

3.7 A Note on JSBML

With the exception of MATLAB and Octave, the language bindings for libSBML are automatically generated.

Unfortunately this limits the platform independence brought by the use of Java because the binding is only a wrapper around the C/C++ core, implemented using the Java Native Interface (JNI). As a consequence, some software developers have experienced difficulties deploying portable libSBML-based Java applications with high concurrent usage. JSBML [19], a Java library for SBML, addresses this issue by providing an API that maps all SBML elements to a flexible and extended Java type hierarchy whilst striving for 100 % compatibility with the libSBML Java API.

4 SBMLToolbox

Many modellers use a mathematical environments such as MATLAB (http://www.mathworks.com) as this provides the computational power they need for the simulation/analysis they wish to perform. Unfortunately models constructed in such an environment do not have a proscribed structure and this makes it difficult, if not impossible, to share and reuse these models. SBMLToolbox, together with the matlab binding for libSBML, provides a means of facilitating the import and export of SBML from both the MATLAB and Octave (http://www.gnu.org/software/octave/) environments. All the functionality of the libSBML binding and SBMLToolbox is compatible with Octave.

SBMLToolbox [20] is not intended to be a fully fledged simulation tool, although some simulation capability is available. It is primarily intended to demonstrate how structures in the MATLAB environment can be used to represent models and how these structures can be compatible with SBML. Here we provide a basic overview of SBMLToolbox. Full documentation is available online and with each SBMLToolbox release.

4.1 The MATLAB SBML Structure

The term *MATLAB SBML Structure* refers to the in-memory data structure used by the libSBML bindings and SBMLToolbox to represent an SBML model in the mathematical environment. A complete model is shown below. This mimics the elements of SBML and how they are contained within each other; again see Fig. 1. In the MATLAB SBML structure the name 'listOfXYZ' has been dropped as the structure contains arrays of the individual elements but the concept of the Model object remains identical to that of SBML and the classes used in libSBML.

```
model =

              typecode: 'SBML_MODEL'
                metaid: ''
                 notes: [1x281 char]
            annotation: ''
            SBML_level: 2
          SBML_version: 1
                  name: ''
                    id: 'Branch'
    functionDefinition: [1x0 struct]
        unitDefinition: [1x0 struct]
           compartment: [1x1 struct]
               species: [1x4 struct]
             parameter: [1x0 struct]
                  rule: [1x0 struct]
              reaction: [1x3 struct]
                 event: [1x0 struct]
           time_symbol: ''
          delay_symbol: ''
            namespaces: [1x1 struct]
```

The structure is a standard MATLAB/Octave structure, with the fieldnames representing SBML attributes or arrays of SBML sub elements. These can be easily accessed using the fieldnames and by indexing into any array.

In addition to fieldnames corresponding to SBML attributes the structures may contain additional fieldnames to assist in determining whether a value has been set or whether a default value is being used. SBML Levels 1 and 2 have inbuilt default values for some attributes whilst SBML Level 3 does not. SBMLToolbox facilitates the use of all SBML Levels and Versions.

```
>> model.species(1)

ans =

                   typecode: 'SBML_SPECIES'
                     metaid: ''
                      notes: ''
                 annotation: ''
                       name: ''
                         id: 'S1'
                compartment: 'compartmentOne'
              initialAmount: NaN
       initialConcentration: 0
             substanceUnits: ''
           spatialSizeUnits: ''
      hasOnlySubstanceUnits: 0
          boundaryCondition: 0
                     charge: 0
                   constant: 0
         isSetInitialAmount: 0
  isSetInitialConcentration: 1
                isSetCharge: 0


>>model.species(1).id

ans =
   'S1'

>>model.species(1).initialConcentration
ans =
   0
```

4.2 Creating SBML

The MATLAB SBML Structures directory of SBMLToolbox contains functions that provide identical functionality to that available in libSBML. Sub elements can be created and attributes values set and queried.

```
% create a model with level 3 and version 1
>> model = Model_create(3,1)

model =

              typecode: 'SBML_MODEL'
                metaid: ''
                 notes: ''
            annotation: ''
            SBML_level: 3
          SBML_version: 1
                  name: ''
                    id: ''
               sboTerm: -1
    functionDefinition: [1x0 struct]
        unitDefinition: [1x0 struct]
           compartment: [1x0 struct]
               species: [1x0 struct]
             parameter: [1x0 struct]
     initialAssignment: [1x0 struct]
                  rule: [1x0 struct]
            constraint: [1x0 struct]
              reaction: [1x0 struct]
                 event: [1x0 struct]
        substanceUnits: ''
             timeUnits: ''
           lengthUnits: ''
             areaUnits: ''
           volumeUnits: ''
           extentUnits: ''
      conversionFactor: ''
           time_symbol: ''
          delay_symbol: ''
       avogadro_symbol: ''
            namespaces: [1x0 struct]


% create a parameter within the model

>> model = Model_createParameter(model)


model =

              typecode: 'SBML_MODEL'
                metaid: ''
                 notes: ''
            annotation: ''
            SBML_level: 3
          SBML_version: 1
                  name: ''
                    id: ''
               sboTerm: -1
```

```
    functionDefinition: [1x0 struct]
        unitDefinition: [1x0 struct]
           compartment: [1x0 struct]
               species: [1x0 struct]
             parameter: [1x1 struct]

             ...


>> Parameter_isSetId(model.parameter)


ans =

     0


>> Parameter_getUnits(model.parameter)


ans =

     ''


>> Parameter_setId(model.parameter, 'p1')


ans =

      typecode: 'SBML_PARAMETER'
        metaid: ''
         notes: ''
    annotation: ''
       sboTerm: -1
          name: ''
            id: 'p1'
         value: NaN
         units: ''
      constant: 0
    isSetValue: 0
         level: 3
       version: 1


>> Parameter_setValue(model.parameter, 3)

ans =

      typecode: 'SBML_PARAMETER'
        metaid: ''
         notes: ''
    annotation: ''
       sboTerm: -1
          name: ''
            id: 'p1'
         value: 3
         units: ''
      constant: 0
    isSetValue: 1
         level: 3
       version: 1
```

4.3 Import and Export of SBML

Import and export of SBML to and from theMATLAB SBML Structure is achieved using the libSBML binding for either MATLAB or Octave. This binding essentially consists of two functions **TranslateSBML** and **OutputSBML**.

4.3.1 TranslateSBML

This function takes a filename (or where the environment allows will browse for a file if no argument is given) and returns the MATLAB SBML structure representing the model. It will optionally perform a full validation of the model, that is, validation with all available validators enabled as described in Subheading 3.4. Any errors reported can be saved to a separate structure.

```
% include the validate flag which is 0 by default
>> [model, errors] = TranslateSBML('test.xml', 1)

model =

              typecode: 'SBML_MODEL'
                metaid: ''
                 notes: [1x281 char]
            annotation: ''
            SBML_level: 2
          SBML_version: 1
                  name: ''
                    id: 'Branch'
    functionDefinition: [1x0 struct]
        unitDefinition: [1x0 struct]
           compartment: [1x1 struct]
               species: [1x4 struct]
             parameter: [1x0 struct]
                  rule: [1x0 struct]
              reaction: [1x3 struct]
                 event: [1x0 struct]
           time_symbol: ''
          delay_symbol: ''
            namespaces: [1x1 struct]

errors =

1x6 struct array with fields:
    line
    errorId
    severity
    message

% the errors structure can be indexed into to retrieve further information
>> errors(1)

ans =

        line: 34
     errorId: 99505
    severity: 'Warning '
     message: [1x405 char]

>> errors(1).message
```

```
ans =

In situations when a mathematical expression contains literal numbers or
parameters whose units have not been declared, it is not possible to verify
accurately the consistency of the units in the expression.  The units of
the <kineticLaw> <math> expression 'k1 * X0' cannot be fully checked. Unit
consistency reported as either no errors or further unit errors related to
this object may not be accurate.
```

4.3.2 OutputSBML

The function OutputSBML is the converse of TranslateSBML: it writes the MATLAB SBML structure to an XMLfile. The structure is checked to ensure it has the all the fieldnames it expects to find. Optionally the function can be configured to restrict this check to ensuring ONLY the expected fields are present. This facilitates the use of other fields by a user

```
% start with a structure that has user fields in addition to SBML fields
model =

               typecode: 'SBML_MODEL'
                 metaid: ''
                  notes: [1x281 char]
             annotation: ''
             SBML_level: 2
           SBML_version: 1
                   name: ''
                     id: 'Branch'
     functionDefinition: [1x0 struct]
         unitDefinition: [1x0 struct]
            compartment: [1x1 struct]
                species: [1x4 struct]
              parameter: [1x0 struct]
                   rule: [1x0 struct]
               reaction: [1x3 struct]
                  event: [1x0 struct]
            time_symbol: ''
           delay_symbol: ''
             namespaces: [1x1 struct]
           anotherfield: 'foobar'

# output
# by default extensions are allowed
>> OutputSBML(model, 'out.xml')
>> Document written.

# output with no extensions allowed
>> OutputSBML(model, 'out.xml', 0)
>> Unexpected field encountered.
```

5 Conclusion

The Systems Biology Markup Language has become the de facto standard for exchanging models in the Systems Biology arena. Supporting SBML in software applications is facilitated by both libSBML and SBMLToolbox enabling developers to concentrate on the functionality of their software and avoid the need to consider parsing, creation, manipulation and validation of the SBML language itself. The variety of language bindings available reduces the need for the developer to move away from their preferred programming language and thus, whether it is for a two line script or a full blown software application, the use of SBML should be accessible to a wide variety of users from many backgrounds.

References

1. Kell DB, Mendes P (2008) The markup is the model: reasoning about systems biology models in the Semantic. Web era. J Theor Biol 252:538–543
2. Hedley W, Nelson MR, Bullivant DP, Nielsen PF (2001) A short introduction to CellML. Philos Transact A Math Phys Eng Sci 359 (5):1073–1079
3. Hucka M, Finney A, Sauro HM, Bolouri H, Doyle JC, Kitano H, Arkin AP, Bornstein BJ, Bray D, Cornish-Bowden A, Cuellar AA, Dronov S, Gilles ED, Ginkel M, Gor V, Goryanin II, Hedley WJ, Hodgman TC, Hofmeyr JH, Hunter PJ, Juty NS, Kasberger JL, Kremling A, Kummer U, Le Novère N, Loew LM, Lucio D, Mendes P, Minch E, Mjolsness ED, Nakayama Y, Nelson MR, Nielsen PF, Sakurada T, Schaff JC, Shapiro BE, Shimizu TS, Spence HD, Stelling J, Takahashi K, Tomita M, Wagner J, Wang J, SBML Forum (2003) The systems biology markup language (SBML): a medium for representation and exchange of biochemical network models. Bioinformatics 19(4):524–31
4. Bray T, Paoli J, Sperberg-McQueen CM, Maler E (2000) Extensible Markup Language (XML) 1.0 (second edition), W3C recommendation 6-October-2000. Available via the World Wide Web at http://www.w3.org/TR/1998/REC-xml-19980210.
5. Stromback L, Lambrix P (2005) Representations of molecular pathways: an evaluation of SBML, PSI ML and BioPAX. Bioinformatics 21(24):4401–4407
6. Le Novère N, Bornstein B, Broicher A, Courtot M, Donizelli M, Dharuri H, Li L, Sauro H, Schilstra M, Shapiro B, Snoep JL, Hucka M (2006) BioModels Database: a free, centralized database of curated, published, quantitative kinetic models of biochemical and cellular systems. Nucleic Acids Res 34: D689–D691
7. Goldbeter A (1991) A minimal cascase model for the mitotic oscillator involving cyclin and cdc2 kinase. Proc Natl Acad Sci U S A 88 (20):9107–9111
8. Curto R, Voit EO, Sorribas A, Cascante M (1998) Mathematical models of purine metabolism in man. Math Biosci 151(1):1–49
9. ElowitzMB LS (2000) A synthetic oscillatory network of transcriptional regulators. Nature 403(6767):335–338
10. Hodgkin AL, Huxley AF (1952) A quantitative description of membrane current and its application to conduction and excitation in nerve. J Physiol 117(5):500–544
11. Izhikevich EM (2004) Simple model of spiking neurons. IEEE Trans Neural Netw 14(6): 1569–1573
12. Tham LS, Wang L, Soo RA, Lee SC, Lee HS, Yong WP, Goh BC, Holford NH (2008) A pharmacodynamic model for the time course of tumor shrinkage by gemcitabine + carboplatin in non-small cell lung cancer patients. Clin Cancer Res 14:4213–4218
13. Munz P, Hudea I, Imad J, Smith RJ (2009) When zombies attack!: Mathematical modelling of an outbreak of zombie infection. In: Tchuenche JM, Chiyaka C (eds) Infectious disease modelling research progress. Nova Science, New York, pp 133–150
14. Hucka M, Bergmann F, Hoops S, Keating SM, Sahle S, Wilkinson DJ (2009) The Systems Biology Markup Language (SBML) language specification for level 3 version 1 core. Available

via the World Wide Web at http://www.sbml.org/Documents/Specifications.

15. Hucka M, Hoops S, Keating SM, Le Nov‘ere N, Sahle S, Wilkinson DJ (2008) Systems Biology Markup Language (SBML) level 2: Structures and facilities for model definitions. Available via the World Wide Web at http://www.sbml.org/Documents/Specifications.
16. Hucka M, Finney A, Sauro HM, Bolouri H (2001) Systems Biology Markup Language (SBML) Level 1: Structures and facilities for basic model definitions. Available via the World Wide Web at http://www.sbml.org/Documents/Specifications.
17. Ausbrooks R, Buswell S, Carlisle D, Dalmas S, Devitt S, Diaz A, Froumentin M, Hunter R, Ion P, Kohlhase M,Miner R, Poppelier N, Smith B, Soiffer N, Sutor R, Watt S (2003) Mathematical Markup Language (MathML) version 2.0 (second edition). Available via the World Wide Web at http://www.w3.org/TR/MathML2/
18. Bornstein BJ, Keating SM, Jouraku A, Hucka M (2008) LibSBML: an API library for SBML. Bioinformatics 24(6):880–881
19. Dräger A, Rodriguez N, Dumousseau M, Dörr A, Wrzodek C, Le Novère N, Zell A, Hucka M (2011) JSBML: a flexible Java library for working with SBML. Bioinformatics 27 (15):2167–2168
20. Keating SM, Bornstein BJ, Finney A, Hucka M (2006) SBMLToolbox: an SBML toolbox for MATLAB users. Bioinformatics 22(10): 1275–1277

Chapter 12

Controlled Annotations for Systems Biology

Nick Juty, Camille Laibe, and Nicolas Le Novère

Abstract

The aim of this chapter is to provide sufficient information to enable a reader, new to the subject of Systems Biology, to create and use effectively controlled annotations, using resolvable Identifiers.org Uniform Resource Identifiers (URIs). The text details the underlying requirements that have led to the development of such an identification scheme and infrastructure, the principles that underpin its syntax and the benefits derived through its use. It also places into context the relationship with other standardization efforts, how it differs from other pre-existing identification schemes, recent improvements to the system, as well as those that are planned in the future. Throughout, the reader is provided with explicit examples of use and directed to supplementary information where necessary.

1 Introduction on MIRIAM Guidelines

Typically models generated in the latter part of the twentieth century were created in isolation, usually by small groups or by individuals. They were frequently encoded in custom formats or were directly written in a programming languages, contained non-standard terminologies to describe model components, and were often simulated or processed with proprietary software applications. Together, these factors resulted in a largely unusable body of work; since models could not be shared with other groups (custom formats), it was not clear what was being modeled (nonstandard nomenclature with insufficient metadata), or the simulation results could not be repeated (software specificity or unavailability).

Over the past decade, a number of standardization efforts have risen to address these deficiencies. There are now a number of description formats, largely based on XML (eXtensible Markup Language) *see* **Note 1**, which are suitable for the representation of models. These include, for instance, Systems Biology Markup Language (SBML) *see* **Note 2**, [1]. In addition, many other formats can be converted into a standardized representation, such as SBML, through the use of community-developed software.

Maria Victoria Schneider (ed.), *In Silico Systems Biology*, Methods in Molecular Biology, vol. 1021,
DOI 10.1007/978-1-62703-450-0_12, © Springer Science+Business Media, LLC 2013

To facilitate the harmonization of terminologies used in mathematical modeling, there now exists a cornucopia of ontologies, which themselves are developed according to shared community guidelines (Open Biomedical Ontologies Foundry (*see* **Note 3,** ref. [2])). These ontologies can be used, for example, to define the roles of various model components and the mathematical equations that describe their behaviors (see Systems Biology Ontology (*see* **Note 4,** ref. [3])), or to describe the algorithms that are needed to reproduce previously demonstrated simulation results (see KiSAO, Kinetic Simulation Algorithm Ontology (*see* **Note 5,** ref. [3])).

Various communities across the biological sciences also define their own Minimum Information checklists (MIs), specifying the key information that should be included with their (experimental) data to aid in their reuse (MIBBI, Minimum Information for Biological and Biomedical Investigations) (*see* **Note 6,** ref. [4])). In the field of Systems Biology, this yielded the Minimum Information Required in the Annotation of Models (MIRIAM; *see* ref. [5])).

The MIRIAM Guidelines are a community-developed effort to define a minimal set of information that should be provided within a model. This information should be sufficient to enable a model to be reused in the manner intended by its creator and is formalized as a set of guidelines to which a model must adhere to be deemed MIRIAM-compliant.

The MIRIAM Guidelines are composed of three sections, each dealing with a different aspect of a model and the manner in which it is encoded: *reference correspondence*, *attribution annotation*, and *external resource annotation*. Briefly, the *reference correspondence* section details information relating to the file format of the model, the accuracy with which it reflects the (biological) process under consideration, and its instantiability in a simulation. The *attribution annotation* section deals with information pertaining to the model creation process, its modification, and the terms under which it can be (re)distributed. The interested reader should consult the original publication for further details regarding these components of the Guidelines [5].

External resource annotation, the final section of the MIRIAM Guidelines, describes how to formalize the relationships between model components, and information about those components that is held externally, for instance on the World Wide Web. The objective of this final section of the Guidelines is to ensure that this information, or metadata, is constructed in such a manner as to prolong its longevity and accuracy. The following section details the considerations that were made in addressing this final part of the Guidelines, providing information on "metadata" and the concepts and existing frameworks that were leveraged to address the highlighted issues. The detailed requirements to comply with this section of the Guidelines, together with examples of use, follow in the section entitled "External resource annotation."

2 Metadata and Annotation

Metadata is often, and vaguely, defined as "data about data" and may refer to some information held elsewhere, perhaps in a repository or database, that relates to or sheds light on the present subject. Annotation is the process by which, in some shorthand notation, one can provide the "reader" with or direct him to this additional information. Annotations can be thought of as supplementary information which can be used to assist in clarification or definition of data components, but are not themselves required in the processing of that data. For example, in the context of modeling, the annotations provided within a model are not necessary to run a simulation.

Annotations can take many forms, many of which are not suitable for formal use. Referred to as "uncontrolled" annotations, they may be expressed as raw text, directly copy/pasted from the information source or web page address, or simply cite an identifier from a database, presented without context. These contribute to many downstream issues such as their unsuitability for computational processing, their unreliability due to fragility and changeability of web pages, and their ambiguities, respectively.

Identifiers assigned to data sets by their providers are almost exclusively composed from a limited pool of characters (alphanumeric). It is therefore often the case that an identifier from one data set is also a legitimate and valid identifier for a completely unrelated piece of information from a different data provider. For instance, the identifier "9606" describes *Homo sapiens* in the NCBI Taxonomy (*see* **Note 7**), a species of bird (*Bombycilla cedrorum*) in the BOLD taxonomy (*see* **Note 8**), and a German article in PubMed (*see* **Note 9**).

Even when care is taken to identify data using stable and established sources, there are some rare instances, in which an identifier scheme can be superseded. Table 1 illustrates the changes implemented in, what was at the time known as "EMBL bank," but is now known as the European Nucleotide Archive (ENA) (*see* **Note 10**).

Table 1
The history of protein identification syntax by release of the European Nucleotide archive

EMBL bank release (month/year)	Protein identification
43 (06/1995)	/note="pid:g2285"
45 (12/1995)	/db_xref="PID:g2285"
58 (03/1999)	/protein_id="CAA03857.1"

Table 2
The web addresses listed all provide alternative means to access exactly the same information from the Enzyme Nomenclature (http://www.chem.qmul.ac.uk/iubmb/enzyme/)

http://www.enzyme-database.org/query.php?ec=1.1.1.1
http://www.genome.jp/dbget-bin/www_bget?ec:1.1.1.1
htttp://www.ebi.ac.uk/intenz/query?cmd=SearchEC&ec=1.1.1.1
http://enzyme.expasy.org/EC/1.1.1.1

In some cases, the change in the syntax used by data providers can be subtle, as highlighted by the change in letter-case of "pid" between release 43 and 45 (Table 1), in ENA. Though not frequent, an entire identifier scheme can itself be deprecated in favor of, or subsumed into, an alternative scheme. Such a transition is shown in release 58 above (PID). In such circumstances, data providers should provide a mapping service to such entries, lest they be lost.

Databases are often accessed through a query-able interface. The resultant web address displayed is usually linkable in that it can be copied and pasted as text. However, web addresses often specify intrinsically an adopted architecture, specify a retrieval system, or direct one to a specific resolving location. Hence, with data being mirrored in various geographical locations, the copying of simple web addresses restricts one to a specific resource. If the specified resource is "down" at the time of query, relevant information cannot be accessed. In addition, over time, some URLs may also become obsolete. Some examples of different web addresses that provide exactly the same information are shown in Table 2.

An additional complication arises from database nomenclature itself. Identifiers provided by the Universal Protein Resource (UniProt (*see* **Note 11**)) have previously been known as "SWISS-PROT," "UniProt/Swiss-Prot," "UniProtKB/Swiss-Prot," and "UNP" identifiers. While self-evident to the reader, particularly one grounded in the biological sciences, the computational processing of such names, together with the diversity of associated web addresses, can be problematic and error-prone.

Clearly the use of "raw" text or any one of a plethora of web addresses, as the basis of an annotation, makes them short-lived, fragile, and difficult to process. Consequently, since the incorporation of annotations within a model has numerous benefits, a "controlled" metadata provision methodology is required.

2.1 Controlled Annotations

Controlled annotations are those which follow a defined structure and syntax. These need to address the major issues highlighted above, namely a way to handle the nomenclature used to identify a set of data (SWISS-PROT vs. UniProtKB/Swiss-Prot), and a way to refer to a piece of data regardless of the architecture through which it is resolved, or of its geographical location (databases query mechanisms and remotely mirrored data). In the field of computer science, such mechanisms already exist: *Uniform Resource Identifiers* (URIs).

2.2 Uniform Resource Identifiers and Namespaces

A Uniform Resource Identifier (URI) (*see* **Note 12**) is a string of characters that is used to identify a resource and comes in two forms: Uniform Resource Name (URN) (*see* **Note 13**) and Uniform Resource Location (URL). Most people will be familiar with URLs; however, there is a key difference between the two which lies in the fact that a URN specifies only a name for a resource, while a URL specifies a name as well as a resolving location.

A namespace is a set of reserved strings of characters that are used to uniquely and unambiguously identify a pool of information. For example, the set of data available from the "Transport Classification Database" (*see* **Note 14**) is assigned the namespace *tcdb*.

By combining the use of a namespace with identifiers supplied by data providers in a URI, it is possible to build unique, robust, and perennial identifiers. To enable such identifiers to be used within any given community, and to ensure that they are used consistently, it is necessary to design a common syntax for encoding identifiers (URIs) and to share a list of legitimate namespaces. In our case, this list of namespaces is the MIRIAM Registry and is central in the creation of resolvable Identifiers.org URIs.

3 MIRIAM Registry

The MIRIAM Registry (*see* **Note 15**) is a product of the MIRIAM Guidelines. Having identified the need for adding metadata to model files, it was necessary to create a suitable repository of approved namespaces: MIRIAM Registry. Importantly, the way in which information is structured in this registry takes into consideration the way information is presented and distributed in the scientific domain (Fig. 1).

For the purposes of simplifying data access and information storage within the Registry, an abstraction is made of a "pool" or "set of data" of interest and is referred to as a "data collection." Each data collection is assigned a namespace, which is human-readable ("taxonomy," Fig. 1). This data collection contains a finite number of data records, each of which exists in this namespace regardless of where the data itself is located. Hence, neither data collection nor individual records are restricted by geolocation or database architecture and are thought to exist as abstract concepts. Each record is of course assigned an identifier

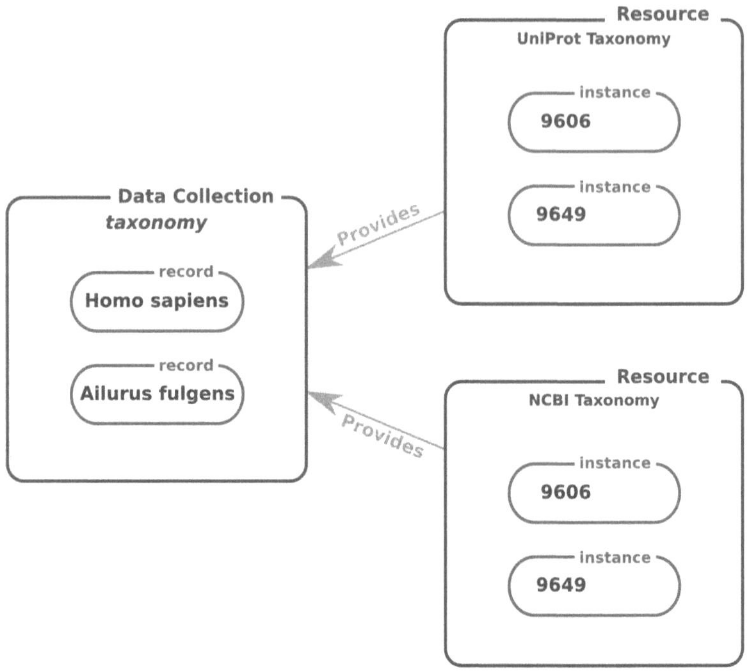

Fig. 1 Structure and nomenclature of information stored in the MIRIAM Registry

by the data providers themselves. For each data collection, there may be one or more "resources" that serve the pertaining information. These "resources," then, are the physical locations where the data itself may actually be accessed, if required. In the example shown (Fig. 1), both the "UniProt" and "NCBI" *resources* provide access to *instances* of *records* from the "Taxonomy" *data collection*.

This simple separation of the data (record) from the locations where the information can be accessed (resources) allows the building of a robust, unambiguous, and perennial identification and cross-referencing system.

The namespace information stored in the Registry can then be used to construct URIs of either URN or URL forms. This requires, besides the namespace assigned and stored in the Registry, a unique collection-specific identifier (generated by the data provider). Since the same namespace is used in both URN and URL forms, and the identifier for a particular record is fixed, it is apparent that both forms are highly related, and indeed it is possible to convert from one form to the other. A typical Registry entry is provided (Enzyme Nomenclature, Fig. 2). It should be stated that the Identifiers.org URLs (cf. below) are the preferred form of identifiers, and that the URN form is becoming largely deprecated, given all the advantages the URL form presents.

Data collection: *Enzyme Nomenclature*

Overview | Categories | Miscellaneous

Overview of the data collection

<table>
<tr><th colspan="3">Name</th></tr>
<tr><td colspan="2">Identifier</td><td>MIR:00000004</td></tr>
<tr><td colspan="2">Name</td><td>Enzyme Nomenclature</td></tr>
<tr><td colspan="2" rowspan="3">Synonyms</td><td>Enzyme Classification</td></tr>
<tr><td>EC code</td></tr>
<tr><td>EC</td></tr>
<tr><th colspan="3">Information</th></tr>
<tr><td colspan="2">Definition</td><td>The Enzyme Classification contains the recommendations of the Nomenclature Committee of the International Union of Biochemistry and Molecular Biology on the nomenclature and classification of enzyme-catalysed reactions.</td></tr>
<tr><td colspan="2">Identifier pattern</td><td>^\d+\.-\.-\.-|\d+\.\d+\.-\.-|\d+\.\d+\.\d+\.-|\d+\.\d+\.\d+\.(n)?\d+$</td></tr>
<tr><th colspan="3">URIs</th></tr>
<tr><td colspan="2">Namespace</td><td>ec-code</td></tr>
<tr><td colspan="2">Root URL</td><td>http://identifiers.org/ec-code/</td></tr>
<tr><td colspan="2">Root URN</td><td>urn:miriam:ec-code:</td></tr>
<tr><th colspan="3">Physical Locations</th></tr>
<tr><td rowspan="4">Resource
MIR:00100308</td><td>Access URL</td><td>http://www.enzyme-database.org/query.php?ec=$id [Example: 1.1.1.1]</td></tr>
<tr><td>Website</td><td>http://www.enzyme-database.org/</td></tr>
<tr><td>Description</td><td>ExploreEnz at Trinity College</td></tr>
<tr><td>Institution</td><td>Trinity College, Dublin, Ireland</td></tr>
<tr><td rowspan="4">Resource
MIR:00100002</td><td>Access URL</td><td>http://www.genome.jp/dbget-bin/www_bget?ec:$id [Example: 1.1.1.1]</td></tr>
<tr><td>Website</td><td>http://www.genome.jp/dbget-bin/www_bfind?enzyme</td></tr>
<tr><td>Description</td><td>KEGG Ligand Database for Enzyme Nomenclature</td></tr>
<tr><td>Institution</td><td>Kyoto University Bioinformatics Center, Japan</td></tr>
<tr><td rowspan="4">Resource
MIR:00100001</td><td>Access URL</td><td>http://www.ebi.ac.uk/intenz/query?cmd=SearchEC&ec=$id [Example: 1.1.1.1]</td></tr>
<tr><td>Website</td><td>http://www.ebi.ac.uk/intenz/</td></tr>
<tr><td>Description</td><td>IntEnZ (Integrated relational Enzyme database)</td></tr>
<tr><td>Institution</td><td>European Bioinformatics Institute, United Kingdom</td></tr>
<tr><td rowspan="4">Resource
MIR:00100003</td><td>Access URL</td><td>http://enzyme.expasy.org/EC/$id [Example: 1.1.1.1]</td></tr>
<tr><td>Website</td><td>http://enzyme.expasy.org/</td></tr>
<tr><td>Description</td><td>Enzyme nomenclature database, ExPASy (Expert Protein Analysis System)</td></tr>
<tr><td>Institution</td><td>Swiss Institute of Bioinformatics, Switzerland</td></tr>
<tr><th colspan="3">References</th></tr>
</table>

Fig. 2 MIRIAM Registry entry for the enzyme nomenclature data collection

A brief description of the variety of information captured for each data collection in the Registry, and its significance, is given in Table 3.

3.1 Registry Accessibility Features and Facilities

A variety of user-centric features have been provided alongside the Registry to facilitate both its efficient use and to encourage its rapid adoption. These include Web Services to allow programmatic access to the Registry [6,7], for example to validate, resolve, or create MIRIAM URIs.

Other useful features:

Collection "tags": A few tags, taken from a defined set of keywords, are associated with each data collection. They describe either the type of information recorded by the collection ("sequence," "phenotype"), the subject of that collection ("gene," "drug"), the domain area to which it relates ("disease," "neuroscience"), or the taxonomic relation of the data ("mammalian," "human"). This allows users to identify collections of interest. The refinements planned for the system are discussed elsewhere [8].

Table 3
Description of the main information components stored for data collections in the MIRIAM Registry

Information field	Description	Significance/comment
Collection name	A human-readable name to refer to the collection	Usually assigned based upon the most commonly associated resource for the collection
Collection identifier	A unique identifier for the collection within the Registry	Not intended for human readability
Collection synonyms	Other names by which the collection may identified	This field is searchable through the web interface, and query-able through web services
Collection identifier pattern	A regular expression pattern that matches all valid identifiers within the collection	Can be used through web services to validate potential identifiers
Collection namespace	The namespace assigned to the data collection	Can be used to construct URIs, and is usually an acronym based upon the collection name, or based upon the most commonly associated resource(s)
Access URLs (resources)	The physical location URL which can be used to access to a given record from the associated collection	URLs can be modified if needed by Registry curators, allowing its seamless use to the community Each URL is attached to a resource (which is also uniquely identified)
References	Reference information for the data collection	Directs to citation information, or user guides

Resource health: A "health status" has been implemented at the level of the individual resources listed in the Registry, whereby a daily health check is automatically performed for each resource. This is summarized on the data collection listing for each resource, where it is depicted by color coding of the "Resource identifier" panel. A calendar view of the uptime and further details are also available. This system is also used by the Registry curators to identify issues with resources.

Registry download: The entire contents of the Registry can be downloaded in XML format, through the "Export" link on the left panel of any MIRIAM Registry page. This is often preferred by users who would otherwise need to perform numerous queries through Web Services.

Submission of new data collections: The Registry aims to provide its services to any domain of the biological sciences. Any users wishing to submit a collection for inclusion can use the "Submit new" feature. In addition, anyone can provide suggestions for modifications/ improvements to the presented information. As a community-driven project, we welcome and encourage all such submissions.

Search facility: It is possible to search the information stored in the Registry using the provided search functionality. This search functions against all textual information stored, such as the collection name, synonyms, and collection description.

Flag system: Over the course of time, the Registry has evolved from collating only "free-to-use" collections and resources, to now accommodating those which may not be free, or have other restrictions. Of course it is useful to present information on such restrictions to the users, since it can affect their choice of collection or resource to use. This is achieved through the use of a "flag" system. Current flags include, for example, "License restriction" (which may preclude access or use for commercial purposes), and "Access restriction" (e.g., requiring registration). Data collections with restrictions are clearly labeled.

3.2 MIRIAM URIs

As stated, the namespace stored in the MIRIAM Registry can be used to construct both URN and URL forms of identifiers. While initially URNs were recommended for use in annotation, an increasing number of users expressed the desire to process these annotations in situ. For instance, given an identifier for a chemical compound in the ChEBI (*see* **Note 16**) data collection, it may be desirable to know if this model component is identical to a component in another model that was annotated with a PubChem (*see* **Note 17**) collection-based annotation. Using the URN form one would need to, for example, perform queries via web services to retrieve resolving locations (resources) for that URN, then to examine any common cross-references and descriptions contained on each target page (one for each CheBI and PubChem record). The provision of URL-based identifiers removes one step in this process, and depending on how such an information was retrieved, provides additional information in various formats (such as RDF/XML).

MIRIAM URIs are composed of four parts partitioned by a separator, "/" for URLs and ":" for URNs. The stem of the construct is constant and is composed of the definition of URI form, e.g., http:/, and the definition of the URI type, e.g., identifiers.org. The next part specifies the data collection to be identified, using the namespace recorded in the MIRIAM Registry, for example pubmed. Finally, the record identifier, which is unique and assigned by the data provider, for example 16333295, is appended to construct the full identifiers.org URL, http://identifiers.org/pubmed/16333295.

3.3 Identifiers.org URLs

Identifiers.org (http://identifiers.org, *see* ref. ([8])) is a resolving layer built upon the information stored in the MIRIAM Registry and provides resolvable identifiers. Each collection in the Registry has an associated namespace and dictates the syntactic stem that is to be used to construct both URN and URL forms of identifiers.

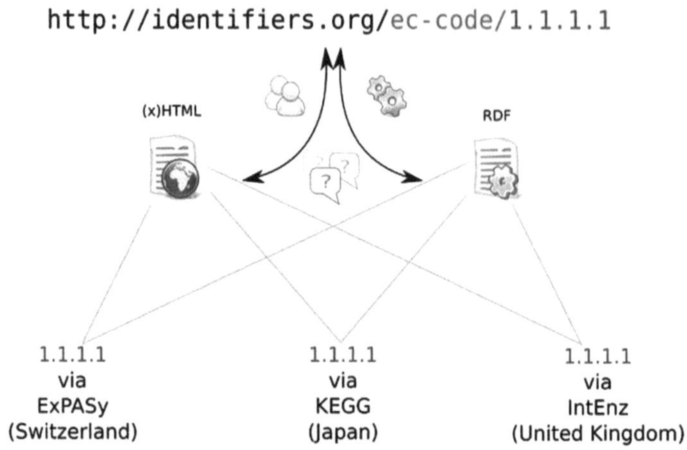

Fig. 3 Illustration of the relationship between the intermediate resolving location and physical locations associated with a specific data collection

A collection record can then be specified using the collection-specific identifier assigned by the data provider. For comparison, both URN (top) and URL (bottom) forms identifying the MIRIAM publication in the PubMed data collection are shown:

urn:miriam:pubmed:16333295

http://identifiers.org/pubmed/16333295

Since the URL above specifies a record, it may be associated with any number of resolving locations (resources). Hence, since it is preferable to provide all of them rather than preselecting a single one, the URL instead resolves to an intermediate page, where all such locations are presented to the user for selection. This behavior is depicted in the illustration below (Fig. 3).

The Identifiers.org URL form also allows various levels of customization in resolving behavior, for example allowing one to request the resolved information to be returned in a specified format, such as RDF.

3.4 Identifiers.org Granularity

Identifiers.org URLs can be used directly to access the information available in the MIRIAM Registry. The following examples illustrate how to build URLs at an appropriate level of granularity.

Identification of a data collection.

The URL below resolves to the entry for the "PubMed" collection in the MIRIAM Registry. http://identifiers.org/pubmed/: Since MIRIAM itself is a collection (of namespaces) and is listed in the MIRIAM Registry, it is also possible to reference the "PubMed" collection using the identifier for "PubMed" in the MIRIAM Registry (MIR:00000015), allowing retrieval of the same entry in the database with the synonymous URL: http://identifiers.org/miriam.collection/MIR:00000015

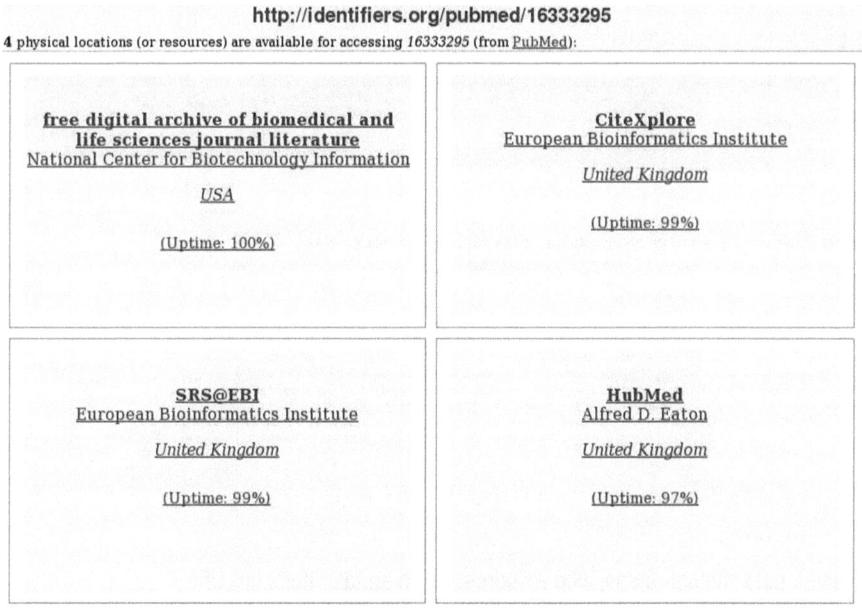

Fig. 4 Illustration of an example intermediate page which is displayed when accessing a "PubMed" data record

Identification of a record within the PubMed data collection.

The URL below resolves to an intermediate page which lists all available resolving locations listed for this collection, in the MIRIAM Registry.

http://identifiers.org/pubmed/16333295: The intermediate page corresponding to this example is shown (Fig. 4).

For convenience, listed alongside each associated resource is its name, geographical location, and its "uptime," summarized from the health check status.

Of course, a user will likely have, or develop over time, a preference for one resource over another, for whatever reason. In such instances, they may wish to directly and repeatedly resolve to that specific resource location. This can be accomplished using the resource identifier associated with each collection resource.

http://identifiers.org/pubmed/16333295?resource=MIR:00100023: In this case, the page corresponds to http://www.ncbi.nlm.nih.gov/pubmed/16333295, which is the NCBI resource location associated with the PubMed data collection (Fig. 5).

Of course, it is not convenient or likely that users will commit individual resource identifiers to memory. This issue necessitated the creation of the "profile" parameter.

3.5 Profiles

Profiles allow one to customize the behavior of the resolving system through pre-selection of the resources to be used in dereferencing Identifiers.org URLs. This means that one can define a set of

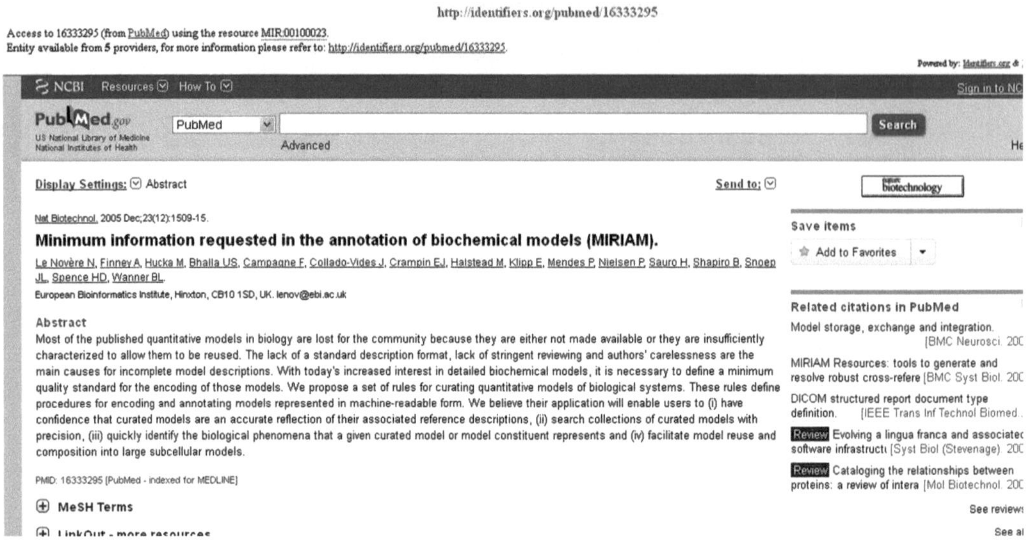

Fig. 5 Accessing data through a specified resource, using an Identifiers.org URI

resolving locations for, potentially, every data collection stored in the Registry. Currently the number of available profiles is limited to those created by the Registry curators. Work is under way to extend this facility and allow users to create and share their own profiles. Profiles will be allowed to be "private," while an interface is being implemented to allow public profiles to be searched. For example, the predefined profile "most_reliable" (below) always returns the instance of a record through the resource with the highest uptime. The "most_reliable" profile is based on the health check history of the resource. http://identifiers.org/pubmed/16333295?profile= most_reliable

The use of the URL form, combined with judicious "parameters," simplifies access to a wealth of information, largely obviating the need for directly querying the Registry through web services. However, it should be noted that the use of the URL form for identification purposes should not incorporate the use of any parameter. Hence, in the unambiguous and perennial identification of data, the identifier should be considered as being the minimal string that specifies a record. From a practical perspective, those users who have in the past used identifiers of the URN form can convert them into identifiers.org URLs if they choose, or indeed vice versa.

3.6 BioModels.net Qualifiers

The purpose of "Qualifiers" is to refine the relationship between, for example, a model component and the resolved target of a cross-reference associated with that component. In the absence of a qualifier, the relationship assumed is an "is" relationship. For instance, given a model component labeled as "Glu" containing

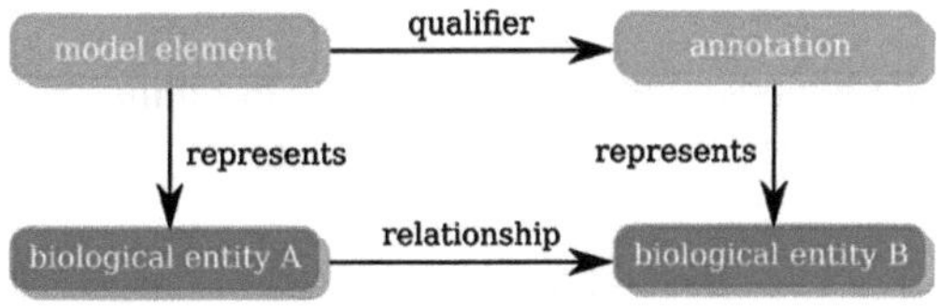

Fig. 6 Schematic representation between model component in a file (model element), the "real-life" entity it seeks to represent (biological entity A), the external resource annotation provided with it (annotation), and the "real-life" target of that external resource annotation (biological entity B)

an unqualified annotation (http://identifiers.org/obo.chebi/CHEBI:17234, which resolves to a page with information about "Glucose"), it should be assumed that the model component (written as "Glu") "is" "Glucose" (the external resource record representing the real-life glucose molecule).

The "is" or "identity" relationship is straightforward to understand, but other qualifiers exist to express more complex relations between model component and an external resource. It should be noted that there are two types of qualifiers, *biological* (in the "bqbiol" namespace) and *modeling* (in the "bqmodel" namespace), which relate either biological/physical objects (genes, proteins, enzymes) or modeling objects/concepts (model files, databases, literature), to model components. Figure 6 illustrates the relationship between model component in a file (model element), the "real-life" entity it seeks to represent (biological entity A), the external resource annotation provided with it (annotation), and the "real-life" target of that external resource annotation (biological entity B).

For example, expanding on the example above, a model of glycolysis may contain a model component labeled "PFK" (model element), representing the "real-life" enzyme phosphofructokinase (biological entity A). The external resource annotation presented alongside it (annotation), when resolved, can be used to represent a database record for the real-life activity of the phosphofructokinase enzyme (biological entity B), which is important with respect to its function in the model, namely its catalysis of a specific reaction. In this case, an appropriate qualifier would be "hasProperty." The qualifier is, in essence, a reflection of the relationship between two representations, one being held in a model, and the other in an external resource. This is necessary since it is not possible to actually attach a PFK molecule to an electronic file, whether it is a model file, or a database record.

Some of the biological relationships that can be represented are shown below, with reference to the figure above. It should be noted that each qualifier is presented in two forms, noun and verb, to allow users to select whichever they are most comfortable with. Both can be used synonymously. The full list of qualifiers is available

Table 4
Examples of the biological qualifiers available to refine the relationships between model component and external resource

Qualifier	Description
bqbiol:hasPart bqbiol:part	The biological entity represented by the model element includes the subject of the referenced resource (biological entity B), either physically or logically. This relation might be used to link complex to the description of its components
bqbiol:isDescribedBy bqbiol:description	The biological entity represented by the model element is described by the subject of the referenced resource (biological entity B). This relation should be used, for instance, to link a species or a parameter to the literature that describes the concentration of that species or the value of that parameter
bqbiol:isEncodedBy bqbiol:encoder	The biological entity represented by the model element is encoded, directly, or transitivity, by the subject of the referenced resource (biological entity B). This relation may be used to express, for example, that a protein is encoded by a specific DNA sequence
bqbiol:isHomologTo bqbiol:homolog	The biological entity represented by the model element is homologous to the subject of the referenced resource (biological entity B). This relation can be used to represent biological entities that share a common ancestor
bqbiol:occrsIn bqbiol:container	The biological entity represented by the model element is physically limited to a location, which is the subject of the referenced resource (biological entity B). This relation may be used to ascribe a compartmental location, within which a reaction takes place
bqbiol:isVersionOf bqbiol:hypernym	The biological entity represented by the model element is a version or an instance of the subject of the referenced resource (biological entity B). This relation may be used to represent, for example, the "superclass" or "parent" form of a particular biological entity

from the BioModels.net website (*see* **Note 18**) and can be expanded and refined upon community request and feedback (Table 4).

3.7 Incorporating Qualifier Relationships

The simplest way to understand qualifiers is to consider them as being the "predicate" in a "subject, object, predicate" sentence, where the subject is the model component, the object is the target of the external resource annotation, and the predicate is the qualifier relationship between them.

Qualifiers are commonly used within metadata in model encoding formats, such as SBML:

1. <rdf:RDF xmlns:rdf="http://www.w3.org/1999/02/22-rdf-syntax-ns#"
2. xmlns:bqbiol="http://biomodels.net/biology-qualifiers/">
3. <rdf:Description rdf:about="#MyModelElement">
4. <bqbiol:hasPart>
5. <rdf:Bag>

6. <rdf:li rdf:resource="http://identifiers.org/uniprot/P04551"/>
7. <rdf:li rdf:resource="http://identifiers.org/uniprot/P10815"/>
8. </rdf:Bag>
9. </bqbiol:hasPart>
10. </rdf:Description>
11. </rdf:RDF>

The use of Identifiers.org URLs does not in itself require any particular format or syntax. It can therefore be incorporated into any structured format relatively easily. However, many structured formats do themselves have a syntactic procedure through which such annotations are to be expressed. For example, within SBML, such annotations are encoded in RDF (*see* **Note 19**) blocks.

A detailed line-by-line description of the example above:

1. The <rdf element open tag and definition of XML namespace declaration for RDF use.
2. Definition of the biology-qualifiers namespace.
3. RDF Description block opened, with the subject being MyModelElement
4. The qualifier for the block is hasPart from the bqbiol namespace.
5. The rdf:Bag construct allows the inclusion of multiple URIs in an annotation.
6. The li line element where a resource is specified.
7. The li line element where a resource is specified.
8. Close tag to end the Bag block.
9. Close tag to end the hasPart block.
10. Close tag to end the Description block.
11. Close tag to end the RDF block.

This annotation block should be interpreted to mean that "MyModelElement" represents a biological object that "has parts" described by the records in the UniProt data collection specified by the identifiers P04551and P10815. In this example, the UniProt specified entries refer to Cyclin-dependent kinase and G2/mitotic-specific cyclin cdc13, which are both involved in the control of the cell cycle at the G2/M (mitosis) transition.

3.8 Alternative Identification Schemes

The Identifiers.org identification scheme offers distinct advantages over some other well-known systems, some of which are described briefly below.

Persistent Uniform Resource Locations (PURLs) (*see* **Note 20**) are subtly different in intent from Identifiers.org URLs. Since this is an open system, in the sense that once registered any individual may create a PURL, there can potentially be a plethora of different PURLs that all identify the same record (have a common end-point). This is an hindrance to data integration. In addition, the focus of PURLs is to permanently identify a record resolved through a specified resource, thus effectively tying a record identifier to a specific instance, within a single URL. This should be contrasted with a record identifier using Identifiers.org URLs, which can be used to resolve information through any number of associated resources.

Digital object identifiers (DOIs) (*see* **Note 21**) are generally associated with online authored publications, and hence may not be as well suited to the referencing of biological entities. In addition, it is a fee-based assignment service, and the identifier designated by DOI does not reuse the identifier assigned by the data provider. Finally, like PURLs, a DOI resolves to a single instance of a record.

Life Science Record Names (LSRN) (*see* **Note 22**) are the closest relative of the MIRIAM identification scheme, in that they use a central database with assigned namespaces and store information on the associated resolving locations. The key differences lie in the extensive curation of the MIRIAM Registry, its broader coverage, and the supporting facilities it offers, including web services, programmatic access to the database, health check, XML download availability, together with an extensive and highly active community of users.

A more complete comparison of these, and other, identification schemes is espoused on the Registry website (*see* **Note 23**). There follows a summary of the key advantages proffered by the MIRIAM system:

- Open submission—Anyone can make a submission to the Registry.
- Curated—The content of the Registry is heavily curated and maintained for accuracy by a dedicated curation team.
- Resolution system—The scheme adopted allows the mapping of records to multiple resolving locations.
- Health check—Daily monitoring of all resources, with curator intervention when necessary.
- Extensive support—A growing community of users to provide software and tools in support of the system (see below).
- Accessibility—A variety of access methods is provided, including web services.
- Export—The entire content of the can be exported as XML, allowing noninteractive processing of Registry content.

- Free—There is no restriction on the use of information in the Registry, and no registration requirement.
- Standardization—MIRIAM is itself a partner in a number of standardization efforts.

3.9 The Registry User Community

There are a number of perspectives that can be taken on the knowledge captured in the MIRIAM Registry. It can be viewed as part of a standardization effort, thus having associated compliant file formats, and supporting software and tools; it can be regarded as a way to identify both data records and a means to standardize namespaces and associated resources; it can also be considered within the landscape of other, sometimes competing, identification schemes. Each of these perspectives is briefly addressed below.

Since many structured formats conform to the MIRIAM Guidelines, they by default should use annotations based on the MIRIAM Registry. Since SBML is one such structured format, all tools that read, write, or manipulate this format will intrinsically handle Identifiers.org URIs. This covers a broad spectrum of activities ranging from the annotation of models, through processing of those models to do novel research, to creating human or machine-readable representations of those models.

Identifiers based on information stored in the Registry are already widely used, most notably within BioModels Database (*see* **Note 24**, ref. [9]). The latest release of the database (22nd release, May 2012) contains over 142,900 models, with over 444,130,000 annotations. These model files are available freely and can be downloaded with either URN or URL annotations, with the latter being the default annotation style.

Since the Registry also assigns and stores namespace information for data collections, as well as associated synonyms, this knowledge itself can also be used to harmonies or standardize resource nomenclature. For example, both the PSI-MI (Proteomics Standards Initiative—Molecular Interactions; *see* ref [10]) and BioPAX (Biological Pathway Exchange; [11]) working groups use this information to assign standard database names in their controlled vocabularies, using the stored synonym information.

As a standardization effort, support for Identifiers.org URIs, and particularly the use of resolvable identifiers, is growing; LSRN, has announced that it will be transitioning its information into the Registry and will cease further support and development of its own identification scheme. This process is already well under way.

Further information on the formats, tools, and software that utilize MIRIAM Registry information is available from the Registry's documentation (*see* **Note 25**).

3.10 Future Perspectives

The MIRIAM Registry is a stable resource which provides an identifier scheme, a perennial URI generation service, and a resolving system. While its foundations lie in Computational Systems Biology, it is by no means restricted to that domain, and indeed data collections from more diverse fields are continually being incorporated. This potential for its use as a universal cross-referencing system, which was noted during the inception of the system, is now being realized. There should be no impediment in its use in any domain.

The user interface and access options are being continually improved, permitting not only the creation of perennial and unambiguous identifiers, but also facilitating the customization of the way the underlying data is addressed. The ability to create "Profiles," for example, will allow the creation of entire sets of resolving preferences, which can potentially be shared at an institutional, community, or group level.

There previously existed various restrictions governing the suitability for inclusion of data collections into the Registry. These have recently been removed in recognition of the referencing needs of the user community at large. For instance, some proprietary data collections were deemed unsuitable since they required either registration or were subject to fee-based access. The provision of the "flag" system discussed above has enabled the incorporation of such historically non-compliant data sources.

When deliberating upon the future of data access on the web, one must also consider the importance of efforts such as the Semantic Web (*see* **Note 26**) and the Linking Open Data (LOD) (*see* **Note 27**) initiative. In providing resolvable URIs the Identifiers.org addresses some of the demands of this growing community.

The MIRIAM efforts (Guidelines, Registry, and Identifiers.org) are all partners in larger community level standardization efforts, such as MIBBI and BioDBCore [12], as well as members of the modeling community, particularly through their involvement in SBML, but also within the COMBINE (*see* **Note 28**) community.

4 Notes

1. http://www.w3.org/TR/REC-xml/
2. http://sbml.org/Documents/Specifications
3. http://obofoundry.org/
4. http://www.ebi.ac.uk/sbo/
5. http://biomodels.net/kisao/
6. http://mibbi.org/
7. http://www.ncbi.nlm.nih.gov/taxonomy
8. http://www.boldsystems.org/views/taxbrowser_root.php
9. http://www.ncbi.nlm.nih.gov/pubmed

10. http://www.ebi.ac.uk/ena
11. http://www.uniprot.org/
12. http://tools.ietf.org/html/rfc3986
13. http://en.wikipedia.org/wiki/Uniform_Resource_Name
14. http://www.tcdb.org/
15. http://www.ebi.ac.uk/miriam/
16. http://www.ebi.ac.uk/chebi/
17. http://www.ncbi.nlm.nih.gov/pccompound
18. http://biomodels.net/qualifiers/
19. http://www.w3.org/TR/REC-rdf-syntax/
20. http://www.purl.org/
21. http://www.doi.org/
22. http://lsrn.org/
23. http://www.ebi.ac.uk/miriam/main/mdb?section=uris
24. http://www.ebi.ac.uk/biomodels/
25. http://www.ebi.ac.uk/miriam/main/mdb?section=use
26. http://www.w3.org/2001/sw/
27. http://linkeddata.org/
28. http://co.mbine.org/

5 Funding

This work has been supported by EMBL, the ELIXIR Preparatory Phase Project, and by BBSRC grants.

References

1. Hucka M et al (2003) The Systems Biology Markup Language (SBML): a medium for representation and exchange of biochemical network models. Bioinformatics 19:524–531
2. Smith B et al (2007) The OBO Foundry: coordinated evolution of ontologies to support biomedical data integration. Nat Biotechnol 25: 1251–1255
3. Courtot M et al (2011) Controlled vocabularies and semantics in systems biology. Mol Syst Biol 7:543
4. Taylor C et al (2008) Promoting coherent minimum reporting guidelines for biological and biomedical investigations: the MIBBI project. Nat Biotechnol 26:889–896
5. Le Novère N et al (2005) Minimum Information Requested in the Annotation of biochemical Models (MIRIAM). Nat Biotechnol 23: 1509–1515
6. Laibe C, Le Novère N (2007) MIRIAM Resources: tools to generate and resolve robust cross-references in Systems Biology. BMC Syst Biol 1:58
7. Li C et al (2010) BioModels.net Web Services, a free and integrated toolkit for computational modelling software. Brief Bioinform 11: 270–277
8. Juty N et al (2012) Identifiers.org and MIRIAM Registry: community resources to provide persistent identification. Nucleic Acids Res 40: D580–D586
9. Li C et al (2010) BioModels Database: an enhanced, curated and annotated resource for published quantitative kinetic models. BMC Systems Biol 4:92
10. Orchard S, Hermjakob H (2008) The HUPO proteomics standards initiative–easing communication and minimizing data loss in a changing world. Brief Bioinform 9:166–173
11. Demir E et al (2010) The BioPAX community standard for pathway data sharing. Nat Biotechnol 28:935–942
12. Gaudet P et al (2011) Towards BioDBcore: a community-defined information specification for biological databases. Nucleic Acids Res 39: D7–D10

Chapter 13

Bayesian Approaches for Mechanistic Ion Channel Modeling

Ben Calderhead, Michael Epstein, Lucia Sivilotti, and Mark Girolami

Abstract

We consider the Bayesian analysis of mechanistic models describing the dynamic behavior of ligand-gated ion channels. The opening of the transmembrane pore in an ion channel is brought about by conformational changes in the protein, which results in a flow of ions through the pore. Remarkably, given the diameter of the pore, the flow of ions from a small number of channels or indeed from a single ion channel molecule can be recorded experimentally. This produces a large time-series of high-resolution experimental data, which can be used to investigate the gating process of these channels. We give a brief overview of the achievements and limitations of alternative maximum-likelihood approaches to this type of modeling, before investigating the statistical issues associated with analyzing stochastic model reaction mechanisms from a Bayesian perspective. Finally, we compare a number of Markov chain Monte Carlo algorithms that may be used to tackle this challenging inference problem.

Key words Ion channels, Mechanistic models, Bayesian inference, Markov chain Monte Carlo

1 Introduction

Cells and subcellular organelles are delimited by membranes that are made of phospholipids, arranged in a bilayer. The lipidic nature of membranes means that they cannot be crossed by hydrophilic compounds, including many that are essential for the cell metabolism and survival. This problem is solved by providing specific active or passive transport mechanisms mediated by appropriate proteins. In the case of ions, their movement in and out of the cell is achieved by the expression of proteins, which either chaperone the ion or form a hydrophilic pore through the membrane. The former are known as ion transporters (and can work against the electrochemical gradient by coupling transport of an ion to hydrolysis of ATP or to countertransport of other ions), and the latter are known as ion channels. The role of these proteins is to allow the selective movement of ions into and between regions of the cell. Ion channels are essential in many biological processes, including cell excitability and

Maria Victoria Schneider (ed.), *In Silico Systems Biology*, Methods in Molecular Biology, vol. 1021, DOI 10.1007/978-1-62703-450-0_13, © Springer Science+Business Media, LLC 2013

fast cell-to-cell communication in the nervous system, and they are the focus of this chapter.

The opening of the transmembrane pore in an ion channel is brought about by conformational changes in the protein, which results in a flow of ions through the pore. Remarkably, given the diameter of the pore, the flow of ions from a small number of channels or indeed from a single ion channel molecule can be recorded experimentally. This produces a large time-series of high-resolution experimental data, which can be used to investigate the gating process of these channels.

Ion channels can be classified into two groups. The first group comprises ligand-gated ion channels, which undergo conformational changes in response to the binding of a chemical, such as a neurotransmitter, to a site on the protein. The energetic perturbation associated with the agonist binding leads to the opening of the pore. The second group consists of voltage-gated channels, in which pore opening is induced in response to local environmental changes in the potential difference between the cell and its extracellular domain. This chapter is concerned with discerning the link between the structure and function of ligand-gated ion channels. Ligand-gated ion channels are essential in synaptic transmission in the nervous system and understanding how they work is consequently of considerable biological and pharmacological interest.

Ligand-gated ion channels can be classified into three different protein superfamilies. For example, the Cys-loop superfamily includes $GABA_A$, glycine, acetylcholine, and serotonin receptors. Subunits that make these receptors all exhibit a characteristic Cys-loop formed by a disulphide bond in the N-terminal extracellular domain and are arranged in a quasi-symmetrical pentamer around a central pore. The most studied receptor in this group is the muscle nicotinic acetylcholine receptor, mainly found in the vertebrate neuromuscular junction, which is responsible for transmitting the signal between the nervous system and muscle.

A second receptor superfamily is the ionotropic Glutamate family, members of which are tetramers of homologous subunits. Glutamate receptors have three broad classes based on their selective agonists—NMDA, kainate, and AMPA. The Glutamate superfamily is believed to be capable of molecular coincidence detection and is therefore hypothesized to play a crucial role in synaptic plasticity and the formation of memory.

The third superfamily is the ATP-gated cationic channels, known as purinergic nucleotide receptors. Their function is not well understood but they may contribute to peripheral signaling in the autonomic nervous system.

Experimental evidence of the gating process of single channels is available at fine time resolution, thanks to patch-clamp techniques. Despite the tiny diameter of ligand-gated channels, long recordings of the currents generated by single channels as they open and close are experimentally achievable.

Electrophysiological recordings alone only tell us when the channel is open or closed and do not provide information of the conformational changes of the channel as it opens and closes. However, the analysis methods we describe here for physical modeling of kinetic states allow us to map the energy landscape for the conformational changes of these proteins, and this in turn can provide a bridge between ion channel structure and function. The use of models combined with high resolution data can therefore serve a number of purposes in studying ion channel gating. Namely they can:

1. Provide a structured way of thinking about the sequence of conformational changes that need to occur in proteins to allow pore formation.
2. Help explain new experimental data observed in lieu of direct structural information for example describing the action of novel agonists on channel gating.
3. Suggest testable hypotheses regarding aspects of the gating process, which may be subsequently verified by novel experimental design.

The analysis of ion channel behavior in this fashion starts with the postulation of model reaction mechanisms that describe the sequence of changes that occur as a single channel opens and closes in a stochastic manner. The use of explicit reaction mechanisms describing the ion channel gating process has a long history dating back to the 1950s, when del-Castillo and Katz [1] sought to explain why some agonists (e.g., partial agonists) were less efficacious than others at opening channels. Their supposition of the existence of both a binding step and a gating step allowed the first explanation of this phenomenon. Although different agonists may bind with the same affinity, i.e., bind with the same "strength" to the receptor, once bound they differ considerably in their ability to produce the conformational change necessary to actually open the receptor.

This mechanistic way of thinking applied to the analysis of single-channel data can also help explain, with appropriate experimental data, the impact of mutations or of novel agonists on receptors, by assessing their impact on the rates of movement between different conformational steps in the reaction mechanism. On the whole, this cannot be achieved using macroscopic cell recordings, as the time course of these responses reflects all the steps in the reaction mechanism.

We therefore see how the modeling of ion channel gating as reaction mechanisms can help our understanding of the structural processes that occur during gating. The benefits of this knowledge that feed back into the biological domain are many. These include the ability to explain the effects of deleterious channel mutations, see for example [2, 3] and the future potential to investigate and assess the impact of novel agonists. Indeed, a fuller understanding of a novel agonist's actions on experimentally distinguishable components of the gating process would have great ramifications for the ultimate aim of rational drug design.

2 Single Channel Experimental Data

Single channel recordings are the most important source of data for evaluating reaction mechanisms. The approach involves observing single ion channels as they open and close by recording the quantum change in current through the channel as ions flow through the pore.

Experimentally, this is achieved through the production of a borosilicate glass pipette with an extremely small diameter, fire-polished to a high resistance. The pipette, in the cell attached configuration, is placed near a target cell which expresses the desired receptor. Suction is then applied in order to seal the pipette against the cell with a high resistance. The current from a small number of ion channels is then recorded as it flows across the tip of the pipette.

The signal recorded from the ion channels is subsequently low-pass filtered to remove high frequency noise generated from the recording equipment, such as the noise resulting from the amplification of the current and the imperfect resistance of the seal made by the pipette against the cell membrane. The signal is then sampled at a high frequency to give an estimate of the conductance of the channel through time.

Although very fast single ion channel gating can be recorded with these techniques, the requirement for filtering and other experimental considerations, such as expression levels of the protein on the surface of the cell, result in a number of well-known inferential complications for statistical modeling. These include:

- *Limited time resolution*: The requirement to filter and sample data restricts the time resolution at which the sample can be recorded. Typical patch-clamp recordings have a best resolution of about 10–20 μs, although this depends greatly on signal size and idealization techniques. This results in a phenomenon known as "missed events," which significantly complicates the modeling and inferential process. See [4–7] for an in-depth discussion.
- *Unknown number of channels in the patch*: Experimental factors such as the level and heterogeneity of channel expression on the surface of the cell mean that openings and closings recorded in the patch can originate from more than one channel.

While statistically we want to consider the gating of *one* channel, the probability of having more than one channel in the patch in experimental recordings needs to be accounted for in order to obtain accurate assessment of reaction mechanisms.

3 Modeling Ion Channel kinetics

3.1 Observable Data

The key summary statistics of biological interest generated from the experimental record are:

- The univariate open and closed dwell time distributions of the channel.
- Bivariate distributions of adjacent open and closed sojourns, which can exhibit correlations for mechanisms with multiple gateway states.
- The dose–response of the receptor across multiple concentrations, typically summarized as the fraction of the total time that the channel is open across an experimental record with a given concentration of agonist.

The modeling challenge is to present a parameterized reaction mechanism that can account for these observations and through which parameters of interest, such as reaction rate constants between the conformational changes, are inferable. Evaluating competing mechanisms is also of great importance, see e.g. [8]. This involves postulating different mechanisms which incorporate varying numbers and connectivity of binding and gating steps to describe the opening trajectory of the channel. The next step is deducing from the data which suggested mechanism is the most plausible.

3.2 The Stochastic Framework and Derivation of the Model Likelihood

For modeling such biological systems we require a mathematical framework within which to specify a reaction mechanism from single recordings. It has been noted that the observable open and closed sojourns typically follow a semi-Markov or renewal framework [6, 7]. This implies that adjacent sojourns are independent and can incorporate a variety of stochastic models.

The most common stochastic models within this generalization are based on discrete time or continuous time Markov processes. Markov models neatly fit the conceptual single-channel biological process. For example, the hypothesis of hidden conformational states, which represent approximate energy minima, translate well into a discrete but countable state space in a Markov model. The gating process of a single channel can reasonably be assumed to be a memoryless stochastic process whose key properties are invariant through time. Deriving likelihoods and statistical inferences is a reasonably tractable process, particularly in the case of perfect time resolution, given the independence of non-overlapping observable sojourns.

Let us formulate a stochastic model for single ion channel kinetics as follows. Consider a continuous time Markov process $\mathbf{S}(t)$, $t \geq 0$ which is parameterized by a generator matrix $\mathbf{Q}$. The $\mathbf{Q}$ matrix defines the reaction mechanism by representing the system as a graph linking different states, and its parameters govern the rate at which transitions between states occur. It is assumed to be homogenous, i.e., invariant through time, and has some initial state distribution π. Each state in the Markov model represents discrete conformational states that the protein can occupy as it moves through the gating process.

Note that the number of states within the model does not necessarily correspond to the number of conductance levels seen in the observable data. If we consider two conductance levels, open and closed, then there may be many Markov states within the open conductance class and within the closed conductance class. These aggregations of states represent the different features of the gating mechanism that are observable experimentally. Models of this type have hidden states within a conductance level, and it is therefore unknown precisely which state the process is in given the observed conductance level. These models are therefore defined as aggregated Markov models.

Given this model structure, there are two main goals of interest. First, can we infer from our data the parameters of the $\mathbf{Q}$ matrix and the initial distribution that governs our model process? From this, we can judge how well our model explains our observable data. Second, if we have competing plausible model hypotheses, can we use the observed data to choose which model structure is most likely to describe the true underlying biological mechanism?

In order to infer the parameters in our models effectively, we need to develop an expression for a likelihood. This quantity tells us the probability of the recorded data being observed for a particular set of parameters and given model. The observed data of interest shows the transitions between the aggregation of closed states and the aggregation of open states. Thus we aim to derive a probability density for the open and closed sojourns, given our parameters in the $\mathbf{Q}$ matrix and the equilibrium distribution of the process. In this instance we assume, naively for now, both perfect time resolution and that there is only one channel in the patch.

Using the approach described in [9, 10], we derive the sojourn density functions as follows. Consider the transition matrix $\mathbf{P}(t)$ of the continuous time process $\mathbf{S}(t)$, $t \geq 0$ considered above. A standard result for solving for $\mathbf{P}(t)$ is by obtaining the solution to the differential equation $\frac{\mathrm{d}\mathbf{p}(t)}{\mathrm{d}t} = \mathbf{P}(t)\mathbf{Q}$, which gives:

$$\mathbf{P}(t) = \mathrm{e}^{\mathbf{Q}t}$$

However, we only observe sojourns in conductance levels, which are represented by aggregations of states. Consider two aggregations, open and closed, where set A represents the open states and set B represents the closed states, such that $A \cup B = S$. We can subpartition our $\mathbf{P}(t)$ and $\mathbf{Q}$ matrices accordingly, such that $\mathbf{Q}_{AA}$ represents the transition rates from open states to other open states, $\mathbf{Q}_{AB}$ from open states to closed states, and so forth.

For the derivation of the observed sojourn times in the open conductance level, we consider the probability of remaining within the open states for a time t before transitioning to one of the closed states in an infinitesimally small time step. We can denote these as a matrix of probabilities, $\mathbf{G}_{AB}(t)$, which is analogous to $\mathbf{P}(t)$ but describes only transitions from open states to closed states. By considering the closed states to be absorbing, such that $\mathbf{Q}_{BA} = \mathbf{Q}_{BB} = \mathbf{0}$, the resulting solution for these probabilities follows as,

$$\mathbf{G}_{AB}(t) = \mathbf{P}_{AA}(t)\mathbf{Q}_{AB} = e^{\mathbf{Q}_{AA}t}\mathbf{Q}_{AB} \qquad (1)$$

The density for closed states $G_{BA}(t)$ can be derived by exchanging the partitioning indices. Given that the process is Markov, non-overlapping time interval events are assumed to be independent. Given that the observed data is a series of closed and open sojourns, then the likelihood of n such sojourns is given by,

$$\phi_A \mathbf{G}_{AB}(t_1)\mathbf{G}_{BA}(t_2)\mathbf{G}_{AB}(t_3)\ldots\mathbf{G}_{BA}(t_n)\mathbf{u}_A \qquad (2)$$

where ϕ_A is the equilibrium probability of starting in each of the open states, and $\mathbf{u}_A$ is a unit vector which sums up the likelihood over the final sojourn entry into the different open states.

The likelihood in Eq. 2 can be maximized to obtain the maximum likelihood estimate for the parameters of the $\mathbf{Q}$ matrix. The maximum likelihood represents the most likely set of rate constants that explain the observed data. Theoretical open and sojourn time distributions, bivariate correlations, and P_{open} curves can then be produced from the parameterized $\mathbf{Q}$ matrix and compared with the experimental distributions to assess the fit of the model. In practice, the fitting is made across a range of agonist concentrations to increase the amount of information that can be extracted for the receptor. Empirical measures such as the burst distribution length and concentration jumps can also provide additional useful information to discriminate between models.

It should be reemphasized that the model describes the hypothetical *underlying* conformational changes of the channel and not necessarily the observed conductance levels of the channel as it opens and closes. This increases the inferential challenges associated with estimating parameters for the Markov process. The aggregated nature of the model process raises issues of parameter identifiability [11, 12], which translates in biological terms as the ability to demarcate transitions between discrete conformational states.

It also raises concerns about model identifiability—the potential to distinguish between different plausible model structures that produce approximately the same observable phenomena, see for example [13].

The method of maximizing the likelihood of an entire sequence of single channel measurements of open and shut times offers many advantages over previously suggested approaches, which involve fitting the model to separate distributions describing summary statistics of the mechanism, such as average open and shut times, see for example [9, 14]. For example, the identification of exponential mixtures in dwell time distributions can be ambiguous, and exact missed event correction impossible without a specified mechanism. The construction of a likelihood based on the entire sequence encapsulates all such underlying properties of the mechanism, including any correlation structures that might arise from the open and shut time intervals.

As highlighted in [15, 16], a likelihood-based approach allows a mechanism's rate constants to be inferred directly from the data and removes any need to subjectively decide which summary statistics to base inference on. Perhaps most importantly it automatically takes into account the cross correlation behavior that may arise in the data, such as when short shut times are often observed to follow long open times; this type of information is missed in the summary statistics and may subsequently impact the inferences made. This is particularly true in the case of mechanisms with more than one gateway path between open and closed states.

The first maximum likelihood approaches for fitting idealized data to continuous time mechanisms were outlined in [17] with subsequent approaches increasing in their generality and correction for missed events in the likelihood calculations [15, 18, 19].

3.3 Fitting Mechanisms in Practice

Experimental data may be interpreted as noisy discrete observations of some continuous underlying process. The true signal must be estimated despite the experimental noise and signal distortion caused by the requirement to filter the raw signal. In addition, the likelihood outlined in Eq. 2 must be corrected for the fact that sampling the data imposes a given resolution on signal detection. This means brief openings and closings that are potentially missed must be corrected for in this record. Also, the fitting procedure must account for the likelihood of having more than one channel in the recording patch. There is a requirement to fit stretches of data where the number of channels in the stretch can be established to be almost certainly one.

A commonly used computational package to fit kinetic mechanisms to single ion channel data is HJCFIT (http://www.ucl.ac.uk/Pharmacology/dcpr95.html). This software, derived from [5, 15], is designed to fit models given an idealized signal, for example using time-course fitting, and exactly correcting sojourn probability densities for the impact of missed events. The fitting process can fit a simultaneous series of recordings at different agonist concentrations to extract the full amount of information across the dose–response curve.

Experimental data is idealized by time course fitting, by superimposing (over the record) a calculated output to a response step input. The fit is adjusted until the signals overlap and from this the idealized sojourn can be inversely inferred. This is incorporated in the SCAN package. The subsequent record can either be fitted to directly using HJCFIT or analyzed heuristically by fitting exponential distributions to the dwell times (using EKDIST) to obtain approximate inference about the number of potential conformational states and bivariate dwell time correlations.

The fitting procedure maximizes the likelihood of the sequence sojourn times and utilizes an exact correction for missed events [4, 5, 20], which calculates the apparent open and closed time probability density functions from a given rate transition matrix and kinetic model. The exact solution is based on a piecewise inversion of a Laplace transform. This transform captures all possible missed transitions within a given open or closed sojourn. It is, however, more computationally expensive to calculate than other approximate corrections such as [21–23] and is numerically stable only for limited multiples of the resolution time. Therefore, the asymptotic form of the missed events correction is used to correct sojourn times that are more than two multiples of the resolution time, and this has proven to be accurate enough in practice.

The modeler must account for the number of channels in the patch by fitting channel records broken into stretches that are only likely to contain one channel using a time interval cutoff. The actual maximization of the corrected likelihood is achieved using a derivative-free simplex method [24]. The properties of its estimators, including the impact of fitting the wrong models, have been thoroughly examined within the maximum likelihood framework [16].

3.4 Alternative Maximum-Likelihood Approaches

There have been several other maximum-likelihood approaches based on fitting mechanisms both to idealized dwell time distributions and by direct fitting to the single channel records using Hidden Markov Models.

For example, another commonly used fitting tool is QuB. The QuB software package incorporates both an idealization procedure and a fitting procedure based on maximizing the joint probability density of dwell times. The idealization procedure [25] is a *k*-means segmentation algorithm based on [26] which tries to restore an ion channel signal given a prior hypothesized mechanism. The discretized idealization is a two step algorithm; first, the most likely hidden state sequence is obtained using the Viterbi algorithm, given the model mechanism, conductance levels, and variance.

In the second step the parameters are re-estimated. These two steps are iterated until the likelihood is maximized.

The idealized signal is then used as the basis for a dwell time fitting algorithm [27, 28] akin in spirit to [29, 30], through the use of forward and backward variables that are used to calculate the likelihood and its analytical derivatives. The derivatives are utilized in a variable metric optimization procedure to search for the maximum likelihood. The procedure employs an approximate first-order missed events correction first developed in [21].

An explicit HMM approach by Qin et al. [31, 32] also employs a direct fitting of the **Q** matrix to the raw data signal. Again, this utilizes the forward–backward procedure and a maximization technique based on the analytical derivatives of the likelihood. The citing procedure allows the specification of rate constraints and mechanism fitting across different experimental sets. Extensions such as [31] improve the modeling of the noise process by the incorporation of correlated noise and the effects of signal filtering.

Additional maximum-likelihood Hidden Markov approaches to fitting and idealizing single channel data account for correlated noise [33–35] and conductance dependent noise, although this increases the computational cost of model fitting. Others have sought to fit to the dwell times directly using either a pseudo-likelihood approach [36], or empirically through simulation [37, 38], particularly for complex models where the number of states makes fitting with HMM approaches computationally prohibitive, see for example [25].

3.5 Achievements and Limitations

Full maximum-likelihood approaches have been successfully employed in a number of investigations into ion channels, for example [8] used the fitting process of [15, 16] to investigate competing mechanisms to explain the gating process of heteromeric glycine receptors using different agonists. The ability to fit rate constants as fast as $130,000\ s^{-1}$ was a feature of the fitting. After considering a wide variety of thirty potential mechanisms, only two mechanisms were considered adequate; specifically, a mechanism adapted from one previously proposed for GABA receptors [39] and an additional model that incorporated the same additional shut states as in [39] in an intermediate closed "flipped" conformation. Both provided a visually reasonable fit to the summary statistics of the experimental data.

Upon further inspection, the kinetic model incorporating flipped conformational states offers a more biologically plausible explanation of the gating process of the receptor, and has fewer rate parameters to estimate. This mechanism explains an increase in binding affinity as more ligand molecules bind to the receptor. This, at face value, implies an ability of distal ligand binding sites to interact. The flipped conformational change accounts for the apparent receptor binding co-operativity by implying that agonist binding stabilized the flipped conformation, which is the higher affinity form of the receptor, rather than the resting conformation. The actual affinities for progressive binding of agonist within each of these two conformations are constant, but the flipped conformation has a 65-fold greater affinity for agonist.

Additional research on the pre-opening flipped conformational change [40, 41] added insights into the mechanics of partial agonism. As described previously in this chapter, the classical view is that partial agonism occurs as a result of the reduced efficacy of some agonists to elicit the final open conformational change from the closed agonist-bound state. However, Lape et al. [40] showed that for glycine and acetylcholine receptors, partial agonism results from differences in an earlier step in the gating process. The authors compared the actions of the full agonist glycine and the partial agonist taurine on the glycine receptor. Maximum likelihood fitting of the flipped mechanism for glycine receptor revealed that kinetically the final conformational step for both agonists was similar. The difference between the two agonists was their ability to elicit the flipped conformational change that occurs when the channel is still shut. Lape et al. [41] investigates the actions of the agonist choline on acetylcholine receptors. Although choline behaves like a partial agonist, it was unknown whether choline is actually a full agonist whose maximal response is limited by channel blocking, or whether it is a genuine partial agonist. Fitting the flip mechanism in conjunction with concentration jump experiments suggests that choline is more likely to be a partial agonist than a full agonist given its reduced ability to elicit the flipped conformation of the receptor.

However, the current maximum-likelihood approaches are reaching their limits, both in terms of parameter identification and model discrimination. For example [42] investigated a mutation in the glycine receptor that slows down the speed of channel gating. This enabled the authors to investigate further the nature of closed intermediate states in order to explain the experimental data produced by mutant receptors. A mechanism was proposed incorporating "primed" intermediate states in which independent subunits of the glycine channel can flip independently. This however required the estimation of a large number of rate parameters, not all of which were identifiable, suggesting that the fully primed model is overparameterized. This required the ideal model to be pruned in order to remove rarely visited states to improve kinetic rate identifiability, although a different subset of the fully primed model was then needed to fit the wild-type control data.

In addition, the existence of modalities in the likelihood surfaces of more complicated models, particularly with the problem of time interval omission, will require more sophisticated techniques to explore fully the likelihood surface. The hope is that the adoption of a Bayesian approach to parameter estimation and systematic model selection may allow informative inferences to be made despite the identifiability issues highlighted above.

4 The Bayesian Approach

Employing a Bayesian approach offers us a consistent way of reasoning about uncertainty, both in the rate parameters and in the mechanism structure. We do so using the language of probability theory. We wish to evaluate the posterior probability distribution, which fully describes the global sensitivities of the rate parameters of a hypothesized ion channel mechanism given some experimental data. The posterior distribution is calculated using Bayes' theorem, which is simply derived from the standard laws for the conditional probability of the model parameters $\boldsymbol{\theta}$ given some observed data $\mathbf{Y}$,

$$p(\boldsymbol{\theta}|\mathbf{Y}) = \frac{p(\mathbf{Y}|\boldsymbol{\theta})p(\boldsymbol{\theta})}{p(\mathbf{Y})} \tag{3}$$

We may interpret this as combining the likelihood $p(\mathbf{Y}|\boldsymbol{\theta})$ with a prior probability distribution $p(\boldsymbol{\theta})$ that characterizes our prior beliefs about the range of values each rate parameter is likely to assume. We note that in a maximum-likelihood approach, we also have to decide on the range of values within which we employ some optimization scheme to search for a solution; a uniform prior plays a similar role, setting the upper and lower bounds on the search space, and ensures we are explicit and clear about any such assumptions that we make. The posterior distribution results from a product of two probability distributions, the likelihood and the prior, and is therefore itself a valid probability distribution. The marginal likelihood $p(\mathbf{Y})$ acts as the normalizing constant, ensuring that the posterior density has unit mass, i.e., the posterior integrates to 1, as is required of a valid normalized probability distribution. In addition this quantity provides us with a method of comparing model hypotheses, which implicitly takes into account the number of parameters each mechanism has. We can see this more clearly by realizing that the posterior distribution is also implicitly conditioned on the model M itself,

$$p(\boldsymbol{\theta}|\mathbf{Y}, M) = \frac{p(\mathbf{Y}|\boldsymbol{\theta}, M)p(\boldsymbol{\theta}|M)}{p(\mathbf{Y}|M)} \tag{4}$$

The marginal likelihood, $p(\mathbf{Y}|M)$, is therefore the probability of the data given a particular model, and this plays an important role as it allows us to compare different models that may have different numbers of parameters. For each model the marginal likelihood is computed by integrating over all possible parameter values,

Table 1
Bayes factors may be interpreted using the following scale

$B_{1,2}$	Evidence in favor of model M_1
1–3	Not worth more than a bare mention
3–10	Substantial
10–100	Strong
>100	Decisive

$$p(\mathbf{Y}|M) = \int p(\mathbf{Y}, \boldsymbol{\theta}|M)\mathrm{d}\boldsymbol{\theta} = \int p(\mathbf{Y}|\boldsymbol{\theta}, M)p(\boldsymbol{\theta}|M)\mathrm{d}\boldsymbol{\theta} \tag{5}$$

The Bayes factor $B_{1,2}$ gives the strength of evidence in favor of model 1 compared to model 2. Given the marginal likelihood for competing hypothesized structures we may then calculate Bayes factors in order to rank the models based on the evidence of the data as follows,

$$B_{1,2} = \frac{p(\mathbf{Y}|M_1)}{p(\mathbf{Y}|M_2)} \tag{6}$$

Bayes factors may be usefully interpreted using the scale in Table 1, as proposed in [2]. These numbers allow us to compare model hypotheses in a quantitative and systematic manner, in comparison to the commonly used subjective approach of visually inspecting the model fit of ion channel models. Bayes factors also automatically take into account the model complexity, since they are defined as an integral (or weighted average) over all model parameters and will naturally pick the least complex model that is still complex enough to describe the data [2].

5 Previous Bayesian Approaches in Ion-Channel Modeling

Subsequent to the maximum-likelihood approaches described earlier in this chapter, there have been a number of Bayesian approaches for modeling ion channels. Compared to the maximum likelihood approach, the use of Bayesian methodology has so far been limited in the field. However, some Bayesian techniques have been implemented to perform signal reconstruction of noisy experimental data, obtain inferences over the rate parameters in specified models, and perform model selection between competing kinetic mechanisms for physiologically different ion channels.

One of the key initial inferential challenges is to recreate idealized signals from noisy data with limited time resolution.

There have been a number of Bayesian approaches to reconstruct a continuous signal from noisy data.

An initial attempt to use HMMs in a Bayesian context to provide a local and global signal restoration is given in [43]. An attempt to reconstruct a simulated dataset by fitting an alternating renewal process to the signal where open and closed sojourn times are gamma distributed is given in [44], in which the noise of the signal is modeled as an autoregressive process. The main technique used for drawing samples of appropriate sojourn times has been the reversible jump algorithm of [45], which proposes changes in the number and placement of observed sojourn times in the record. The authors noted however that their method is inefficient for large numbers of points and therefore is ill-suited for long ion channel recordings in practice.

An alternative restoration procedure was proposed in [46]. Their approach proceeds by sampling the hidden states at the sampling intervals given the noisy signal and the model parameters using a stochastic version of the Baum forward–backward algorithm proposed by Carter and Kohn [3] and a Gibbs sampling scheme. The restored signal is then obtained from the aggregation of state sequences along with the noise parameters for the observed recording, which is assumed to be a white noise process. The advantage of this approach is the low computational cost, with the method outperforming thresholding after low-band pass filtering.

Bayesian methods have also been employed in parameter estimation and model selection. Early work by Ball et al. [47] approximated the continuous time process using discrete sampling intervals and constructed Gibbs and Metropolis–Hastings samplers to extract a transition rate matrix for the hidden states across the given sampling interval. The **Q** matrix could then be inferred from the transition matrix. Subsequent extensions for modeling the gating process in continuous time followed [48, 49] using reversible jump techniques for sampling sojourn intervals of the observed process. This method also performs signal reconstruction as well as estimating transition rates and means and variances of the conductance states from the raw noisy data.

Modeling the noise process is considered crucial if a model is being fit directly to the experimental data without an idealization step. An early attempt to provide a complete noise model within a Bayesian framework is illustrated by de Gunst et al. [50]. This approach models the underlying signal as a discrete HMM where the signal is corrupted by the conductance level of the state, noise correlation from the recording equipment, and the smoothing effects of low-pass filtering on the raw signal. This noise modeling approach is computationally more expensive but its benefits were seen in the subsequent analysis of the K^+ outward rectifier channel in Barley Leaf [51]. The particular physiology of this receptor exhibits "flickering" behavior—rapid closings in long open intervals which can be missed if the filter effects are not taken into account and which makes accurate signal idealization difficult. This subsequent study also used reversible jump techniques for model selection. The flat posterior distribution across the analyzed models highlighted the requirements for large stretches of data to distinguish between models, increasing the computational burden of the long sampling times given the required full noise model.

Other discrete time approximations include [52] which also incorporates a noise model accounting for correlated noise and low-pass filtering. The Markov chain Monte Carlo (MCMC) approach samples a hidden realization of the unobservable Markov model using stochastic methods introduced by Carter and Kohn [3], which was extended to incorporate the required features of the noise model. This framework was applied to Ryanodine receptors, which play a significant role in intracellular calcium dynamics in the heart. This Bayesian approach was used to parameterize and evaluate 16 competing models of Ryanodine ion channel behavior.

Further recent approaches adopt fitting a continuous time Markov model directly to the noisy signal, akin to [49]. Such an approach was adopted in [53] to fit models directly to single ion channel data from IP3 receptors, which control intracellular calcium dynamics in the endoplasmic reticulum. This approach assumes that the signal is corrupted by Gaussian white noise and adopts the approximate effective rate constant correction for missed events described in [22]. The approach reveals that the main effect of Ca^{2+} is to modulate the probability that the receptor is in a state that is able to open directly, rather than to modulate the open probability directly. This sampling approach is extended in [54] to build a "Metropolised-Gibbs" sampler for the **Q** matrix with a generalized number of aggregated states. The sampler is used to perform Bayesian inference for model discrimination for the IP3 receptor.

The brief review above highlights the still nascent use of Bayesian methods for tackling the various challenges of signal detection, model parameterization, and model selection. As is usual with Bayesian methods, there is a trade off between model complexity and computational cost. However, the key balance is perhaps to be found in adapting the Bayesian approach to the physiology at hand, as investigations of different ion channels present different modeling challenges. For example, given the rapid openings of the glycine channel detected in previous investigations, the most useful Bayesian approaches should prioritize missed event correction and signal restoration. In addition, any approach must be scalable enough for performing inference over the large state spaces of the Markov models that are now commonplace.

6 MCMC for Ion Channel Modeling

It is advantageous to move from a point estimate using maximum likelihood to one characterized by a probability distribution that fully describes the identifiability and sensitivities of the model parameters. MCMC methods offer us a way of drawing samples from arbitrary probability distributions, and advances in this area have fueled the rise of Bayesian approaches to statistical inference over the last 20 years.

The idea is to simulate a Markov chain such that its stationary distribution is the density of interest; in the case of Bayesian statistics this is the posterior distribution. The position of a Markov chain corresponds to a set of model parameter values $\boldsymbol{\theta}$, and the next position of the Markov chain, $\boldsymbol{\theta}^*$, depends only on the values at the previous position, as demonstrated in Algorithm 1.

Algorithm 1. Standard Metropolis–Hastings Algorithm

1. Given current position $\boldsymbol{\theta}$, draw proposed position $\boldsymbol{\theta}^*$ from the proposal distribution $q(\boldsymbol{\theta}^* |\boldsymbol{\theta})$
2. Calculate the acceptance ratio $\mathrm{R}(\boldsymbol{\theta}^*|\boldsymbol{\theta}) = min\left[1, \frac{p(\boldsymbol{\theta}^*)q(\boldsymbol{\theta}|\boldsymbol{\theta}^*)}{p(\boldsymbol{\theta})q(\boldsymbol{\theta}^*|\boldsymbol{\theta})}\right]$
3. Draw $U \sim \mathrm{Uniform}[0, 1]$
4. Let $\boldsymbol{\theta} = \begin{cases} \boldsymbol{\theta}^* & \textit{if}\ \mathrm{U} < R(\boldsymbol{\theta}^*|\boldsymbol{\theta}) \\ \boldsymbol{\theta} & \textit{otherwise} \end{cases}$

At each iteration, a new position is proposed conditional on the current position of the chain. This is then accepted with some probability, such that as the number of moves becomes large, the collected points constitute correlated samples from the desired probability distribution. This property is ensured by the *balance condition*,

$$p(\boldsymbol{\theta}^*) = \int p(\boldsymbol{\theta})A(\boldsymbol{\theta}^*|\boldsymbol{\theta})\mathrm{d}\boldsymbol{\theta} \tag{7}$$

which states that the average probability of moving from any point to the current point is equal to the probability of the current point itself. The term $A(\boldsymbol{\theta}^*|\boldsymbol{\theta})$ defines the total probability of firstly a proposed point in the parameter space being sampled from the proposal distribution $q(\boldsymbol{\theta}^*|\boldsymbol{\theta})$, and secondly this proposed point actually being accepted with probability $R(\boldsymbol{\theta}^*|\boldsymbol{\theta})$, such that

$$A(\boldsymbol{\theta}^*|\boldsymbol{\theta}) = q(\boldsymbol{\theta}^*|\boldsymbol{\theta})R(\boldsymbol{\theta}^*|\boldsymbol{\theta}) \tag{8}$$

If a Markov chain is *reversible*, then the probability of moving from $\boldsymbol{\theta}$ to $\boldsymbol{\theta}^*$ is the same as the probability of the reverse move, such that,

$$p(\boldsymbol{\theta})A(\boldsymbol{\theta}^*|\boldsymbol{\theta}) = p(\boldsymbol{\theta}^*)A(\boldsymbol{\theta}|\boldsymbol{\theta}^*) \tag{9}$$

This is known as the *detailed balance condition* and it provides an easy way of satisfying the balance condition in Eq. 7. Finally, it has been shown that employing the following acceptance probability results in a reversible Markov chain that satisfies detailed balance, and therefore converges to the correct target distribution,

$$R(\boldsymbol{\theta}^*|\boldsymbol{\theta}) = \min\left\{1, \frac{p(\boldsymbol{\theta}^*)q(\boldsymbol{\theta}|\boldsymbol{\theta}^*)}{p(\boldsymbol{\theta})q(\boldsymbol{\theta}^*|\boldsymbol{\theta})}\right\} \tag{10}$$

In practice it sometimes only requires a few thousand samples to obtain information about the structure of the target distribution, although this depends on the dimensionality of the problem and the level of correlation in the samples. Initially, the Markov chain is likely to be in a region of negligible probability and there will be a transient "burn-in" period during which the chain must converge to the stationary distribution and find regions of high probability mass. After the burn-in period, the samples of the Markov chain are assumed to be truly representative of the target distribution.

Given these samples we may calculate summary statistics, such as means and covariances, and plot the posterior samples to see the global correlation structure. The algorithm proceeds by sampling new points from the proposal distribution $q(\boldsymbol{\theta}^*|\boldsymbol{\theta})$, and the choice of this density can drastically affect the performance and efficiency of the MCMC algorithm. Often, geometric information may be employed to propose better moves; for example the MALA sampler [55] utilizes gradient information to move the mean of the proposals to points of higher probability. However such geometric information is not always available, and even when it is it may be computationally costly to obtain. In this chapter we shall compare standard Metropolis–Hastings, as described above, with Adaptive MCMC, another MCMC variant that also requires only the pointwise evaluation of the likelihood and is designed to offer improved sampling from strongly correlated probability distributions.

We employ the Adaptive MCMC algorithm described in [56, 57], whereby the mean and covariance of the proposal distribution at the *i*th iteration are given by $\boldsymbol{\mu}_i$ and $\boldsymbol{\Sigma}_i$ and are updated taking into account the next sampled values of the chain $\boldsymbol{\theta}_{i+1}$ as follows,

$$\boldsymbol{\mu}_{i+1} = \boldsymbol{\mu}_i + \gamma(\boldsymbol{\theta}_{i+1} - \boldsymbol{\mu}_i) \tag{11}$$

$$\boldsymbol{\Sigma}_{i+1} = \boldsymbol{\Sigma}_i + \gamma((\boldsymbol{\theta}_{i+1} - \boldsymbol{\mu}_i)(\boldsymbol{\theta}_{i+1} - \boldsymbol{\mu}_i)^T - \boldsymbol{\Sigma}_i) \tag{12}$$

The parameter γ controls the rate of adaptation and may be set to $\gamma = 0.05$. For further details and discussion of the properties and implementation of Adaptive MCMC algorithms see [56, 57].

We note that an alternative method is required for sampling from multimodal posterior distributions. One approach is to use a population MCMC method based on parallel tempering. This proceeds in a similar manner as described in [58] by defining a collection of tempered distributions, given by $p(\boldsymbol{\theta}|\mathbf{Y}, \beta_i) \propto p(\mathbf{Y}|\boldsymbol{\theta})^{\beta_i} p(\boldsymbol{\theta})$, for $i = 1 : N$. The N points are spaced between 0 and 1 inclusively, such that the first tempered distribution is proportional to the prior, $p(\boldsymbol{\theta}|\mathbf{Y}, \beta_1 = 0) \propto p(\boldsymbol{\theta})$, and the last tempered distribution is proportional to the true posterior, $p(\boldsymbol{\theta}|\mathbf{Y}, \beta_N = 1) \propto p(\mathbf{Y}|\boldsymbol{\theta})p(\boldsymbol{\theta})$. The tempered distributions therefore form a smooth path of probability distributions from the prior to the posterior. A Markov chain can then be run in each distribution simultaneously and at each iteration swapping moves between adjacent temperatures can be proposed, which satisfy detailed balance across the overall product distribution [58]. In this manner we may draw samples from all tempered distributions, which results in better sampling of a multimodal posterior distribution of interest as chains caught in local maxima may escape by moving up and down the temperature ladder via the swapping moves. We note that the use of such a tempered scheme also allows for efficient estimation of the marginal likelihood, which may be employed for model comparison [58], as mentioned previously in the chapter. In the simulations that follow, however, we found that the posterior distribution was unimodal and we therefore employed the standard single chain versions of the Metropolis–Hastings and adaptive MCMC algorithms.

7 Bayesian Analysis of Ion Channel Models

We consider two aggregated Markov models of differing complexity that may be used to describe ion channel dynamics. We begin with the simplest possible Markov process model that can describe a ligand-activated ion channel. The following model was suggested by del-Castillo and Katz [1] and comprises two closed states and an open state. Each of the closed states corresponds to the ligand being unbound and bound; Fig. 1 depicts this simple model structure. For the purpose of investigating Bayesian inference over such mechanisms, we generate synthetic data from the model, such that we know the true parameter values to determine the quality of inferences we may obtain given different quantities of data. We assume perfect resolution with no missed events, although the

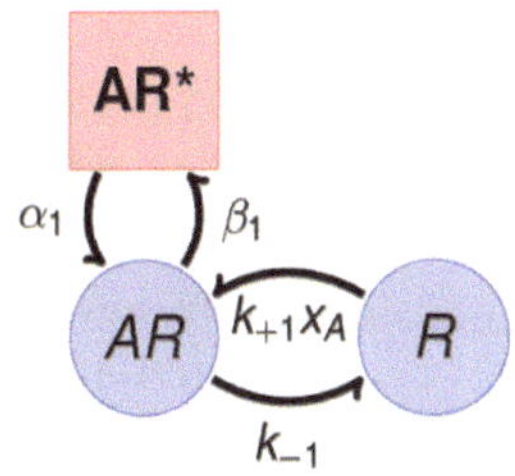

Fig. 1 Diagram showing the del-Castillo Katz model with two closed states (represented by *circles*) and one open state (represented by a *square*)

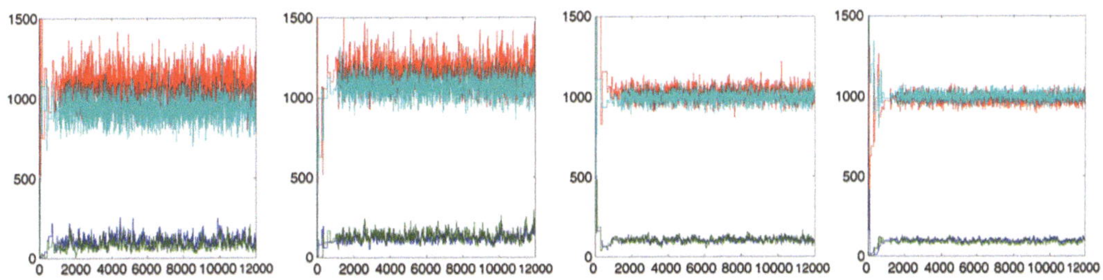

Fig. 2 Trace plots of a Markov chain using standard Metropolis–Hastings to explore the posterior distributions induced by using [0.5 s, 1 s, 3 s, 5 s] of data (from *left* to *right*). The first 2,000 samples are regarded as burn-in and the final 10,000 samples are considered correlated samples from the true posterior distribution

methodology is straightforwardly applied in the case of missed events by employing the appropriate likelihood function.

We employ the parameters $\alpha = 1{,}000$, $\beta = 1{,}000$, $k_{-1} = 100$ and $k_{+1}X_A = 100$, and generate [0.5, 1, 3, 5] seconds of data, corresponding to [370, 719, 1,992, 3,402] observations respectively. We infer the parameters with the standard Metropolis–Hastings algorithm using wide uniform priors and observe that the Markov chain converges to the same region of posterior probability mass from a wide variety of starting positions (Fig. 2). We employed 2,000 "burn-in" iterations to allow the chains to converge, after which 10,000 samples were collected to estimate summary statistics for the stationary posterior distribution. As we increase the length of the recording, and hence the number of observations, we observe that the uncertainty in the estimated parameter values decreases, as detailed in Table 2. For this relatively simple example, all four parameters are identifiable and the posterior is unimodal, from which a standard Metropolis–Hastings algorithm is able to draw samples efficiently.

Let us now consider a more realistic and complex model with five hidden states and nine rate parameters to be inferred from the

Table 2
Comparison of summary statistics of log posterior distributions inferred for the del Castillo–Katz model based on 5 runs of 10,000 samples with varying lengths of simulated recordings

Length of recording (s)	Mean $(\alpha, \beta, k_{-1}, k_{+1}X_A)$	Standard deviation $(\alpha, \beta, k_{-1}, k_{+1}X_A)$
0.5	(1.99, 1.86, 3.03, 2.97)	(0.149, 0.166, 0.036, 0.032)
1	(2.06, 2.10, 3.06, 3.03)	(0.091, 0.111, 0.030, 0.023)
3	(1.99, 2.01, 3.01, 2.99)	(0.056, 0.065, 0.017, 0.014)
5	(1.98, 1.94, 2.99, 2.99)	(0.045, 0.052, 0.013, 0.011)

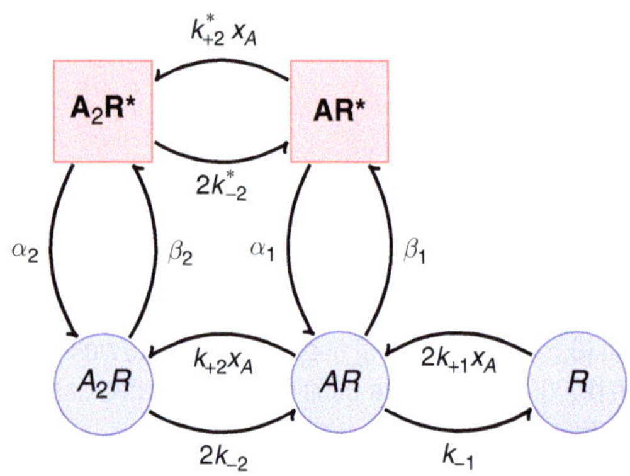

Fig. 3 A five state model with three closed states (represented by *circles*) and two open states (represented by *squares*)

data, shown in Fig. 3. This model represents the simplest possible mechanism that may describe a (muscle-type) nicotinic acetylcholine receptor, yet we will see that the induced posterior distribution exhibits a much more complicated structure.

This time we assume a fixed agonist concentration $X_A = 1e - 7$, and employ parameter values $k_{+1} = 5e7$, $k_{-1} = 2{,}000$, $k_{+2} = 5e8$, $k_{-2} = 2{,}000$, $k^*_{+2} = 15$, $k^*_{-2} = 3{,}000$, $\beta_1 = 15{,}000$, $\alpha_1 = 500$, $\beta_2 = 5e8$, to generate 2 s of data, resulting in 4,419 observations. Parameter α_2 was then calculated automatically based on the other parameter values by assuming detailed balance. From this dataset, we generate 2 shorter datasets of 1,130 and 2,946 observations. In each case, starting the chain from multiple initial values results in convergence to the same region of the posterior, suggesting that the induced posterior distribution is unimodal, although there appears to be a very strong correlation structure

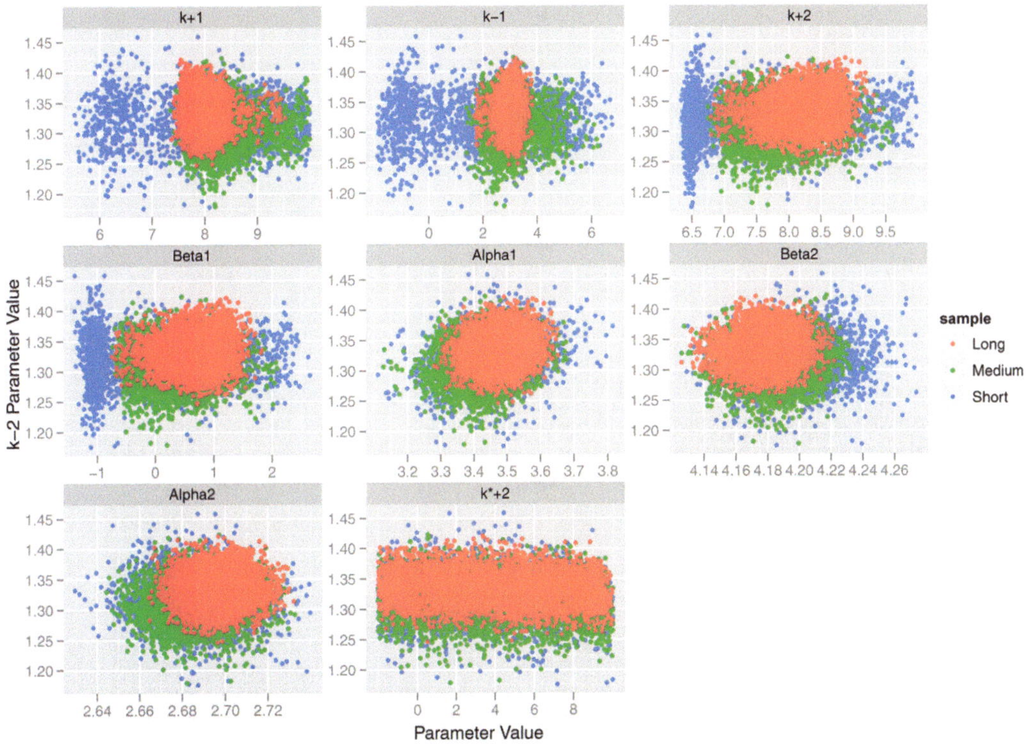

Fig. 4 Scatter plots comparing the posterior samples drawn with the five state model using Adaptive MCMC and the short, medium, and long datasets. Parameter 4 is plotted against each of the other eight parameters to display the marginal pairwise correlation structure that is present. All parameters are plotted in log space. Adaptive MCMC is able to sample from all regions of high density in the parameter space, despite the long tails, strong correlation structure, and unidentifiability of parameter K_{+2}^{*}

between the parameters, which results in poor sampling when using the standard Metropolis–Hastings algorithm. We therefore require a more advanced MCMC sampler to help ensure we fully explore all areas of high probability mass. We employ an Adaptive MCMC algorithm [56], in which the proposal distribution is chosen dependent on the previously sampled values of the chain via an adaptation procedure that attempts to estimate the covariance structure of the target distribution.

Figure 4 shows the overlaid scatter plots of the samples obtained using each of the three datasets, and Fig. 5 shows the overlaid kernel density estimates of the marginal posterior distributions, both using Adaptive MCMC. Adaptive MCMC offers far better sampling than standard Metropolis–Hastings, as it allows the underlying correlations to be taken into account when proposing moves through the parameter space. This is clear from Fig. 6, which compares the trace plots using the two methods. The standard Metropolis–Hastings algorithm would therefore need to be

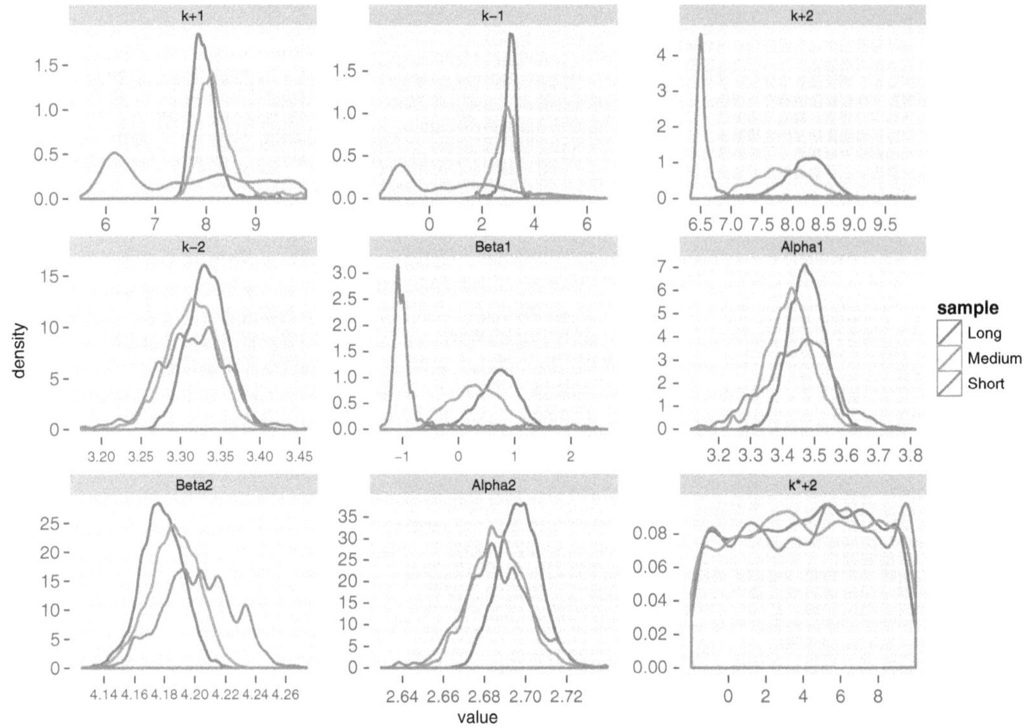

Fig. 5 Kernel density estimate plots comparing the marginal log posterior distributions of each of the parameters of the five state model using varying amounts of data and an adaptive MCMC sampling scheme. We observe that the marginal posterior distributions become less diffuse, as the amount of observed data increased, with the exception of parameter K_{+2}^{*}, which remains unidentifiable

run for a much larger number of iterations to obtain same number of effectively independent samples as the Adaptive MCMC algorithm produces.

8 Conclusions

The maximum likelihood methods introduced at the start of this chapter attempt to optimize the probability of the data given the parameters of a particular model; however this approach may pose problems. As we observed from the simulation study, there may be multiple sets of parameters that have similar likelihoods and ideally we wish to characterize them all. We have shown that using a Bayesian approach allows the full uncertainty in the parameters to be quantified, including any unidentifiable and strongly correlated structures that might be present.

Such a Bayesian approach however requires careful application of MCMC algorithms to obtain estimates of the posterior distribution. We have compared two approaches and discovered that for models with a relatively simple structure, consisting of five hidden

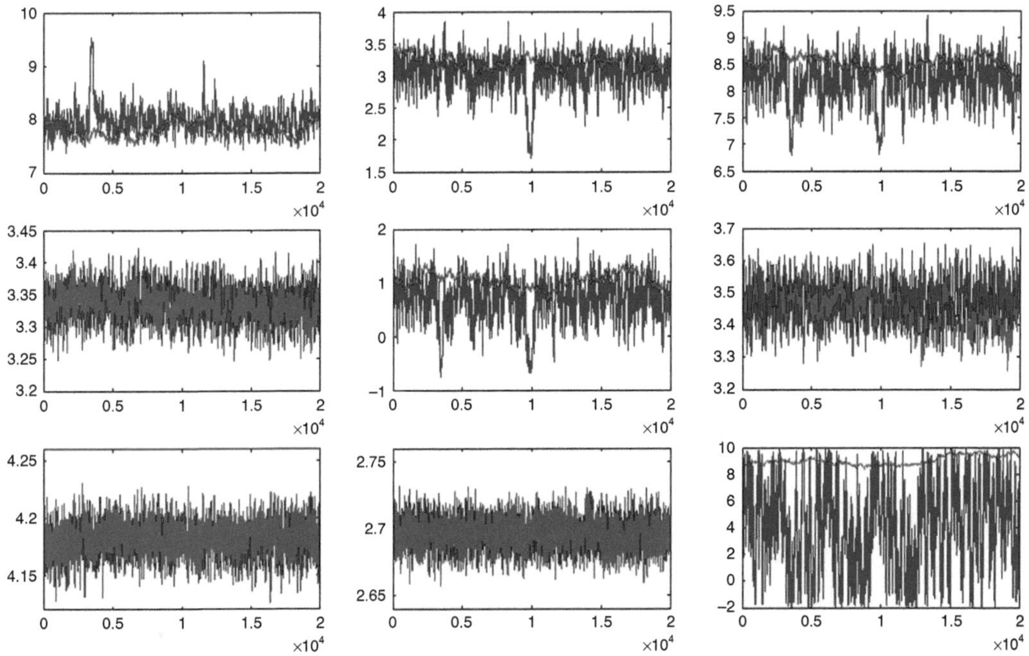

Fig. 6 A comparison of the log trace plots for each parameter using Metropolis–Hastings (*red*) and Adaptive MCMC (*blue*). The parameters plotted from *left* to *right* are K_{+1}, K_{-1}, K_{+2} on the *top line*, K_{-2}, Beta_1, Alpha_1 on the *middle line*, and Beta_2, Alpha_2, K_{+2}^* on the *bottom line*. It is clear from these plots that the Metropolis–Hastings sampler struggles to fully explore the posterior distribution efficiently due to the strong correlations between the parameters. Adaptive MCMC on the other hand takes these correlations into account and efficiently generates samples from the target distribution (Color figure online)

states, Adaptive MCMC was necessary to sample efficiently from the entire posterior distribution, and thus fully characterize the uncertainty in the parameter estimates.

One of the challenges of employing a Bayesian approach is the computational effort required; the current simulations took 15 min on a laptop; however larger models with more hidden states would take considerably longer, particularly if there are multiple modes necessitating the use of population MCMC. The implementation of such algorithms in a low level language such as C instead of Matlab should improve performance, although more realistically we require highly parallelized code on a computer cluster for performing inference on more complex models within a reasonable amount of time.

In future work we shall investigate the use of an appropriate likelihood to account for missed events in the data, and also tackle the problem of estimating marginal likelihoods. This kind of approach to modeling ion channels offers us a way of comparing hypothesized model structures in a systematic and quantitative manner, which should ultimately help us more rapidly advance our knowledge of these fundamental biological mechanisms.

References

1. Del Castillo J, Katz B (1957) Interaction at end-plate receptors between different choline derivatives. Proc R Soc Lond Ser B Biol Sci 146 (924):369–381
2. Kass R, Raftery AE (1995) Bayes factors. J Am Stat Assoc 90:773–795
3. Carter CK, Kohn R (1994) On Gibbs sampling for state space models. Biometrika 81 (3):541–553
4. Hawkes AG, Jalali A, Colquhoun D (1990) The distributions of the apparent open times and shut times in a single channel record when brief events cannot be detected. Phil Trans Phys Sci Eng 332:511–538
5. Hawkes AG, Jalali A, Colquhoun D, Hawkes AG, Jalali A, Colquhoun D (1992) Asymptotic distributions of apparent open times and shut times in a single channel record allowing for the omission of brief events. Phil Trans R Soc Lond Ser B Biol Sci 337 (1282):383–404
6. Ball F, Sansom M (1988) Aggregated Markov processes incorporating time interval omission. Adv Appl Prob 20:546–572
7. Ball FG, Sansom MS (1988) Single-channel autocorrelation functions: the effects of time interval omission. Biophys J 53(5):819–832
8. Burzomato V, Beato M, Groot-Kormelink PJ, Colquhoun D, Sivilotti LG (2004) Single-channel behavior of heteromeric $\alpha 1\beta$ glycine receptors: an attempt to detect a conformational change before the channel opens. J Neurosci 24(48):10924–10940
9. Colquhoun D, Hawkes AG (1981) On the stochastic properties of single ion channels. Proc R Soc Lond Ser B Biol Sci 211 (1183):205–235
10. Colquhoun D, Hawkes AG (1982) On the stochastic properties of bursts of single ion channel openings and of clusters of bursts. Phil Trans R Soc Lond B Biol Sci 300 (1098):1–59
11. Fredkin DR, Montal M, Rice JA (1985) Identification of aggregated Markovian models: application to the nicotinic acetylcholine receptor. In: Proceedings of the Berkeley conference in honor of Jerzy Neyman and Jack Kiefer, vol 1, pp 269–289
12. Colquhoun D, Hawkes AG (1987) A note on correlations in single ion channel records. Proc R Soc Lond Ser B Biol Sci 230(1258):15–52
13. Kienker P (1989) Equivalence of aggregated Markov models of ion-channel gating. Proc R Soc Lond B Biol Sci 236(1284):269–309
14. Blatz AL, Magleby KL (1986) Quantitative description of three modes of activity of fast chloride channels from rat skeletal muscle. J Physiol 378(1):141–174
15. Colquhoun D, Hawkes AG, Srodzinski K (1996) Joint distributions of apparent open and shut times of single-ion channels and maximum likelihood fitting of mechanisms. Phil Trans R Soc Lond Ser A Math Phys Eng Sci 354(1718):2555–2590
16. Colquhoun D, Hatton CJ, Hawkes AG (2003) The quality of maximum likelihood estimates of ion channel rate constants. J Physiol 547 (3):699–728
17. Horn R, Lange K (1983) Estimating kinetic constants from single channel data. Biophys J 43(2):207–223
18. Fredkin DR, Rice JA (1986) On aggregated Markov processes. J Appl Prob 23(1):208
19. Ball FG, Sansom MSP (1989) Ion-channel gating mechanisms: model identification and parameter estimation from single channel recordings. Proc R Soc Lond B Biol Sci 236 (1285):385–416
20. Ball FG, Rice JA (1989) A note on single-channel autocorrelation functions. Math Biosci 97(1):17–26
21. Roux B, Sauve R (1985) A general solution to the time interval omission problem applied to single channel analysis. Biophys J 48 (1):149–158
22. Blatz AL, Magleby KL (1986) Correcting single channel data for missed events. Biophys J 49(5):967–980
23. Crouzy SC, Sigworth FJ (1990) Yet another approach to the dwell-time omission problem of single-channel analysis. Biophys J 58 (3):731–743
24. Nelder JA, Mead R (1965) A simplex method for function minimization. Comput J 7 (4):308–313
25. Qin F, Li L (2004) Model-based fitting of single-channel dwell-time distributions. Biophys J 87(3):1657–1671
26. Rabiner L, Juang B (1986) An introduction to hidden Markov models. ASSP Mag IEEE 3 (1):4–16
27. Qin F, Auerbach A, Sachs F (1996) Estimating single-channel kinetic parameters from idealized patch-clamp data containing missed events. Biophys J 70(1):264–280
28. Qin F, Auerbach A, Sachs F (1997) Maximum likelihood estimation of aggregated Markov

processes. Proc R Soc Lond Ser B Biol Sci 264 (1380):375–383
29. Rabiner LR (1989) A tutorial on hidden Markov models and selected applications in speech recognition. Proc IEEE 77(2):257–286
30. Baum LE, Petrie T, Soules G, Weiss N (1970) A maximization technique occurring in the statistical analysis of probabilistic functions of Markov chains. Ann Math Stat 41(1):164–171
31. Qin F, Auerbach A, Sachs F (2000) Hidden Markov modeling for single channel kinetics with filtering and correlated noise. Biophys J 79(4):1928–1944
32. Qin F, Auerbach A, Sachs F (2000) A direct optimization approach to hidden Markov modeling for single channel kinetics. Biophys J 79(4):1915–1927
33. Venkataramanan L, Walsh JL, Kuc R, Sigworth FJ (1998) Identification of hidden Markov models for ion channel currents. I. Colored background noise. IEEE Trans Signal Process 46(7):1901–1915
34. Venkataramanan L, Kuc R, Sigworth FJ (1998) Identification of hidden Markov models for ion channel currents. II. State-dependent excess noise. IEEE Trans Signal Process 46 (7):1916–1929
35. Fredkin DR, Rice JA (2001) Fast evaluation of the likelihood of an HMM: ion channel currents with filtering and colored noise. IEEE Trans Signal Process 49(3):625–633
36. Colquhoun D, Sakmann B (1985) Fast events in single-channel currents activated by acetylcholine and its analogues at the frog muscle end-plate. J Physiol 369(1):501–557
37. Magleby KL, Weiss DS (1990) Identifying kinetic gating mechanisms for ion channels by using two-dimensional distributions of simulated dwell times. Proc R Soc Lond Ser B Biol Sci 241(1302):220–228
38. Magleby KL, Weiss DS (1990) Estimating kinetic parameters for single channels with simulation. A general method that resolves the missed event problem and accounts for noise. Biophys J 58(6):1411–1426
39. Jones MV, Westbrook GL (1995) Desensitized states prolong GABA A channel responses to brief agonist pulses. Neuron 15(1):181–191
40. Lape R, Colquhoun D, Sivilotti LG (2008) On the nature of partial agonism in the nicotinic receptor superfamily. Nature 454 (7205):722–727
41. Lape R, Krashia P, Colquhoun D, Sivilotti LG (2009) Agonist and blocking actions of choline and tetramethylammonium on human muscle acetylcholine receptors. J Physiol 587 (21):5045–5072
42. Lape R, Plested AJR, Moroni M, Colquhoun D, Sivilotti LG (2012) newblock The α1K276E startle disease mutation reveals multiple intermediate states in the gating of glycine receptors. J Neurosci 32(4):1336–1352
43. Fredkin DR, Rice JA (1992) Bayesian restoration of single-channel patch clamp recordings. Biometrics 48(2):427–448
44. Hodgson MEA (1999) A Bayesian restoration of an ion channel signal. J R Stat Soc Ser B (Stat Methodol) 61(1):95–114
45. Green PJ (1995) Reversible jump Markov chain Monte Carlo computation and Bayesian model determination. Biometrika 82 (4):711–732
46. Rosales R, Stark JA, Fitzgerald WJ, Hladky SB (2001) Bayesian restoration of ion channel records using hidden Markov models. Biophys J 80(3):1088–1103
47. Ball FG, Cai Y, Kadane JB, O'Hagan A (1996) MCMC methods for discrete sojourn time ion channel data. Nottingham Stat Group Res Rep 96:12
48. Ball FG, Cai Y, O'Hagan A (1997) MCMC for hidden continuous-time Markov chains. University of Nottingham, Nottingham
49. Ball FG, Cai Y, Kadane JB, O'Hagan A (1999) Bayesian inference for ion–channel gating mechanisms directly from single-channel recordings, using Markov chain Monte Carlo. Proc R Soc Lond Ser A Math Phys Eng Sci 455 (1988):2879–2932
50. de Gunst MCM, Ünsch HR, Schouten JG (2001) Statistical analysis of ion channel data using hidden Markov models with correlated state-dependent noise and filtering. J Am Stat Assoc 96(455):805–815
51. Gunst MCM, Schouten JG (2005) Model selection and parameter estimation for ion channel recordings with an application to the K+ outward-rectifier in barley leaf. J Math Biol 50(3):233–256
52. Rosales RA, Fill M, Escobar AL (2004) Calcium regulation of single ryanodine receptor channel gating analyzed using HMM/ MCMC statistical methods. J Gen Physiol 123(5):533–553
53. Gin E, Falcke M, Wagner LE, Yule DI, Sneyd J (2009) Markov chain Monte Carlo fitting of single-channel data from inositol trisphosphate receptors. J Theor Biol 257(3):460–474
54. Siekmann I, Wagner II LE, Yule D, Fox C, Bryant D, Crampin EJ, Sneyd J (2011) MCMC estimation of Markov models for ion channels. Biophys J 100(8):1919–1929
55. Stramer O, Tweedie RL (1999) Langevin-type models. I: Diffusions with given stationary

distributions and their discretizations. Methodol Comput Appl Prob 1(3):283–306
56. Haario H, Saksman E, Tamminen J (2001) An adaptive Metropolis algorithm. Bernoulli 7:223–242
57. Andrieu C, Thoms J (2008) A tutorial on adaptive MCMC. Stat Comput 18:343–373
58. Calderhead B, Girolami M (2008) Estimating Bayes factors via thermodynamic integration and population MCMC. Comput Stat Data Anal 53(12):4028–4045

Chapter 14

Building Models Using Reactome Pathways as Templates

David Croft

Abstract

The first steps of building a new model can be very time-consuming, involving consulting many research papers and then assembling a plausible network of reactions. In this chapter, tools for speeding up this process will be discussed. Reactome is a database containing extensive coverage of pathways in *Homo sapiens* and numerous reference species. It offers researchers wishing to create new models from scratch various tools for extracting the relevant reactions, complete with layout information. In this chapter, two use cases will be described, in which a modeller provides certain essential pieces of information and Reactome automatically constructs the basic models and then dumps them in SBML-ML format.

Key words Reactome, Pathway, Reaction, Model, SBML, Data

1 Introduction

In order to construct a new model, systems biologists often start out by consulting the literature in the field of interest and pulling relevant reactions out of the papers they have read. These reactions then need to be spliced together and laid out, probably using a tool such as CellDesigner [5], which can then export the model as an SBML file or run simulations of the model. The user has to live with the uncertainty of whether they have found all of the relevant literature and whether it is reliable or not. Copying over the reactions by hand to a pathway editing tool is error-prone and time-consuming. This chapter will show how Reactome tools can be used to partially automate this process and increase its reliability.

Reactome is a curated, peer-reviewed pathway database. It focuses on pathways in *Homo sapiens* but also infers pathways to 20 other reference species, using gene orthologies. Curators record pathways, reactions, proteins, small molecules and subcellular compartment, as well as literature references backing up all of the reactions present in a pathway. Comprehensive literature searches

Maria Victoria Schneider (ed.), *In Silico Systems Biology*, Methods in Molecular Biology, vol. 1021,
DOI 10.1007/978-1-62703-450-0_14, © Springer Science+Business Media, LLC 2013

are made for each pathway, and only the literature deemed most reliable is actually used. A sophisticated data model allows all of this information to be stored in a computationally accessible and searchable form. In addition, diagrams are hand-drawn for all pathways, and the layout of these diagrams is stored as coordinates for all reactions and participating molecules.

The scope of Reactome is very broad, encompassing the traditional metabolic pathways but also extending to a wide range of signalling pathways, the cell cycle, apoptosis and mechanisms of bacterial and viral infection, amongst other things. At the time of writing, Reactome covered 187,219 proteins, taking part in 52,818 distinct reactions, of which 5,568 were from *H. sapiens.*

Reactome has tools that can accept as input lists of molecule identifiers, e.g., UniProt IDs, and then determine which reactions or pathways these molecules take part in. From there, it is possible to export entire pathways, or sets of individual reactions, as models in SBML files. If required, layout data can also be incorporated into these models, based on the hand-drawn diagrams produced by Reactome curators.

SBML files can be imported by a wide range of simulation and model editing tools, where they can then be further enriched with extra information, such as reaction kinetics.

So the general concept of model generation using Reactome works as follows:

1. The user puts together a list of proteins, genes and possibly small molecules that they expect to see in their model.
2. The pathways or reactions in which these entities are involved are determined using Reactome tools.
3. An export to SBML creates the model.
4. The SBML is imported into an external tool for further enrichment and modelling.

2 Materials

2.1 General

In order to obtain the best performance from Reactome, a computer with a processor speed of at least 1.5 GHz and memory of at least 2 GB is recommended. Reactome will work with Windows, MacOS and Linux. Reactome is a browser-based tool, and a broadband connection will be needed. It is known to work in Internet Explorer 7 and 8, Chrome, Firefox, Safari and Opera (*see* **Note 1**). Firefox is the browser that is recommended. To start a Reactome session, the URL http://www.reactome.org should be entered into the browser. This will open the Reactome home page.

In the following sections, a number of Reactome tools relevant to model construction will be described in detail.

2.2 Analyze Expression Data

From the Reactome home page, the button on the left-hand side, labelled *Analyze Expression Data*, should be clicked (*see* **Note 2**). This tool can also be used to analyze simple lists of identifiers, and it will be used in this way in the Methods section.

Identifiers should be stored in a simple text file, with each identifier separated by a newline (*see* **Note 3**). The following identifier types are known to Reactome:

- Reactome
- KEGG COMPOUND [6]
- PubChem Substance [7]
- ChEBI [8]
- GO [9]
- UniProt [10]
- RefSeq [11]
- Ensembl [12]
- Affymetrix [13]
- NCBI gene [14]
- IPI [15]
- Illumina [16]
- OMIM [17]
- EC [18]
- MGI [19]
- PDB [20]
- EMBL [21]
- miRBase [22]

Identifiers will be automatically recognized by Reactome in most cases. If one has purely numerical identifiers, they will by default be assumed to be NCBI gene, but the user will be provided with the opportunity to select the identifier type after the analysis is complete.

Once an identifier list has been constructed, it can be either copied and pasted or uploaded as a file, as indicated in Fig. 1.

Clicking the *Analyze* button will, after a short delay, produce a list of the pathways known to Reactome and, for each one, an indication of the number of matching proteins from the supplied identifier list, as seen in Fig. 2 (*see also* **Note 4**).

2.3 BioMart

BioMart [23, 24] is a querying infrastructure that allows very general queries to be specified and which delivers the answers in tabular format. From the Reactome home page, it can be accessed by mousing over the *Tools* menu and clicking on *BioMart: query, link* [25]. *see* Fig. 3.

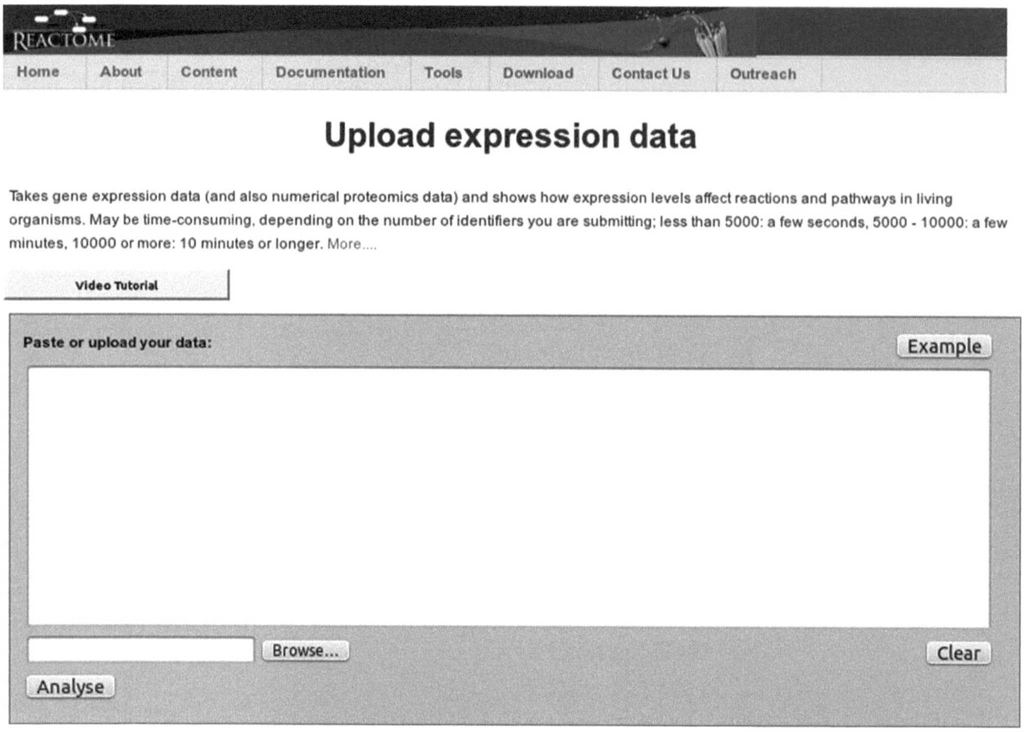

Fig. 1 Launch page for expression analysis, with functionality for uploading identifier lists

To start using it, a database, and dataset must be selected. Clicking on the dropdown labelled *CHOOSE DATABASE* will show the available databases. Once a database has been chosen, a new dropdown will appear, labelled *CHOOSE DATASET.* Selecting one of these datasets then causes the *Filters* and *Attributes* to appear on the left-hand side of the page (*see* **Note 5**).

Clicking on *Attributes* reveals the full range of attributes available for the given dataset. Clicking in any of the checkboxes causes the associated attribute to be added as a column to the output table. Clicking once on an already checked attribute will cause it to be deselected.

Clicking on *Filters* shows the filters relevant to the current dataset; these allow the user to put constraints on the results of the query (*see* **Note 6**). Only the first of these filters is significant for this chapter. This contains a text area, which can be used to copy and paste a newline-separated list of identifiers, plus a dropdown that allows the type of the identifier to be specified. The following identifier types are recognized: ChEBI compound, KEGG gene, NCBI gene, UniProt and Ensembl gene. If gene identifiers are used, Reactome will find the corresponding translated proteins and use those. This filter is not able to automatically determine the identifier type, so it must be explicitly specified.

Clicking the *Results* button causes the query to be performed. Results are returned in tabular form. By default, only the first ten

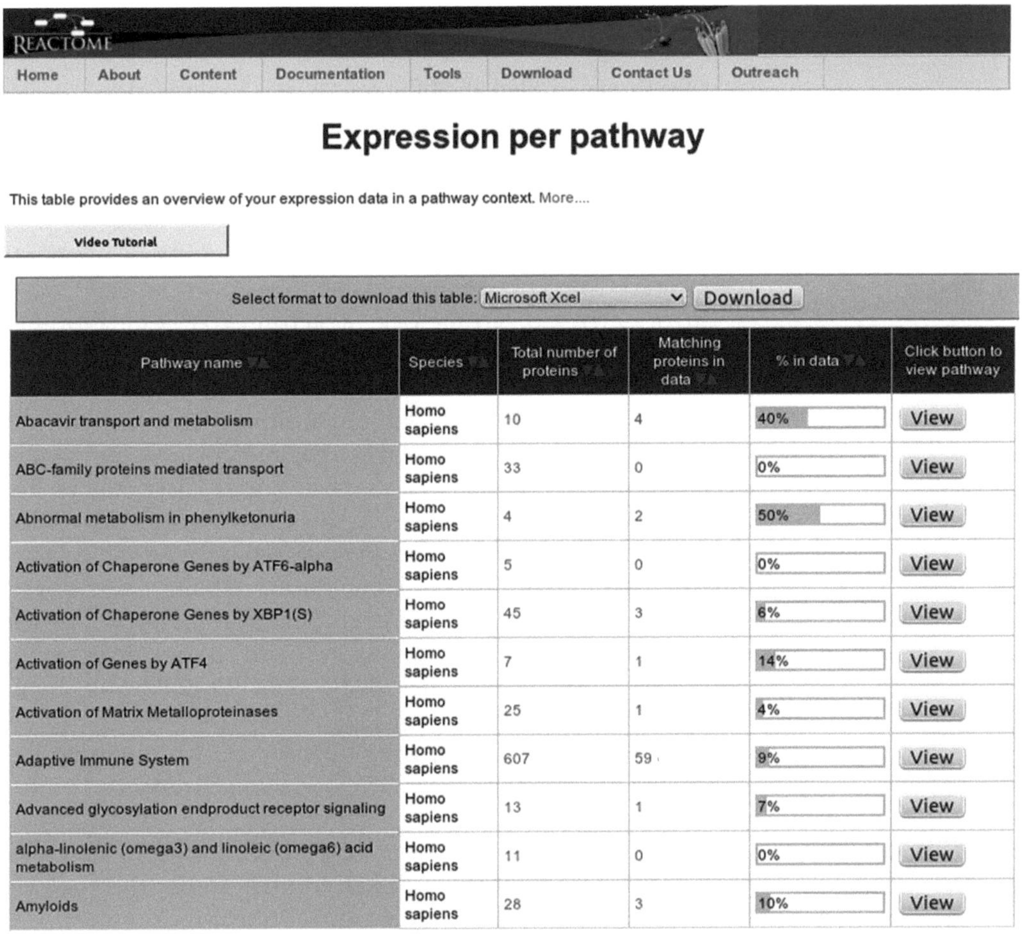

Pathway name	Species	Total number of proteins	Matching proteins in data	% in data	Click button to view pathway
Abacavir transport and metabolism	Homo sapiens	10	4	40%	View
ABC-family proteins mediated transport	Homo sapiens	33	0	0%	View
Abnormal metabolism in phenylketonuria	Homo sapiens	4	2	50%	View
Activation of Chaperone Genes by ATF6-alpha	Homo sapiens	5	0	0%	View
Activation of Chaperone Genes by XBP1(S)	Homo sapiens	45	3	6%	View
Activation of Genes by ATF4	Homo sapiens	7	1	14%	View
Activation of Matrix Metalloproteinases	Homo sapiens	25	1	4%	View
Adaptive Immune System	Homo sapiens	607	59	9%	View
Advanced glycosylation endproduct receptor signaling	Homo sapiens	13	1	7%	View
alpha-linolenic (omega3) and linoleic (omega6) acid metabolism	Homo sapiens	11	0	0%	View
Amyloids	Homo sapiens	28	3	10%	View

Fig. 2 Results page for expression analysis, showing the overlap with user data on a per-pathway basis

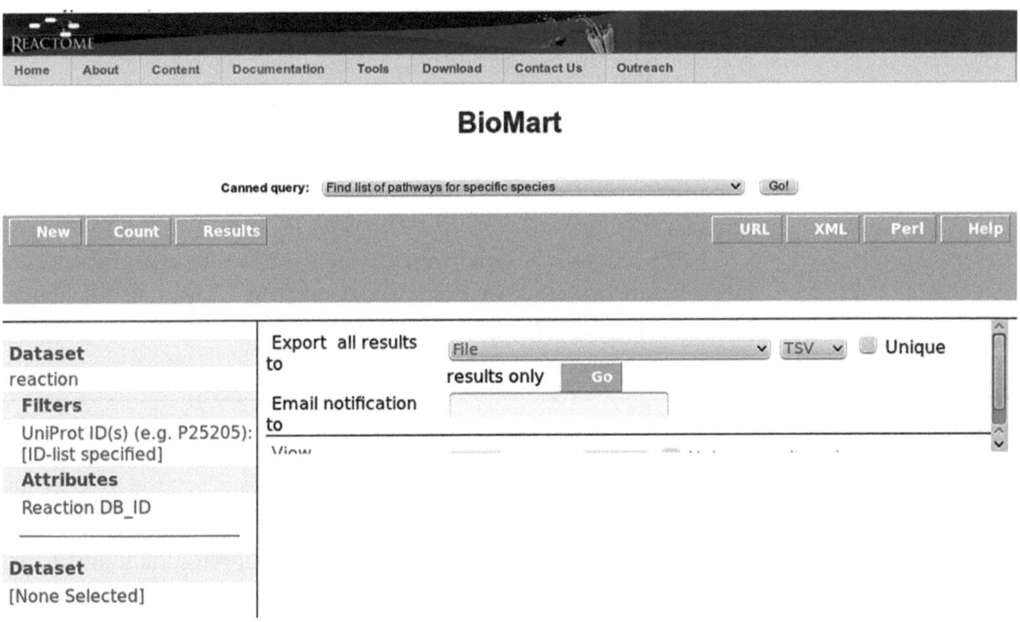

Fig. 3 BioMart start page

rows are shown. To get all of the results, it is recommended to click the *Go* button in the *Export all results to* section. This will allow the results to be deposited into a file in TSV (tab-separated value) format.

2.4 SBML Exporter: URL Version

The URL version of the exporter requires that a URL be typed into the browser. The URL stem looks like this:

http://www.reactome.org/ReactomeGWT/entrypoint/sbmlRetrieval?

To this can be added parameters that will determine the content of the SBML produced. These should be separated by & symbols. Parameters are specified as a parameter name, followed by an equals symbol, followed by a comma-separated list of values. The two parameters that are relevant to this chapter are *LAYOUT* and *ID*, which are used to specify the type of layout to use and pathway IDs, respectively. For example, if one wished to build a model based on the two pathways 109607 (Extrinsic Pathway for Apoptosis) and 169911 (Regulation of Apoptosis), with layout information embedded as SBGN into the SBML, then the following should be supplied to the browser:

http://www.reactome.org/ReactomeGWT/entrypoint/sbmlRetrieval?LAYOUT=SBGN&ID=109607,169911

2.5 Interactive SBML Exporter

The interactive SBML generator is used to turn a list of Reactome reaction IDs into a model in SBML format. To use this tool, the following URL should be entered into the browser:

http://www.reactome.org/ReactomeGWT/entrypoint.html#SBMLRetrievalPage

The layout of this page is illustrated in Figs. 4 and 5. The tool needs to be supplied with a list of newline-separated Reactome reaction identifiers (*see* **Note 7**).

The panel below the data entry area provides extensive possibilities for customizing the exported SBML. It is possible to set the desired SBML level and version, choose the format of the layout included with the SBML (if any), and filter according to various criteria.

Once the user is satisfied with the data format and the selected customizations, the *Generate SBML* button should be clicked to perform the export.

3 Methods

In this section, two use cases will be described, providing step-by-step instructions that lead users through the process of creating models based on the combination of user-supplied data and Reactome pathways. In the first use case, the pathways that are richest in proteins or genes found in the user's data will be used to generate a

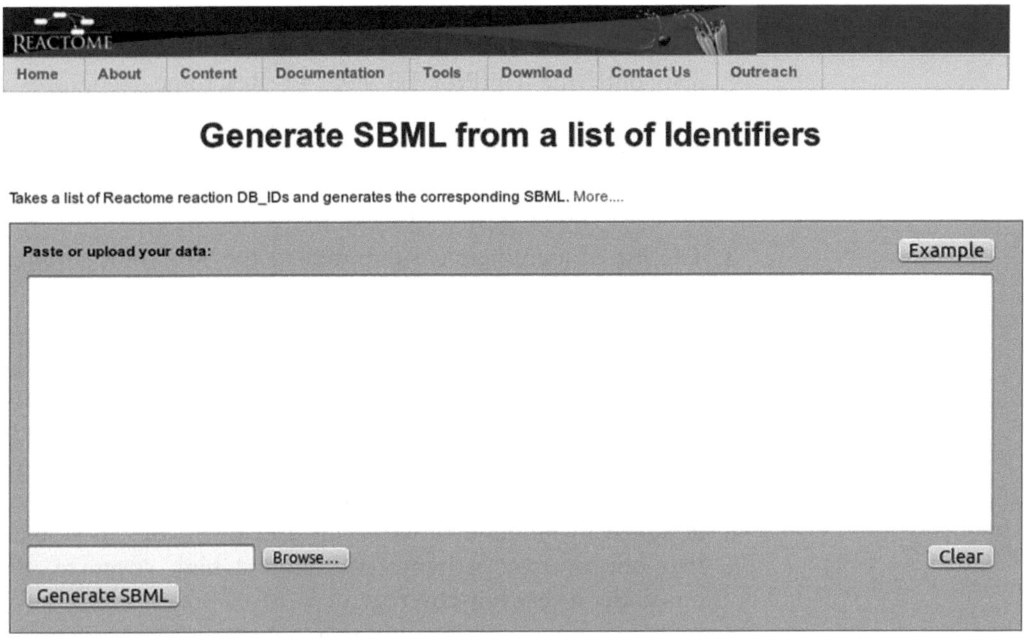

Fig. 4 Top of interactive SBML generator page, showing data entry panel

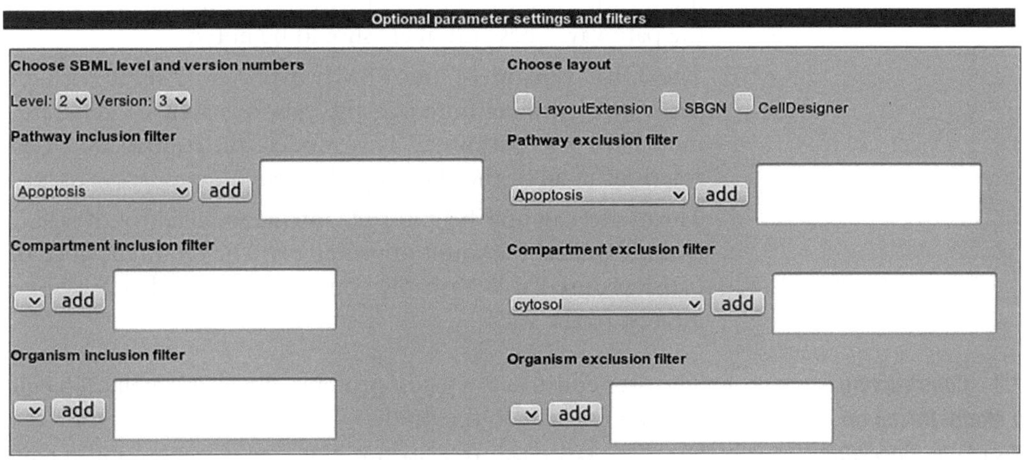

Fig. 5 Bottom of interactive SBML generator page, showing parameter setting options

model. In the second use case, all reactions in Reactome that utilize proteins, genes or small molecules from the user's data will be put together into a model.

3.1 Constructing a Model Based on Pathways Enriched in User Data

1. The user constructs a list of proteins and/or genes that are relevant to the model. It is not necessary for the list to be complete, since Reactome will try to fill in any gaps, but the more complete the list is, the more likely it is that the correct pathways will be found.

2. The user translates this list into a list of corresponding identifiers. For example, *serum amyloid P-component* might map onto the UniProt ID *P02743*. Reactome recognizes a wide variety of identifier types, which gives the user a significant amount of freedom. If gene identifiers are used, Reactome will find the corresponding translated proteins and use those. Identifiers should be separated by newlines.
3. The list of identifiers is submitted to the Reactome *Analyze Expression Data* tool.
4. Once the analysis has completed and the results table has been loaded, the user clicks on the top of the *% in data* column. The rows in the table will be reordered according to the percentage of the proteins in the Reactome pathway that are also found in the user's data, with those having the highest coverage appearing at the top of the table. Assuming that the dataset was designed to model a fairly specific area of biology, it is likely that one or two pathways will show a high coverage, the remainder a very low coverage.
5. For each of the pathways with high coverage, the user clicks on the *View* button. A new tab opens when this button is clicked. The user examines the URL and makes a note of the very last number in the URL. This is the Reactome internal identifier for the pathway. These numbers should be noted.
6. The URL version of the SBML exporter is used with the pathway identifiers found in the previous step to generate a model for the pathways. This model will contain all of the reactions in all of the chosen pathways.
7. The model can now be imported into a model editor. It is likely that a significant amount of pruning will be required, since the pathways used will probably contain reactions that are not of interest to the user.

3.2 Constructing a Model Based on Reactions Operating on Molecules in User Data

1. The user constructs a list of proteins, genes or small molecules that are relevant to the model.
2. The user translates this list into a list of corresponding identifiers. For example, *serum amyloid P-component* might map onto the UniProt ID *P02743*. Identifiers should be separated by newlines. This list needs to be as complete as possible, since the only reactions that will be found are those which contain identifiers in the user's data.
3. From the BioMart page, the *REACTOME* database is chosen and the *reaction* dataset selected.
4. Under *Filters*, the user pastes the identifier list into the text field of *Limit to reactions containing these IDs* and chooses the appropriate identifier type.

5. Under *Attributes*, the user must deselect *Reaction stable ID* (*see* **Note 8**).
6. The *Results* button is clicked.
7. The query results are exported to a file (*see* **Note 9**). This file must be edited and the first line deleted, since this simply contains the column name.
8. This file is then imported into the interactive SBML generator, using the *Browse* button (*see* **Note 10**). Clicking on *Generate SBML* produces the model.
9. The model can now be imported into a model editor. It is likely that a significant amount of pruning will be required, since the pathways used will probably contain reactions that are not of interest to the user.

4 Notes

1. The Reactome website is fairly robust, and it is likely that it will also work without problems in other browsers.
2. The name of this tool is a bit misleading in this context.
3. Clicking on the *Example* button will illustrate the required format, but note that the example provided also contains numerical expression values, which are not needed for model generation. Also, the example shows a table with column names. These are not mandatory.
4. The column labelled *Matching proteins in data* in this table contains, for each pathway, a count of the number of identifiers from the user's data that have been found in the pathway. These numbers are also links, and clicking on one of them provides a more detailed analysis of the identifiers hitting that particular pathway.
5. The interactions dataset is a little bit different from the others, in both the available attributes and filters, but it will not be used in any of the methods presented in this chapter.
6. Probably the best way to understand filters is as a form of query. The terms that are supplied to the filter are the ones that will be queried against.
7. Clicking on the *Example* button will illustrate the required format.
8. The idea is to produce an output table with only a single column for reaction identifiers.
9. The default file name suggested by BioMart may be used, but the file name and path should be noted.

10. At this point, the user can also select the desired level and version for the generated SBML. The defaults are level 2, version 3. If a layouter was selected, then reaction layout will also be incorporated into the SBML file. Note that the *CellDesigner* layouter currently does not work. Filters can be used for doing things like selecting only those reactions that occur within a given cellular compartment.

Acknowledgements

The development of Reactome is supported by a grant from the US National Institutes of Health (P41 HG003751), EU grant LSHG-CT-2005-518254 "ENFIN," Ontario Research Fund, and the EBI Industry Programme.

References

1. Matthews L, Gopinath G, Gillespie M et al (2009) Reactome knowledgebase of human biological pathways and processes. Nucleic Acids Res 37:D619–D622
2. Matthews L, D'Eustachio P, Gillespie M et al (2007) An Introduction to the Reactome Knowledgebase of Human Biological Pathways and Processes. Bioinformatics Primer, NCI/Nature Pathway Interaction Database. doi:10.1038/pid.2007.3
3. Croft D, O'Kelly G, Wu G et al (2011) Reactome: a database of reactions, pathways and biological processes. Nucleic Acids Res 39 (Database Issue):D691–D697
4. Hucka M, Finney A, Sauro HM et al (2003) The Systems Biology Markup Language (SBML): a medium for representation and exchange of biochemical network models. Bioinformatics 9(4):524–531
5. Funahashi A, Tanimura N, Morohashi M et al (2003) Cell Designer: a process diagram editor for gene-regulatory and biochemical networks. BIOSILICO 1:159–162
6. Kanehisa M, Goto S, Kawashima S et al (2004) The KEGG resource for deciphering the genome. Nucleic Acids Res 32(Suppl 1): D277–D280
7. Bolton E, Wang Y, Thiessen PA et al (2008) PubChem: integrated platform of small molecules and biological activities (Chapter 12). In: Wheeler RA, Spellmeyer DC (eds) Annual reports in computational chemistry, vol 4. American Chemical Society, Washington, DC
8. de Matos P, Alcántara R, Dekker A et al (2009) Chemical entities of biological interest: an update. Nucleic Acids Res 38(Database Issue):D249–D254
9. Ashburner M, Ball CA, Blake JA et al (2000) Gene ontology: tool for the unification of biology. The Gene Ontology Consortium. Nat Genet 25(1):25–29
10. The UniProt Consortium (2012) Reorganizing the protein space at the Universal Protein Resource (UniProt). Nucleic Acids Res 40: D71–D75
11. Pruitt KD, Tatusova T, Maglott DR (2007) NCBI reference sequences (RefSeq): a curated non-redundant sequence database of genomes, transcripts and proteins. Nucleic Acids Res 35(Database Issue):D61–D65
12. Paul Flicek M, Amode R, Barrell D et al (2011) Ensembl 2012. Nucleic Acids Res 40(Database Issue):D84–D90
13. Lockhart DJ et al (1996) Expression monitoring by hybridization to high-density oligonucleotide arrays. Nat Biotechnol 14 (13):1675–1680
14. Maglott D, Ostell J, Pruitt KD et al (2005) Entrez Gene: gene-centered information at NCBI. Nucleic Acids Res 33(Database Issue): D54–D58
15. Kersey PJ, Duarte J, Williams A et al (2004) The International Protein Index: an integrated database for proteomics experiments. Proteomics 4(7):1985–1988

16. Wang C, Krishnakumar S, Wilhelmy J et al (2012) High-throughput, high-fidelity HLA genotyping with deep sequencing. Proc Natl Acad Sci 109(22):8676–8681

17. Hamosh A, Scott A, Amberger J et al (2004) Online Mendelian Inheritance in Man (OMIM), a knowledgebase of human genes and genetic disorders. Nucleic Acids Res 33(Database Issue): D514–D517. doi:10.1093/nar/gki033

18. Webb EC (1992) Enzyme nomenclature 1992: recommendations of the Nomenclature Committee of the International Union of Biochemistry and Molecular Biology on the nomenclature and classification of enzymes. Published for the International Union of Biochemistry and Molecular Biology by Academic Press, San Diego. ISBN 0-12-227164-5

19. Eppig JT, Blake JA, Bult CJ et al (2012) The Mouse Genome Database (MGD): comprehensive resource for genetics and genomics of the laboratory mouse. Nucleic Acids Res 40(1): D881–D886

20. Berman HM, Henrick K, Nakamura H (2003) Announcing the worldwide Protein Data Bank. Nat Struct Biol 10:980

21. Cochrane G, Akhtar R, Bonfield J et al (2008) Petabyte-scale innovations at the European Nucleotide Archive. Nucleic Acids Res 37 (Database Issue):D19–D25. doi:10.1093/nar/gkn765

22. Griffiths-Jones S (2004) The microRNA Registry. Nucleic Acids Res 32(Suppl 1): D109–D111

23. Smedley D, Haider S, Ballester B et al (2009) BioMart—biological queries made easy. BMC Genomics 10:22

24. Zhang J, Haider S, Baran J et al (2011) BioMart: a data federation framework for large collaborative projects. Database (Oxford) 2011:bar038. doi:10.1093/database/bar038

25. Haw RA, Croft D, Yung CK et al (2011) The Reactome BioMart. Database (Oxford) 2011: bar031

Chapter 15

Uniform Curation Protocol of Metazoan Signaling Pathways to Predict Novel Signaling Components

Máté Pálfy, Illés J. Farkas, Tibor Vellai, and Tamás Korcsmáros

Abstract

A relatively large number of signaling databases available today have strongly contributed to our understanding of signaling pathway properties. However, pathway comparisons both within and across databases are currently severely hampered by the large variety of data sources and the different levels of detail of their information content (on proteins and interactions). In this chapter, we present a protocol for a uniform curation method of signaling pathways, which intends to overcome this insufficiency. This uniformly curated database called SignaLink (http://signalink.org) allows us to systematically transfer pathway annotations between different species, based on orthology, and thereby to predict novel signaling pathway components. Thus, this method enables the compilation of a comprehensive signaling map of a given species and identification of new potential drug targets in humans.

We strongly believe that the strict curation protocol we have established to compile a signaling pathway database can also be applied for the compilation of other (e.g., metabolic) databases. Similarly, the detailed guide to the orthology-based prediction of novel signaling components across species may also be utilized for predicting components of other biological processes.

Key words Literature curation, Signaling database, Signalogs, Orthology-based prediction

1 Introduction

Signal transduction pathways, functional building blocks of intracellular signaling, control various cellular processes, including cell growth, proliferation, differentiation, and stress response in divergent animal phyla [1]. In humans, defects in intracellular signaling can cause various diseases, *such as* cancer, neurodegeneration, muscle atrophy, immune deficiency, or diabetes. Therefore, a better understanding of the structure, function, and evolution of signal transduction is important for both basic research and medicine. This requires the construction of a comprehensive signaling map, which would (ideally) contain all components of distinct signaling pathways and their genetic and physical interactions. Genome programs and high-throughput (HTP) protein–protein interaction

Maria Victoria Schneider (ed.), *In Silico Systems Biology*, Methods in Molecular Biology, vol. 1021,
DOI 10.1007/978-1-62703-450-0_15, © Springer Science+Business Media, LLC 2013

analyses have greatly contributed to the construction of signaling maps in various model organisms, ranging from invertebrates to mammals. Accordingly, the effort to map novel signaling components and interactions has largely benefited from network alignment techniques and other widely used functional genomics methods, allowing the integration of functional data among and within species [2, 3].

Most of these methods predict new gene or protein properties (annotations) on the basis of sequence homology and similarities between known functions. Similar annotation transfer approaches have been applied to predict structural properties (e.g., domain composition), expression profiles, and physical interactions of proteins [4–6]. For predicting interactions, several techniques have been suggested, out of which one of the most widely used is the method of "interologs": two proteins are predicted to physically interact with each other, if their orthologs in another organism also interact [7]. Interologs, however, are found to be less conserved than orthologs [8] and also less reliable than interactions generated by HTP approaches [9].

Despite a great wealth of protein interaction data obtained from HTP experiments, such as yeast two-hybrid screens, the low abundance of extracellular, membrane-bound, and nuclear signaling components (e.g., ligands, receptors, and transcription factors) make these experimental techniques only partially efficient for identifying signaling interactions [10]. Accordingly, several signaling pathway databases have been generated manually by collecting relevant data from the literature [11]. However, so far most of them lack those key features (e.g., uniform pathway curation across more than one species) that would be necessary for transferring signaling pathway membership information between species [10]. Reliable and detailed signaling pathway databases are crucial for predicting novel signaling components because they are needed (1) as sources of known pathway information from which prediction can be performed (i.e., seed data) and (2) as reference data sets against which the novelty of predictions can be tested (i.e., those predicted signaling pathway member proteins that are already known pathway members should be removed from the list of predictions, while others can be regarded as predicted components).

A comprehensive pathway resource, SignaLink, developed in our lab, applies uniform curation rules to keep the levels of detail identical in all examined pathways for *Caenorhabditis elegans*, *Drosophila melanogaster*, and humans [12]. Compared to three widely used pathway databases (KEGG, Reactome and NetPath), SignaLink contains the (1) highest numbers of signaling proteins and interactions; (2) highest numbers of signaling cross-talks and multi-pathway proteins; (3) and above the average number of publications used per pathway [12]. Moreover, the uniform curation protocol and data structure of the SignaLink database allow systematic transfer of pathway annotation between two species on the basis of sequence orthology.

The topology of signaling pathways is crucial for selecting possible novel drug target candidates [13]. As an example, drugs used for inhibiting a specific signaling protein in order to affect proliferation may actually activate the corresponding pathway by triggering an unknown negative feedback loop [14]. Transferring signaling pathway annotations across species may alleviate such difficulties and can provide a more comprehensive signaling network. Identification of novel signaling components may help to discover novel drug targets as (1) these signaling components can increase the applicability of model organisms for testing drugs and drug target candidates, (2) in humans, they can serve as potential novel drug targets, and (3) in the case of already used target proteins they can help to uncover possible side effects.

2 Materials

1. The data serving as a basis for building the SignaLink database were obtained from both review papers and primary research articles (*see* Table 1).
2. These were complemented with data derived from species-specific databases for *Drosophila* and *C. elegans* (Flybase and Wormbase, respectively) that contain information from different sources—ranging from large-scale experiments to primary research articles (*see* Table 1).
3. We collected Ensembl IDs for human proteins from the genome browser Ensembl and ORFs for worms and flies from species-specific databases (Flybase and Wormbase), while UniProt IDs were collected from UniProt for all three species (*see* Table 1).

Table 1
Sources of the manually curated SignaLink database

Source	Protein	Signaling interaction	Link	Reference
170 Review papers	✓	✓		
771 Research articles	✓	✓		
Wormbase	✓	✓	http://www.wormbase.org/	[37]
Flybase	✓	✓	http://flybase.org/	[32]
UniProt	✓		http://www.uniprot.org/	[29]
Ensembl	✓		http://www.ensembl.org/	[24]

Table 2
Search engines used for the compilation of SignaLink

Search engines	Protein	Signaling interaction	Link	Reference
iHOP	✓	✓	http://www.ihop-net.org/	(18)
Chilibot	✓	✓	http://www.chilibot.net/	(15)
PubMed		✓	http://www.ncbi.nlm.nih.gov/pubmed/	
InParanoid	✓		http://inparanoid.sbc.su.se/	(17)

4. We searched directly for suggested interactions between two selected proteins with iHOP and ChiliBot [15, 16] (*see* Table 2). iHOP uses genes and proteins as hyperlinks between sentences and abstracts, meaning that information of a single protein and its interaction is given as a sentence retrieved from source abstracts [16].
5. We also used the synonym identification tool of iHOP for collecting protein synonyms.

3 Methods

In this section, we describe a unified curation protocol for assigning signaling proteins to signaling pathways and for compiling signaling interactions within a pathway. This standardized curation protocol in three different organisms is a prerequisite for enabling systematic transfer of pathway annotations between different species to predict new signaling components based on orthology.

3.1 Creating a Signaling Database (SignaLink) by a Uniform Manual Curation Protocol

The following section describes our workflow for the construction of a signaling database, which contains eight pathways in three species (*see* Fig. 1). The main steps involve listing signaling proteins of the given pathways, collecting information on the proteins, assigning each protein to the region/section of a given pathway, and collecting protein interaction information of the proteins, thereby also compiling additional proteins to the pathway.

3.1.1 Collecting Pathway Information for Signaling Proteins

All pathways examined from three species (*C. elegans*, *D. melanogaster*, and *Homo sapiens*) were compiled (i.e., manually curated) separately. For the challenges and importance of pathway definitions, *see* **Note 1**. For each pathway, three main steps were performed:

1. A search for pathway-specific review articles and databases using PubMed, Google Scholar, and Google.
2. The assignment of signaling proteins to signaling pathways based on the full text of reviews.

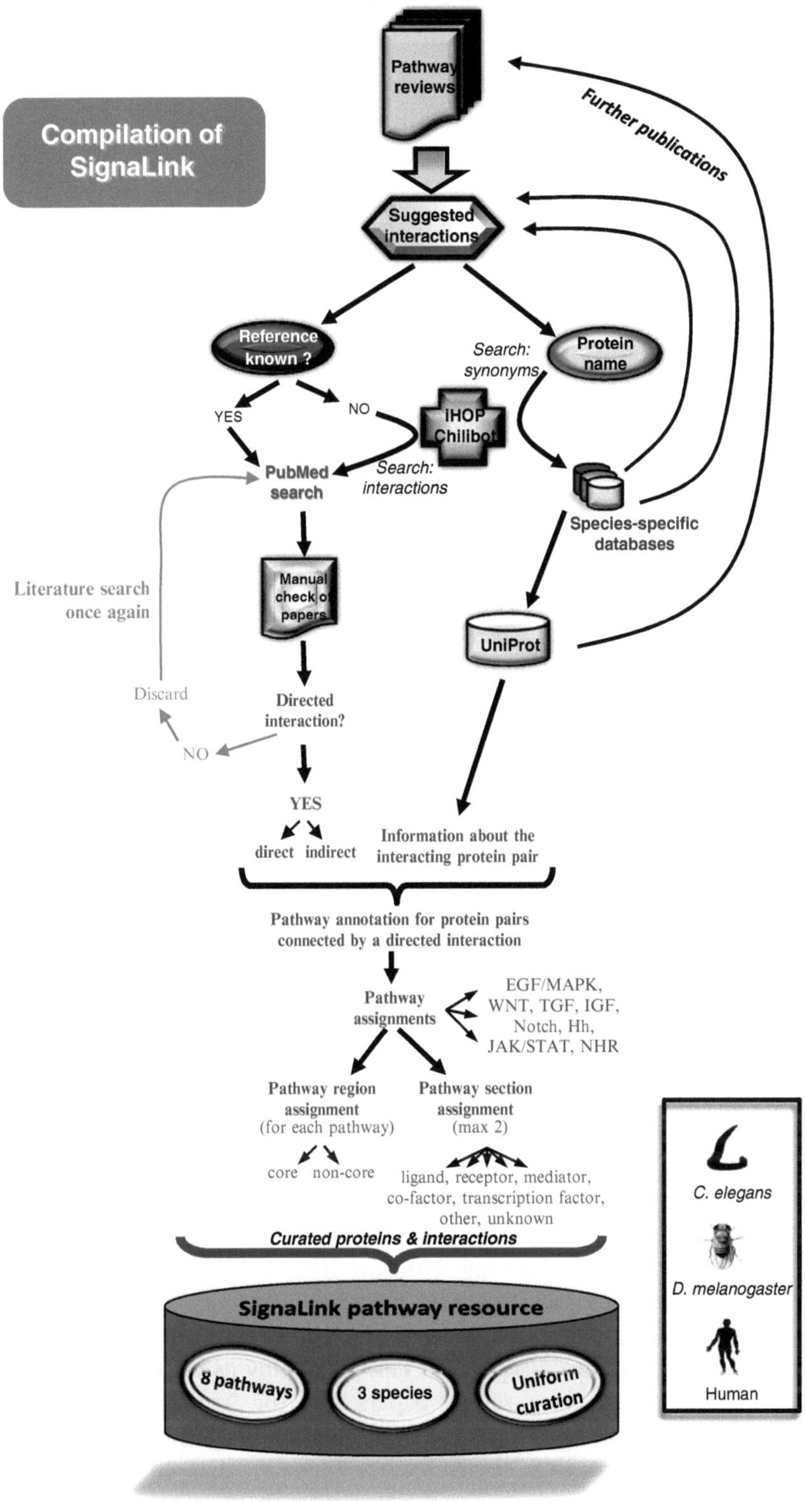

Fig. 1 Manual curation process of SignaLink. To compile the SignaLink pathway resource [12] (http:/signalink.org), signaling interactions were collected from pathway reviews, species-specific databases, and UniProt. Only interactions with references were included after manual checks via PubMed. The iHOP and ChiliBot search engines were used for finding references for suggested interactions lacking a reference in the reviews, and these search results were also manually checked. Synonyms for the interacting proteins were obtained with the help of the synonym finder tool of iHop. Finally, curated signaling proteins were assigned to pathway regions and pathway sections

3. An extended search for additional pathway proteins using iHOP and ChiliBot [15, 16].
4. When inserting a protein into SignaLink we assigned it to one pathway and—within this pathway—to one pathway region. Later, further pathways and pathway regions were added for this protein, if necessary. We marked a protein as a "core" component of a pathway, if it is essential for transmitting the signal of its pathway and has at least one of the pathway's biochemical characteristics, e.g., "Ser/Tyr-kinase activity". A "non-core" (or "peripheral") component modulates the pathway's core proteins, but it does not participate directly in the transduction of the signaling flow.
5. Additionally, the pathway section(s) of each protein was determined separately and a maximum of two sections per protein were allowed. The pathway position *ligand* indicates that the given protein initiates the signal of its pathway. A *receptor* is the direct receiver of this signal. A *mediator* is a member of the pathway that transduces the signal from the receptor towards downstream transcription factors. A *co-factor* modulates the function of any other protein from the pathway. Notably, co-factors often reside in the peripheral (non-core) region of their pathways. A *transcription factor* (1) activates another transcription factor (TF) after receiving the signal from its pathway, or (2) forms a complex with other TF proteins, or (3) binds to a specific promoter region (i.e., a specific binding site) on the DNA. Non-signaling proteins with roles in cellular motion, transport, and membrane anchoring were marked as *other*. When information on the position of a signaling protein in its pathway was lacking, the protein was marked *unknown*.

3.1.2 Collecting Signaling Protein Information

After listing pathway proteins from review and research papers, information on the signaling proteins were collected from different databases (*see* Table 1). For each protein, we also listed its orthologs in the other two species with the help of the ortholog clusters of the InParanoid database [17]. During collecting UniProt IDs, if more than one UniProt ID were available for the same protein, then the ID(s) of the protein(s) with the longest amino acid sequence was (were) used. To make the database more comprehensive, we assigned all known synonyms of the proteins. These were listed from review papers, and the "synonym" field of the iHOP database [18]. For the conversion of protein IDs, the Protein Identifier Cross-Reference Service (PICR) [19] and Synergizer [20] were used.

3.1.3 Collecting Signaling Interaction Information

A key feature of a signal transduction network is that the direction of an interaction is well distinguishable (e.g., protein A activates or negatively regulates protein B). Accordingly, all interactions inserted into SignaLink had to be directed. Each interaction had to be documented with the PubMed ID of the publication reporting the verifying experiment(s). Signaling interactions of a protein were collected from primary research articles, listed in review papers, species-specific databases (FlyBase, WormBase), and UniProt, iHOP, ChiliBot, and PubMed search results (*see* Tables 1 and 2). All research articles were manually examined, and in terms of biochemical experimental evidence, we marked every protein interaction as either *direct* or *indirect*. Direct experimental evidence indicates that there is a published biochemical evidence for signaling interaction between two given proteins, whereas indirect experimental evidence indicates that there is no direct biochemical evidence for interaction, but published experimental results suggest that interaction is very likely possible. Evidence types accepted here involve (1) changes in mRNA/protein levels, enzyme activities, concentrations of the products of catalyzed reactions, and (2) docking domain structures.

Importantly, not only the direction, but also the effect of an interaction is highly relevant to a signaling database. All interactions can be characterized as *activating* or *inhibitory.*

For interactions with indirect evidence, we marked activating interactions as ++ and −−, while inhibitory interactions were marked +− and −+. A unidirectional interaction (A and B interact as either A→B or B→A) has only one type of effect, but for the few bidirectional interactions (A→B and B→A are both present) more than one type of effects are possible between the two proteins. Two signaling interactions between the same two proteins in opposing directions are listed separately in SignaLink. For the challenges and limitations of manual curation, *see* **Note 2**.

3.1.4 Curation Process Example: The Notch Signaling Pathway and the NOTCH1 Protein

As an example, we present here the human Notch pathway and one of its components, the human NOTCH1 receptor protein. We describe the process of (1) obtaining information for the protein NOTCH1 and (2) obtaining protein interaction information for NOTCH1.

According to pathway-based reviews [21], there are 4 members of Notch receptor family proteins in humans: NOTCH1, NOTCH2, NOTCH3, and NOTCH4. Notch proteins have a specific role in transmitting signals [22] between ligands and transcription factors, as well as several additional proteins which influence the function of Notch proteins [23].

Alternative splicing can generate functionally different proteins from the same coding region; however, in the majority of proteins functional significance of different splice variants remains unknown. Despite their potentially different roles, databases and review papers do not differentiate between splice variants. For the human NOTCH1 protein, Ensembl [24] contains two splice variants: ENSP00000277541 and ENSP00000360765. From these two, the InParanoid database [17] contains only the first, ENSP00000277541. Therefore, we inserted only this splice variant into SignaLink. For proteins that have more than one splice variant, but none of them is present in the InParanoid database, we inserted into SignaLink the splice variant that has a primary UniProt accession (AC), as listed by Ensembl version 49.

From Ensembl, we included into SignaLink the UniProt accession(s) of a protein, and from UniProt we used the following data fields of the protein: description, reference—if it contained interaction data,—and cellular component. In addition, data from protein description and interaction fields were manually tested in primary publications for further information.

Regarding the region/section, NOTCH1 is a core protein of its pathway and functions as a receptor, mediator, or transcription factor, according to ref. [25]. However, within the Notch pathway, NOTCH1 functions either as a receptor or a transcription factor. Thus, we included only these two pathway sections for NOTCH1-into SignaLink.

To make SignaLink as complete as possible, we searched for orthologs of the human NOTCH1 protein. Orthologs without known signaling interactions became predicted pathway proteins in SignaLink. From the InParanoid database we identified the *C. elegans* and *D. melanogaster* orthologs of human NOTCH1 (ENSP00000277541). (In several cases we searched by both the UniProt and Ensembl protein IDs in InParanoid to find the protein.) Interestingly, the human NOTCH1 has two worm orthologs (LIN-12 and GLP-1), but only one fly ortholog (the protein N). We inserted all three orthologs into SignaLink. We listed species-specific protein IDs and UniProt ACs of the orthologs from WormBase and FlyBase. For ligands and transcription factors interacting with NOTCH1, we followed the same steps.

Next, we listed articles describing signaling interactions between NOTCH1 and other proteins by browsing through the references of the above mentioned review papers and by using the search engines iHOP [18] and ChiliBot [15]. iHOP allows users to search for all abstracts with interactions containing NOTCH1. With ChiliBot the interaction between two selected proteins can be directly searched for. As an example, interaction between NOTCH1 and TACE/ADAM17 has been described in an experimental article [26]. After reading the article, we found that it describes (1) a putative cleavage site for TACE on NOTCH1 and (2) a correlation between the in vitro enzymatic activity of TACE and the activity of NOTCH1. Thus, this article provides evidence for the activation of NOTCH1 by TACE. In addition, we directly searched for interactions between the orthologs of TACE and NOTCH1 in the other two species.

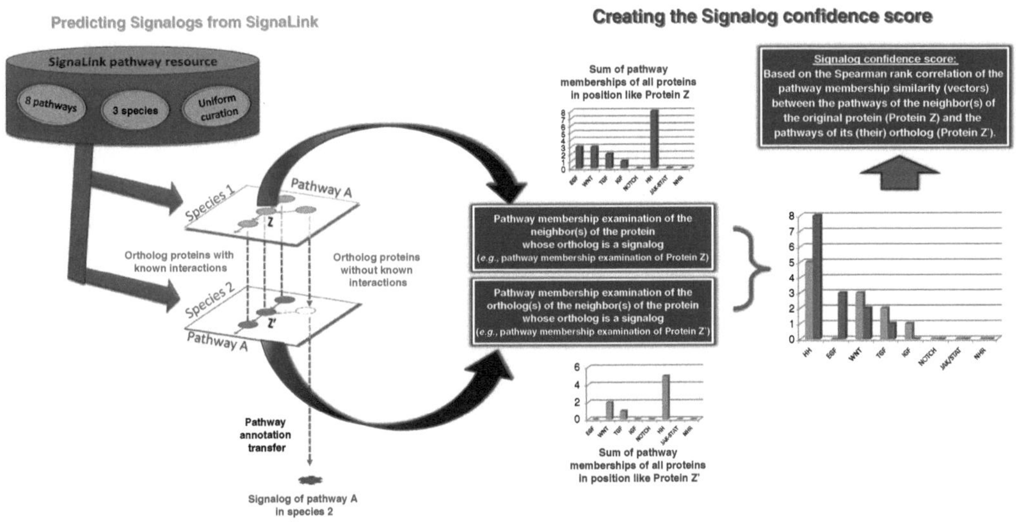

Fig. 2 Prediction of signalogs and calculation of the signalog confidence score. Based on the SignaLink resource [12] orthology assignment was performed between each pair of the three species. Proteins were predicted to be members of the same signaling pathway(s) where their orthologs belong. An interaction with a signaling protein Z′ was predicted for a protein, if the ortholog of the protein interacted with Z (the ortholog of Z′) in the same pathway A in a different species. A confidence score was calculated based on the pathway membership similarity between the neighbors of Z and its ortholog Z′. See main text for details

3.2 Signalog Prediction Based on Orthologous Signaling Components

Despite the conservation of many biological processes (e.g., developmental signaling pathways) throughout evolution, there is a poor overlap in protein–protein interactions between species in different databases [27]. Furthermore, the catalogue of proteins annotated with signaling function is incomplete even in highly studied model organisms. Therefore, the prediction of new potential signaling interactions and also new signaling proteins based on orthology is an important task.

3.2.1 Prediction of Signalogs

We started with creating a list from the three species examined in SignaLink (*C. elegans*, *D. melanogaster*, and *H. sapiens*) by collecting those proteins that have no known signaling interactions, but have at least one signaling pathway member ortholog in the other two species. Similarly to the concept of functional orthology [28], for each of these proteins we assumed that their pathway annotations (i.e., signaling role) can be transferred between species. Thus, we predicted that a protein is a member of the same signaling pathway(s) in which its ortholog(s) belong(s) (*see* Fig. 2). These proteins were termed as signalog proteins (signalogs). Because in SignaLink a protein can belong to more than one pathway [12], a signalog can also be annotated to more than one pathway. Using this approach we were able to predict 88, 92, and 73 novel signaling proteins in worms, flies, and humans, respectively [10]. For the limitations of orthology-based pathway annotation transfer, *see* **Note 3**.

3.2.2 Defining the Novelty of Signaling Protein Predictions Based on Orthology

To verify the novelty of the predicted signaling roles which have not been featured in other resources yet, we searched the literature with semiautomated methods for already known annotations. Next, we compared the list of signalogs and their predicted pathway memberships to pathway annotations found in pathway databases, as well as the list of ortholog predictions to previously published interolog predictions. To assess the novelty of signalogs and quantify the confidence level of each prediction, we performed semiautomated searches using PubMed, UniProt, GO, Wormbase, FlyBase, iHOP, and Chilibot web services [15, 18, 29–32]. During this process, direct manual curation and Python scripts checking multiple proteins in one webservice were used. In each of the three species examined, we classified the predicted signalogs into five groups on the basis of their known properties in the literature: (1) no orthology information and/or no biochemical function is available; (2) there are known orthologs with unknown biochemical function; (3) only biochemical function is available, but orthology information is lacking; (4) data on orthology as well as biochemical function(s) exist; (5) orthologs, biochemical function(s), and pathway annotation(s) are all known. Categories 1–5 denote a decreasing level of novelty. However, even category (5) contains signalogs for which at least one novel signaling pathway membership is predicted. Additionally, to check the novelty of the predicted signaling pathway memberships, we compared the list of signalogs and their predicted pathway memberships to known pathway membership annotations from Reactome and KEGG [33, 34]. We next applied interologs to verify the novelty of our ortholog predictions (an interolog is a pair of proteins predicted to interact based on the interaction of the two proteins' orthologs in at least one other organism) [7]. To reveal the presence of signalogs in current orthology-based prediction databases, we compared already identified interologs in worms, flies, and humans using three species-specific datasets (WI8, DroID, and HomoMINT) [8, 35, 36] with interologs generated from SignaLink data. Since neither SignaLink [12] nor the current signalog identification approach identify interologs directly, we used an indirect method by first deducing interologs from SignaLink data: we linked two proteins in an organism, if their orthologs interacted in at least one of the other three organisms. After generating all possible interologs from SignaLink, we examined only those (predicted interactions) in which at least one of the interactors is a signalog protein (predicted signaling pathway member).

3.2.3 Creating a Confidence Score for Signalogs

To assess the reliability of a signalog, a confidence score was calculated in each case (*see* Fig. 2). For the signalog Z′ that was predicted to be a component of Pathway A′ (PA′) in Species 2, we examined pathway membership of each neighbors (protein interactors) of Z′

in Species 2 and the known signaling component, Z in Species 1. For each Z and Z′ proteins, we summed pathway memberships as 2 pathway vectors (Vector_Z and Vector_Z′). Vectors have components indexed with the name of their signaling pathways. Finally, we computed the Spearman rank correlation of vectors computed for Z and Z′, and based on this correlation we defined the Signalog confidence score: [(Spearman_corr + 1)/2] * 100. This confidence score quantifies similarity between the signaling pathway membership profile of the possible interactors of a signalog protein and the original signaling protein (i.e., the orthologs of the signalog protein). Predictions above 50% can be considered as confident predictions.

4 Notes

1. Pathway definition is a critical task when compiling a pathway database. Pathway databases tend to use different pathway definitions, such as:
   - Canonical (e.g., MAPK)
   - Functional (e.g., inflammation)
   - Inferred (e.g., from gene expression data)
   - Cellular process regulating (e.g., autophagy induction)
   - Organ-related (e.g., vulva development)
   - Disease-related (e.g., list of connected proteins affected by mutations in breast cancer; Alzheimer's disease)
   - Drug-related (e.g., pharmacologically affected list of connected proteins)

   To develop a database for comparative purposes or systems-level examinations, pathway definitions must be the same in the whole database. For SignaLink, we applied a biochemically based, well-documented, and clear pathway definition. For example, the EGF/MAPK pathway in SignaLink contains (with evolutionary and biochemical reasoning) the pathway from the EGF ligand to the terminal MAPK kinases. In several other databases this pathway is scattered across many separate (sub)pathways (e.g., EGFR, RAS, p38, JNK, ERK, ASK). An important consequence of precise pathway definitions is the reduced number of examined pathways. An appropriate and precise grouping can be important to avoid artificial pathway constructs [12].
2. Despite recent advances in the technology of manual pathway curation, this technology still does have several limitations. First, curation highly depends on the knowledge and background of the curator as well as on the quality of the protocol

used for the curation [9]. Second, all data are based on the actual knowledge from the literature. Therefore, these databases have to be updated regularly (e.g., annually or bi-anually).

3. According to our current knowledge, the limitations of systematical pathway annotation transfer between species are the following. Interactions of membrane-bound and nuclear proteins are still underrepresented in most databases, thus predictions involving these proteins are less reliable. Furthermore, interactions between signaling proteins have been shown to be generally more unique to their species than PPIs in most biological processes [7].

Acknowledgement

Authors were supported by the European Union and the European Social Fund [TAMOP-4.2.1/B-09/1/KMR-2010-0003], the Hungarian Scientific Research Fund [OTKA K75334, NK78012], and a János Bolyai Scholarship to TK and TV.

References

1. Pires-daSilva A, Sommer RJ (2003) The evolution of signalling pathways in animal development. Nat Rev Genet 4:39–49
2. Gabaldon T, Huynen MA (2004) Prediction of protein function and pathways in the genome era. Cell Mol Life Sci 61:930–944
3. Kuzniar A, van Ham RC, Pongor S et al (2008) The quest for orthologs: finding the corresponding gene across genomes. Trends Genet 24:539–551
4. Yellaboina S, Dudekula DB, Ko MS (2008) Prediction of evolutionarily conserved interologs in Mus musculus. BMC Genomics 9:465
5. Storm CE, Sonnhammer EL (2003) Comprehensive analysis of orthologous protein domains using the HOPS database. Genome Res 13:2353–2362
6. Salgado D, Gimenez G, Coulier F et al (2008) COMPARE, a multi-organism system for cross-species data comparison and transfer of information. Bioinformatics 24:447–449
7. Yu H, Luscombe NM, Lu HX et al (2004) Annotation transfer between genomes: protein-protein interologs and protein-DNA regulogs. Genome Res 14:1107–1118
8. Persico M, Ceol A, Gavrila C et al (2005) HomoMINT: an inferred human network based on orthology mapping of protein interactions discovered in model organisms. BMC Bioinformatics 6(Suppl 4):S21
9. Cusick ME, Yu H, Smolyar A et al (2009) Literature-curated protein interaction datasets. Nat Methods 6:39–46
10. Korcsmaros T, Szalay MS, Rovo P et al (2011) Signalogs: orthology-based identification of novel signaling pathway components in three metazoans. PLoS One 6:e19240
11. Bauer-Mehren A, Furlong LI, Sanz F (2009) Pathway databases and tools for their exploitation: benefits, current limitations and challenges. Mol Syst Biol 5:290
12. Korcsmaros T, Farkas IJ, Szalay MS et al (2010) Uniformly curated signaling pathways reveal tissue-specific cross-talks and support drug target discovery. Bioinformatics 26:2042–2050
13. Chaudhuri A, Chant J (2005) Protein-interaction mapping in search of effective drug targets. Bioessays 27:958–969
14. Sergina NV, Rausch M, Wang D et al (2007) Escape from HER-family tyrosine kinase inhibitor therapy by the kinase-inactive HER3. Nature 445:437–441
15. Chen H, Sharp BM (2004) Content-rich biological network constructed by mining PubMed abstracts. BMC Bioinformatics 5:147

16. Hoffmann R, Valencia A (2004) A gene network for navigating the literature. Nat Genet 36:664
17. Berglund AC, Sjolund E, Ostlund G et al (2008) InParanoid 6: eukaryotic ortholog clusters with inparalogs. Nucleic Acids Res 36: D263–D266
18. Fernandez JM, Hoffmann R, Valencia A (2007) iHOP web services. Nucleic Acids Res 35:W21–W26
19. Cote RG, Jones P, Martens L et al (2007) The Protein Identifier Cross-Referencing (PICR) service: reconciling protein identifiers across multiple source databases. BMC Bioinformatics 8:401
20. Berriz GF, Roth FP (2008) The Synergizer service for translating gene, protein and other biological identifiers. Bioinformatics 24:2272–2273
21. Baron M (2003) An overview of the Notch signalling pathway. Semin Cell Dev Biol 14:113–119
22. Weinmaster G (1997) The ins and outs of notch signaling. Mol Cell Neurosci 9:91–102
23. Bray SJ (2006) Notch signalling: a simple pathway becomes complex. Nat Rev Mol Cell Biol 7:678–689
24. Flicek P, Aken BL, Beal K et al (2008) Ensembl 2008. Nucleic Acids Res 36:D707–D714
25. Ilagan MX, Kopan R (2007) SnapShot: notch signaling pathway. Cell 128:1246
26. Brou C, Logeat F, Gupta N et al (2000) A novel proteolytic cleavage involved in Notch signaling: the role of the disintegrin-metalloprotease TACE. Mol Cell 5:207–216
27. Gandhi TK, Zhong J, Mathivanan S et al (2006) Analysis of the human protein interactome and comparison with yeast, worm and fly interaction datasets. Nat Genet 38:285–293
28. Bandyopadhyay S, Sharan R, Ideker T (2006) Systematic identification of functional orthologs based on protein network comparison. Genome Res 16:428–435
29. Boutet E, Lieberherr D, Tognolli M et al (2007) UniProtKB/Swiss-Prot: the manually annotated section of the UniProt KnowledgeBase. Methods Mol Biol 406:89–112
30. Ashburner M, Ball CA, Blake JA et al (2000) Gene ontology: tool for the unification of biology. The Gene Ontology Consortium. Nat Genet 25:25–29
31. Harris TW, Antoshechkin I, Bieri T et al (2010) WormBase: a comprehensive resource for nematode research. Nucleic Acids Res 38: D463–D467
32. Drysdale R (2008) FlyBase: a database for the *Drosophila* research community. Methods Mol Biol 420:45–59
33. Ogata H, Goto S, Sato K et al (1999) KEGG: Kyoto encyclopedia of genes and genomes. Nucleic Acids Res 27:29–34
34. Joshi-Tope G, Gillespie M, Vastrik I et al (2005) Reactome: a knowledgebase of biological pathways. Nucleic Acids Res 33:D428–D432
35. Yu J, Pacifico S, Liu G et al (2008) DroID: the *Drosophila* Interactions Database, a comprehensive resource for annotated gene and protein interactions. BMC Genomics 9:461
36. Simonis N, Rual JF, Carvunis AR et al (2009) Empirically controlled mapping of the Caenorhabditis elegans protein-protein interactome network. Nat Methods 6:47–54
37. Rogers A, Antoshechkin I, Bieri T et al (2008) WormBase 2007. Nucleic Acids Res 36: D612–D617

Chapter 16

Bioinformatics Workflows and Web Services in Systems Biology Made Easy for Experimentalists

Rafael C. Jimenez and Manuel Corpas

Abstract

Workflows are useful to perform data analysis and integration in systems biology. Workflow management systems can help users create workflows without any previous knowledge in programming and web services. However the computational skills required to build such workflows are usually above the level most biological experimentalists are comfortable with. In this chapter we introduce workflow management systems that reuse existing workflows instead of creating them, making it easier for experimentalists to perform computational tasks.

Key words Workflows, Web services, SBML, BioModels, Gene Ontology, Taverna, Biocatalogue, myExperiment

1 Introduction

Bioinformatics is perhaps similar to experimental biology in that many repetitive tasks are performed following standard protocols. Many of these tasks can be automated creating computational workflows from existing software components and algorithms. The traditional approach is to glue components together using scripts written in a programming language like Perl. This method is not, however, practical for many wet lab biologists lacking programming skills. A more user-friendly approach to create these pipelines can be established by using workflow management systems (WMS). A WMS is a type of software aimed to manage and define a series of computational tasks that produce an outcome. WMS are thus really useful to facilitate multistep data analysis and data integration. In bioinformatics, this type of software usually presents a graphical user interface that allows users to easily create a concatenation of automated scientific tasks, even without any previous knowledge of programming.

To date working with workflows has been difficult because for the average experimentalist, it can be hard to discover the

Maria Victoria Schneider (ed.), *In Silico Systems Biology*, Methods in Molecular Biology, vol. 1021, DOI 10.1007/978-1-62703-450-0_16, appropriate functionality or it may not be intuitive enough to be easily run. This is particularly the case when the output of a specific task is not compatible as input to the following task, perhaps due to different format standards. A way round this problem is to use those workflows that have already been created to automatically run standard pipelines. Reutilization of existing workflows is easily achievable using myExperiment [1], to date the most comprehensive catalogue of workflows in bioinformatics. In what follows we demonstrate and provide several workflow examples from myExperiment that demonstrate the ease of use and capability of reutilization of existing workflows. For that we use the Taverna [2] WMS. We hope that by the end of this chapter the user becomes familiar with the concept of reusing and reconfiguring existing workflows.

2 Materials

2.1 Web Services

We will focus on workflows that use remote software functions (web services) for this chapter, as they do not require any installation and are readily available for public use. A web service could be equated to a "task" in our workflow and technically is a piece of software that runs remotely being accessible via the Internet. Web services can provide access to data or analysis tools. Web services are designed for programmatic use (not directly by the user). Web services are independent from programming languages and can be operated following specific rules. For example, a rule for a web service can be that a particular web address (URL) is a request.

2.2 Taverna

Taverna (*see* **Note 1**) is a WMS for the life sciences, an application that eases the use and integration of software tools and databases. Taverna is especially suited for web services. Taverna's functionality is offered via a desktop application that provides a graphical user interface for designing and executing workflows. Taverna's main advantage is that it hides the technical aspect of web services facilitating their use for the nonexpert [2]. At the time of writing, Taverna offers access to more than 3,500 web services and allows users to add new ones (*see* **Note 2**).

Herein we used Taverna Workbench 2.4. When the application is executed, it shows by default the "Design" section with three main panels. On the top left corner is a "service panel" containing a list of services that can be used to build workflows. On the right, a "workflow panel" offers a visual representation of steps, inputs, outputs, and services involved in a workflow. At the bottom left corner, an "information panel" provides general details about the workflow and its components (*see* Fig. 1). Although such panels are mostly used to create new workflows, we will focus on prebuilt workflows, skipping the need for understanding how they work (for more information on their functionality, please *see* **Note 3**).

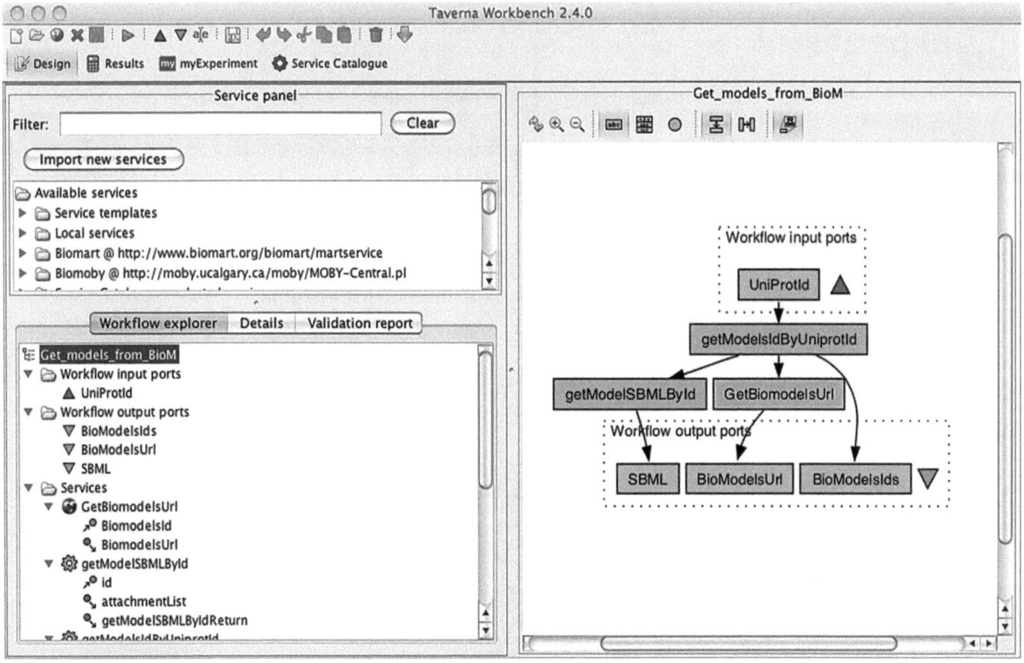

Fig. 1 Screenshot for the "Get models from BioModels including the input protein" workflow in Taverna. This workflow takes as input a protein identifier and then it runs the BioModels service to retrieve model identifiers from the BioModels database. The model identifiers are used by another service of BioModels to retrieve the names of the models. This workflow is available at http:/www.myexperiment.org/workflows/3112.html

2.3 myExperiment

myExperiment (*see* **Note 4**) is an online catalogue of workflows for public access and sharing (*see* Fig. 2). It is accessible through Taverna, which offers a graphical interface to browse and import workflows (*see* Fig. 3). myExperiment allows the researcher submission of workflows and the capability of searching for those relevant to his/her desired functionality. These can then be reused and repurposed to the user's specific requirements, encouraging reproducible research [1]. At the time of this writing (October 2012), myExperiment contained around 2,000 workflows. Some of them have been designed to support the analysis of microarray data [3]; others include the integration of gene expression levels with systems biology [4], the extraction and structuring of knowledge from text [5], or the identification of genes associated with diseases [6]. "Pathways and Gene annotations for QTL region" is probably the most accessed workflow (*see* **Note 5**) and its purpose is to identify genes residing in quantitative trait loci (QTL) regions. This workflow was initially designed for the identification of genes in mouse although it has also been successfully used for different species and QTL regions [7].

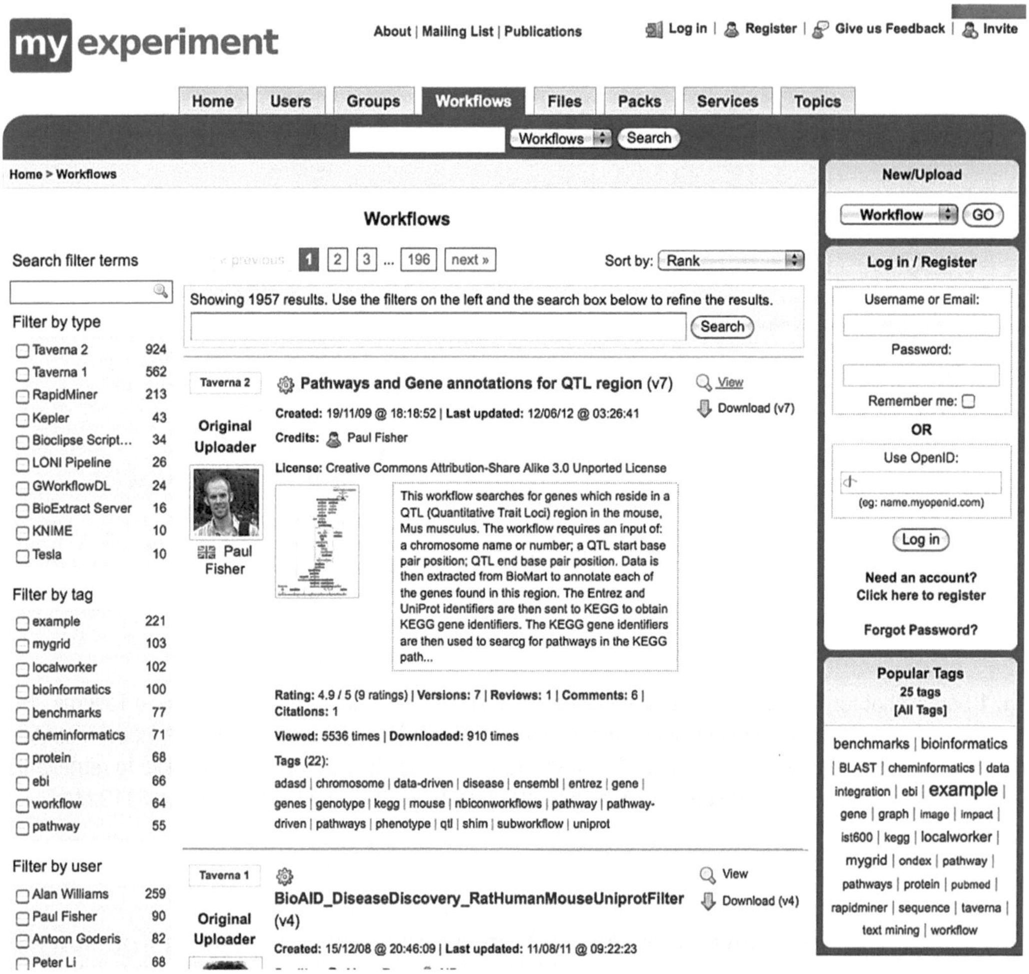

Fig. 2 myExperiment website displaying workflows sorted by "Rank." On the left it shows a search and filter functionality to narrow down the number of workflows to explore. On the center is a list of workflows, showing on the top a workflow entry with the title "Pathway and gene annotations for QTL region"

2.4 Biocatalogue

We will also use Biocatalogue (*see* **Note 6**), an interface for registering, browsing, and annotating web services (*see* Fig. 4). The Biocatalogue repository contains web services, regularly monitored to notify users of service problems and changes. At the time of this writing, Biocatalogue offers more than 2,300 services from over 170 service providers. Like myExperiment, Biocatalogue is conveniently integrated within Taverna, offering a graphical interface to browse and import web services (*see* Fig. 5).

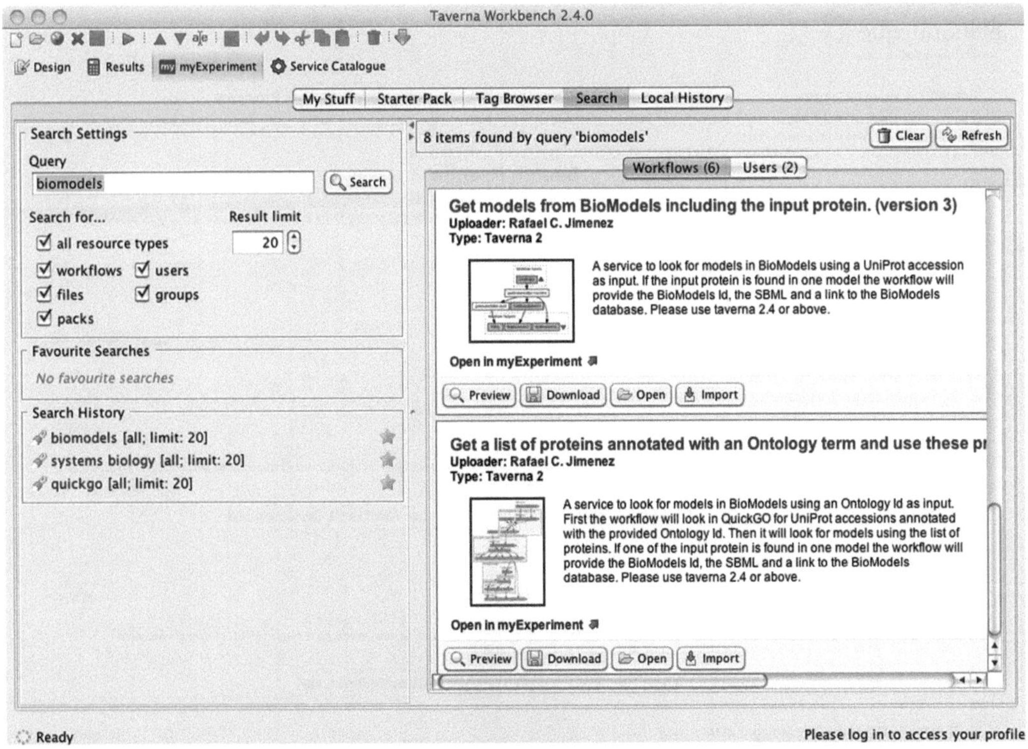

Fig. 3 Graphical user interface included in Taverna to search and import workflows from myExperiment. In this example a search was performed to find workflows using the query "biomodels." This screenshot shows two BioModels-related workflows of six found in myExperiment

3 Examples

In this section we present two biological examples that show how to reuse and reconfigure workflows. The first example explores the SMAD2 signal transducer and transcriptional modulator. Expression of SMAD2 has been associated to colorectal carcinoma [8]. Our aim here is to find all the instances of the SMAD protein found in the BioModels database [9], which contains published mathematical models of biological interest. To do so, we will be querying existing workflows available in myExperiment through Taverna. The second workflow shows how to retrieve information available in BioModels searching proteins involved in the "transforming growth factor beta receptor signaling pathway" biological process to find the association of proteins involved in this biological process with SMAD2.

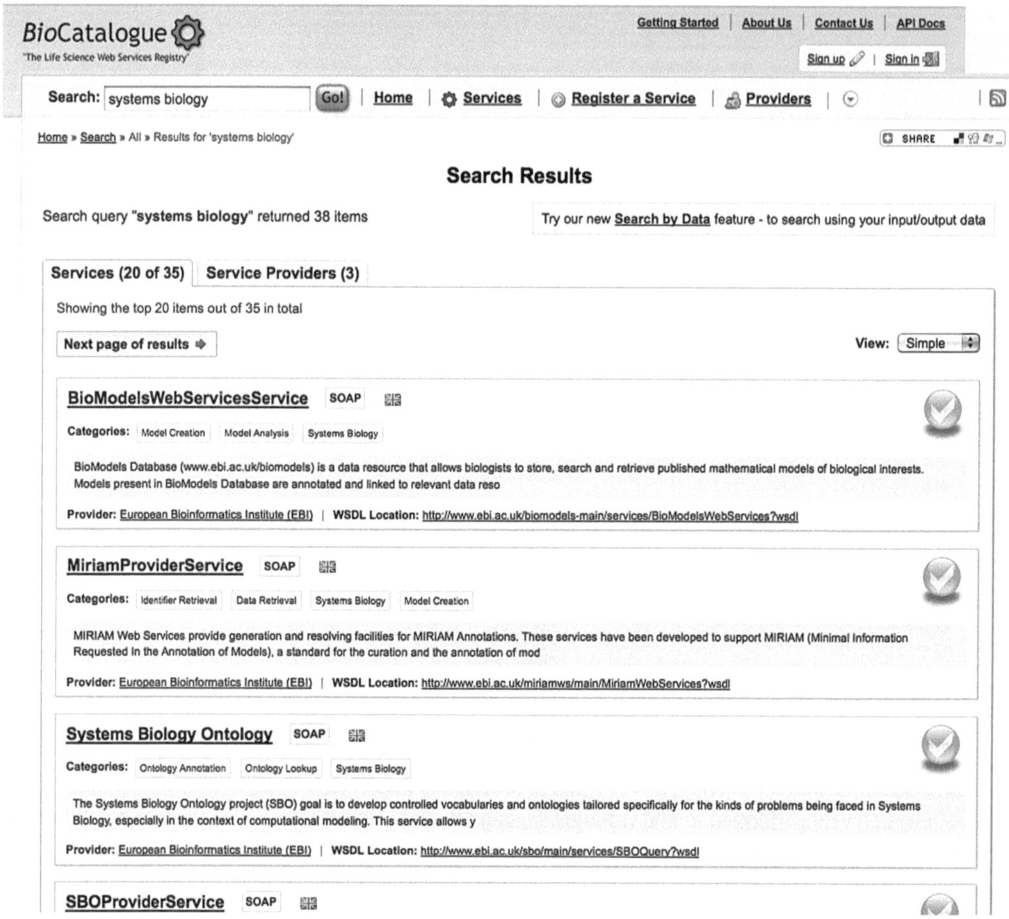

Fig. 4 The Biocatalogue website displaying a list of web services after searching for the query "systems biology." This screenshot shows four of the thirty-five services found in Biocatalogue

3.1 Querying BioModels Using Protein Accessions

In the first example we will use a workflow from myExperiment to retrieve models from BioModels involving SMAD2.

3.1.1 Searching and Loading a Workflow

Click on the "myExperiment" button located on the top left corner below the main menu. This will open a new view. On the "Search" tab, find a "Search Settings" panel where you can make a new search by typing your query in the "Query" field. Then click the "Search" button. Next, search for "BioModels" to find publicly available workflows suiting our needs. As a result of this query, the right panel now displays a list of workflows (*see* **Note 7**). Look for the workflow named "Get models from BioModels including the input protein" (*see* Fig. 3). Read the description and click on the "Open" button to visualize it in Taverna (*see* **Note 8**) (*see* Fig. 1).

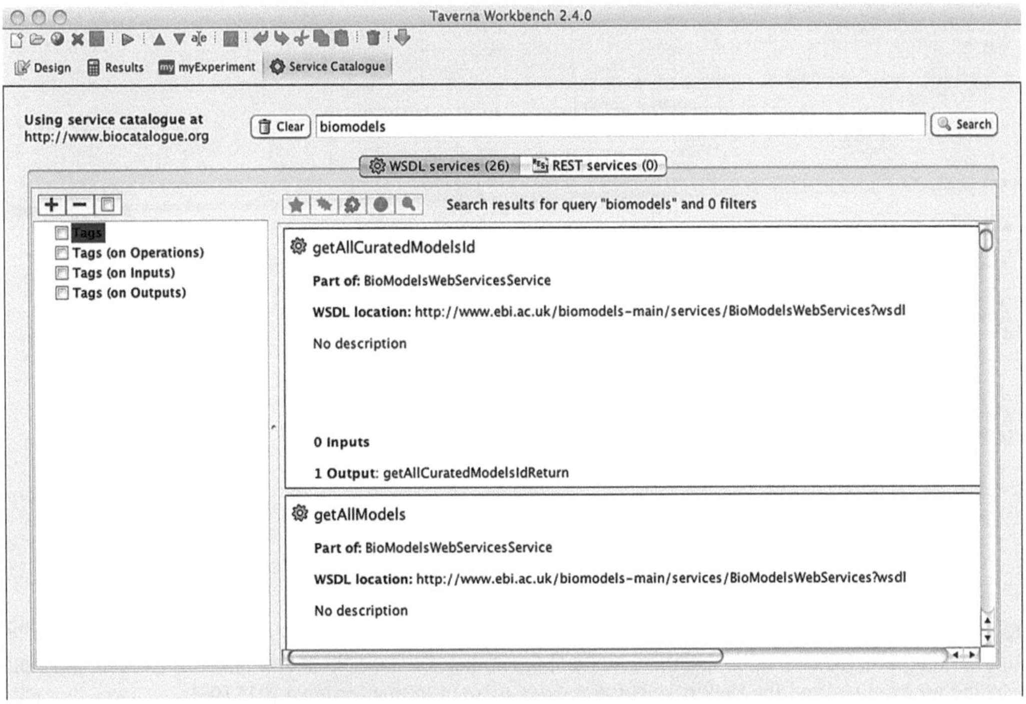

Fig. 5 Graphical user interface included in Taverna to search and import web services from "Biocatalogue." In this example a search was performed to find workflows using the query "biomodels." This screenshot shows two related web service of the twenty-six found in myExperiment

3.1.2 Running the Workflow

In the menu, click on "File" and "Run workflow." A new window is then opened in Taverna. This window serves input values to the workflow. The left side shows some general information about the workflow: workflow diagram, workflow description, and author. On the top right side, information can be found about the different input parameters used to run this workflow (e.g., "port description" and "example values" (*see* **Note 9**)). To set a value for the "Protein Accession" input parameter, make sure you are on the "ProteinAccession" tab and click on the "Set value" button. By default the value will be set using the example value. Change this value to "Q15796," the protein accession number for the example protein "SMAD family member 2" (*see* Fig. 6). Now that the input parameter has been defined, run the workflow by clicking on the "Run workflow" button.

3.1.3 Checking Results and Saving Them

After running the workflow, Taverna points the user to the "Results" section. On the top left panel, a history of the workflows with its results is shown. On the top right panel, the progress and the state of the workflow are shown. To run the workflow should take only a few seconds. On the bottom panel, in "Workflow results," several header tabs appear. Tabs preceded with a red triangle contain

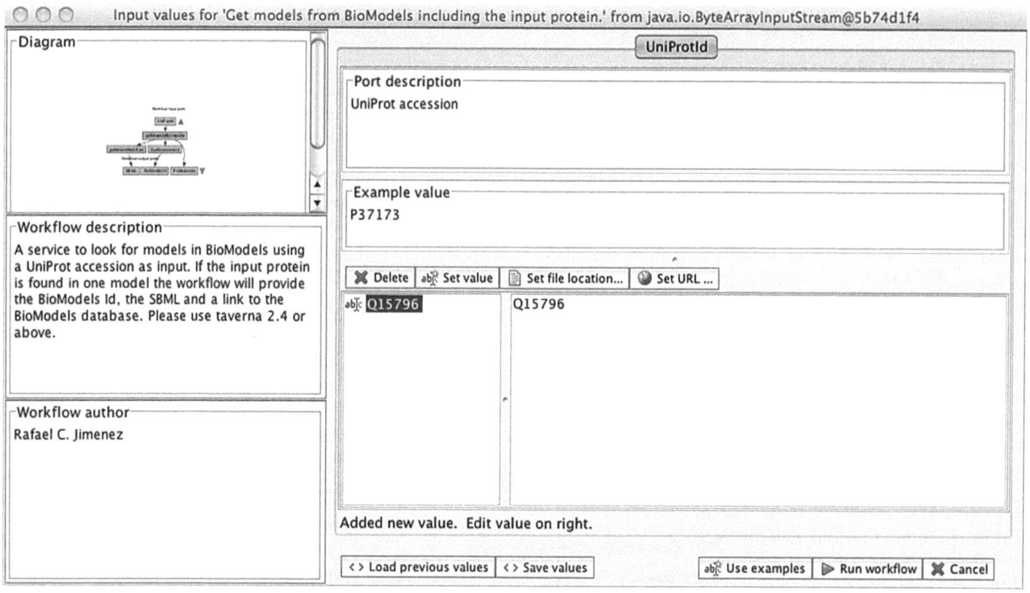

Fig. 6 Screenshot of Taverna to configure input parameters before running a workflow. On the left shows the workflow diagram, the description, and the name of the author. On the right shows the description of the input including an example and the UniProt accession used as input in this example (Q15796)

input information. Tabs preceded with a green triangle contain the results. Clicking on the "BioModelsIds" tab shows four values corresponding to four BioModel ids (BIOMD0000000163, BIOMD0000000342, BIOMD0000000173, and BIOMD00000 00112). The values for the "BioModelsUrl" tab provide links to browse the models in the BioModels database. The values of the "SBML" tab provide the models in SBML format including detail information about the model. One can optionally save this information by clicking the "Save all values" button or saving one specific value (*see* **Note 10**).

3.2 Querying BioModels Using a List of Proteins Annotated with a Gene Ontology Term

Here we retrieve information present in BioModels for human proteins annotated with the Gene Ontology term for the "transforming growth factor beta receptor signaling pathway" cellular component: GO:0007179. GO stands for Gene Ontology (http://www.geneontology.org/).

3.2.1 Search and Load a Workflow

Click on the "myExperiment" button. On the "Search" tab, search for "biomodels" to find publicly available workflows. Look for the workflow named "Get a list of proteins annotated with an Ontology term and use these proteins to query BioModels" (*see* Fig. 3). Click on the button "Open" to visualize it in Taverna (*see* Fig. 7).

3.2.2 Running the Workflow

In the menu, click on "File" and "Run workflow." Click on the "GeneOntologyId" tab and set the value to the example: "GO:0007179." Now click on the "TaxonomyId" tab and set

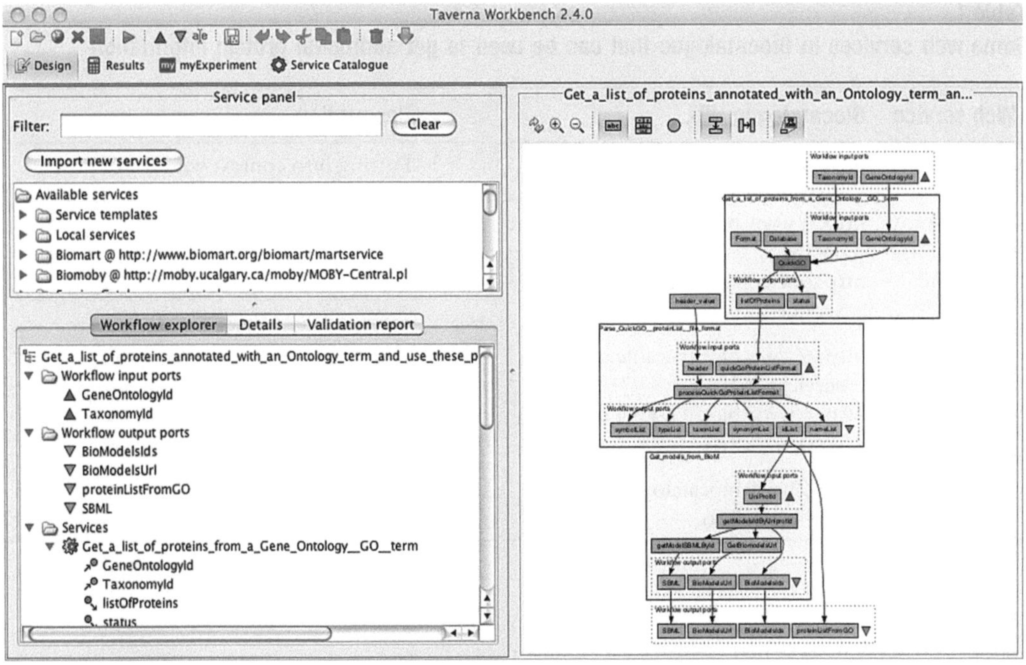

Fig. 7 Screenshot of the "Get a list of proteins annotated with an Ontology term and use these proteins to query BioModels" workflow. This workflow recycles three other workflows. The first workflow gets a list of proteins for a Gene Ontology term from QuickGO (19744993). The second workflow helps to shape the output of the QuickGO service to be able to use it as input in the following workflow (see **Notes 15** and **16**). The third workflow takes a list of proteins from the preceding workflow and for each protein it looks for models in BioModels. This workflow is available at http:/www.myexperiment.org/workflows/3113.html

the value to "9606" to limit the results to human proteins (*see* **Note 11**). Now that your input parameters are defined, run the workflow by clicking on the "Run workflow" button. After the workflow retrieves and transforms results from QuickGO (24), it will iterate over the BioModels service, once per protein. Find a count of the iterations as they are processed (*see* **Note 12**).

3.2.3 Checking Results

In the "Results" section check the output tabs on the "Workflow results" panel. Click on the values available in the output tabs to see the list of results for each protein accession (*see* **Note 13**). At the time of processing this experiment, 217 proteins are found categorized as "transforming growth factor beta receptor signaling pathway" Gene Ontology term. Fifty two of these proteins were annotated in models from the BioModels database. The four most represented models (hit by other 6, 4, 3, and 3 proteins, respectively) are the models found in the first workflow ("BIOMD0000000163, BIOMD0000000342, BIOMD0000000173," and "BIOMD0000000112"), showing the association of SMAD2 with proteins involved in the aforementioned biological process.

Table 1
Some web services in Biocatalogue that can be used to get additional protein information

Web service	Biocatalogue URL	Description
KEGG	http://www.biocatalogue.org/services/11	Putting into context your list of protein with pathway information
WikiPathways	http://www.biocatalogue.org/services/1935	
Reactome	http://www.biocatalogue.org/services/2033	
UniProt	http://www.biocatalogue.org/services/610	Retrieving protein sequences and annotations
InterPro	http://www.biocatalogue.org/services/2753	
PDB	http://www.biocatalogue.org/services/1766	Finding structures

3.3 Adding More Functionality by Extending a Workflow

BioModels is not the only type of service that can be used to query SMAD2. Additional functionality can be extended for previous two workflows using any of the web services described in Biocatalogue (*see* Table 1 for a representative list).

Workflows themselves can be run as services, taking an input and producing an output; Taverna allows this functionality.

4 Notes

1. http://www.taverna.org.uk
2. Taverna is freely available for Linux, Macintosh, and Windows and can be downloaded from http://www.taverna.org.uk/download/.
3. For more information Taverna offers comprehensive documentation including a glossary of terms, a quick start guide, a user manual, tutorials, videos, and a frequently asked questions section at http://www.taverna.org.uk/documentation/.
4. http://www.myexperiment.org
5. http://www.myexperiment.org/workflows/16.html
6. http://www.biocatalogue.org/
7. Workflows from "myExperiment" can be browsed in its website and also from Taverna itself. For this example, we use Taverna

to search, retrieve, and execute workflows directly from "myExperiment."

8. The workflow panel has several options for displaying the workflow diagram: diagram orientation, zooming, service details, etc.
9. The information provided in "port description" and "example values" is worth reading since most of the time it includes tips to define the input required by the workflow.
10. In Taverna save all the results with the "Save all values" buttons or individual values by clicking on the "Save value" button available inside of each "Workflow results" tab. Results can be saved in a file in text, XML, or Microsoft Excel/Open Office format.
11. Find more information about organisms and their classification using the UniProt taxonomy website (http://www.uniprot.org/taxonomy/). You can search there for species or browse organisms using a hierarchical tree view. The input used in the workflow is called "Taxon identifier."
12. Explore intermediate inputs and outputs by clicking on a service on the graph on the progress table of the "results perspective."
13. Checking the results will find some values that do not have BioModels results (values in red). This is because some proteins are not found in the BioModels database. The values on the BioModels tabs are associated to the "Values" on the "proteinListFRomGO" tab.
14. In Taverna this is considered to be a "Shim service": a service that does not perform a scientific function but acts as an intermediary to transform an output from one service into a suitable input for the next service.
15. Access the individual workflows used to work with GO annotations from "myExperiment," which are the "Get a list of proteins from a Gene Ontology (GO) term" workflow at http://www.myexperiment.org/workflows/2742 and the "Parse QuickGO proteinList file format" workflow at http://www.myexperiment.org/workflows/2744.

Acknowledgements

R.C.J. is supported by the NHLBI Proteomics Center Award HHSN268201000035C.

References

1. Goble CA, Bhagat J, Aleksejevs S, Cruickshank D, Michaelides D, Newman D, Borkum M, Bechhofer S, Roos M, Li P, De Roure D (2010) myExperiment: a repository and social network for the sharing of bioinformatics workflows. Nucleic Acids Res 38(Web Server Issue):W677–W682
2. Hull D, Wolstencroft K, Stevens R, Goble C, Pocock MR, Li P, Oinn T (2006) Taverna: a tool for building and running workflows of services. Nucleic Acids Res 34(Web Server Issue):W729–W732
3. Li P, Castrillo JI, Velarde G, Wassink I, Soiland-Reyes S, Owen S, Withers D, Oinn T, Pocock MR, Goble CA, Oliver SG, Kell DB (2008) Performing statistical analyses on quantitative data in Taverna workflows: an example using R and maxdBrowse to identify differentially-expressed genes from microarray data. BMC Bioinformatics 9:334
4. Li P, Oinn T, Soiland S, Kell DB (2008) Automated manipulation of systems biology models using libSBML within Taverna workflows. Bioinformatics 24(2):287–289
5. Roos M, Marshall MS, Gibson AP, Schuemie M, Meij E, Katrenko S, van Hage WR, Krommydas K, Adriaans PW (2009) Structuring and extracting knowledge for the support of hypothesis generation in molecular biology. BMC Bioinformatics 10(Suppl 10):S9
6. Fisher P, Hedeler C, Wolstencroft K, Hulme H, Noyes H, Kemp S, Stevens R, Brass A (2007) A systematic strategy for large-scale analysis of genotype phenotype correlations: identification of candidate genes involved in African trypanosomiasis. Nucleic Acids Res 35 (16):5625–5633
7. Fisher P, Noyes H, Kemp S, Stevens R, Brass A (2009) A systematic strategy for the discovery of candidate genes responsible for phenotypic variation. Methods Mol Biol 573:329–345
8. Moustakas A, Souchelnytskyi S, Heldin CH (2001) Smad regulation in TGF-beta signal transduction. J Cell Sci 114(Pt 24):4359–4369
9. Li C, Donizelli M, Rodriguez N, Dharuri H, Endler L, Chelliah V, Li L, He E, Henry A, Stefan MI, Snoep JL, Hucka M, Le Novère N, Laibe C (2010) BioModels Database: an enhanced, curated and annotated resource for published quantitative kinetic models. BMC Syst Biol 4:92
10. Binns D, Dimmer E, Huntley R, Barrell D, O'Donovan C, Apweiler R (2009) QuickGO: a web-based tool for Gene Ontology searching. Bioinformatics 25(22):3045–3046

INDEX

Maria Victoria Schneider (ed.), *In Silico Systems Biology*, Methods in Molecular Biology, vol. 1021,
DOI 10.1007/978-1-62703-450-0,

MIX
Papier aus verantwortungsvollen Quellen
Paper from responsible sources
FSC® C105338

If you have any concerns about our products,
you can contact us on
ProductSafety@springernature.com

In case Publisher is established outside the EU,
the EU authorized representative is:
Springer Nature Customer Service Center GmbH
Europaplatz 3, 69115 Heidelberg, Germany

Printed by Libri Plureos GmbH
in Hamburg, Germany